Quantitative *Eco*–nomics

Quantitative *Eco*–nomics

How sustainable are our economies?

Peter Bartelmus

Springer

Peter Bartelmus

ISBN: 978-1-4020-6965-9 e-ISBN: 978-1-4020-6966-6

Library of Congress Control Number: 2007943326

Printed on acid-free paper

9 8 7 6 5 4 3 2 1

springer.com

To
Liza, Arik and Claude

Preface

Inspired by participating in the 1992 Earth Summit I set out to explore the new paradigm of sustainable development from an economist's point of view. In the preface of the resulting book I conceded: 'the meaning of sustainable development and its implications for policy-making are not yet clearly understood' (Bartelmus, 1994a). Further scrutiny revealed the vagueness of the popular human-needs-based definition of sustainable development. Most likely such vagueness is responsible for the wide acceptance of the paradigm by government, business and environmentalists. More than a decade later I have to admit that little has changed. There is still no agreement on a blueprint or policy framework for sustainable development. The concept remains rhetoric, embroidering national and international laws such as those of the European Union and the World Trade Organization. Hardly any publication on environment and/or development can resist summoning the concept for authoritative support.

This book shows some resistance. The last question of its introductory part dares to ask whether the paradigm has run its course. At the end of the book the answer is a 'guarded yes' – guarded because there is considerable goodwill attached to the concept. The answer does not come lightly. It is based on a thorough examination of empirical analysis, which makes up the main parts of the book. Rejecting the practical value of a cornucopian paradigm also justifies concentrating on what is measurable, comprehensively and comparably. The conclusion is to focus on the immediate interaction between economic activity and the natural environment. I do not claim that only the measurable is manageable, but I do believe that facts and figures make for *better* management.

Assessing the environmental sustainability of economic performance and growth is one aspect of *eco*–nomics.* The other is a persistent dichotomy in dealing with the environment-economy interface. Environmentalists and ecological economists warn us about looming environmental disaster. Control and regulation of the physical scale of economic activity is their policy response. In contrast, environmental economists seek to change the behaviour of producers and consumers by making them accountable for their environmental impacts. *Eco*–nomics stands for analysing

*To my knowledge Postel (1990) was the first to use the term 'eco'-nomics.

environmental sustainability with the tools of both ecological and environmental economics. The book compares the purpose and practicality of these tools with a view to bridging the environmental-economic dichotomy. Integrative data and accounting systems provide the structure and material for building the bridge.

There has been progress in measuring, accounting and empirical modelling of sustainability. Measurement by statistics, indicators and accounts may be less thrilling than doomsday predictions from environmentalists and anti-globalization movements; but the new measurement tools, and in particular green accounting, generated operational concepts that can make environmental and economic policies more integrative and accountable. The objective of this book is to bring this progress to the critical attention of classrooms, boardrooms and offices.

The infiltration of analysis and policy from the bottom of 'data crunchers' has met with resistance from both, data users and – conservative – data producers. Conventional economic indicators still dominate much of official statistics. National accountants feel threatened by the greening of their established economic accounts. For instance, China's much-heralded greening of its GDP was recently halted to the 'delight' of the national statistical bureau (R. Spencer in *The Daily Telegraph*, 23 July 2007). Surprisingly environmentalists seem to concur. Their warnings about any pricing of a 'priceless' ecological heritage show apprehension about being 'colonized' by mainstream economics. As an alternative, they offer short and long indicator lists, which in their view represent environmental quality and human well-being.

Policymakers, public media and stakeholders prefer, however, information in a nutshell – like GDP for the economy. This might be one reason for the current preoccupation with global warming as a convenient surrogate for environmental degradation and as an indicator of unsustainable consumption and production. The critical evaluation of all these claims and counterclaims should help building another bridge of transparency and understanding between data users and producers.

Part I sets out with identifying environmental concerns and asks what economics has to do with them. The practicality of the broad concept of sustainable development is also discussed in this part. In line with the general focus on quantitative analysis, Parts II and III deal with measuring the environmental sustainability of economic activity. Part II presents the physical assessment tools, favoured by environmentalists looking for evidence of environmental impacts on ecosystems. Material flow balances, in particular, measure the (lack of) dematerialization of the economy. Reducing the flow of materials from the environment is expected to decrease pressures on nature's carrying capacities for attaining *ecological sustainability*. Part III extends the conventional national accounts to include environmental assets and their source and sink services to the economy. The objective is to assess *economic sustainability* in terms of produced and natural capital maintenance. Part IV reviews the modelling of environmental trends, limits and policy scenarios. The question is whether decision makers should rely on more objective (observed) data or use the filters of assumption-laden, but policy-oriented models.

The concluding Part V reviews strategies and evaluates policy instruments as to their practicality in attaining economic and ecological sustainability. It also devotes one chapter to the heated debate of globalization. The generic anti-globalization movement leads us back to development concerns of equity, culture and global governance. Some of the initial questions raised in Part I re-emerge thus in a global context. Admittedly partial or non-conclusive answers in the last chapter should not discourage further debate and research. Rather, they should prompt further exploration of a subject that is frequently obscured by fuzzy vision, anecdotal reporting and rhetoric.

The book seeks to provide a concise and systematic guide through this fuzziness. Its focus on comparative quantitative analysis provides structure and perspective to a subject characterized by political agendas, media hype and dramatization of selected social and environmental issues. It might also help open a dialogue between mainstream and ecological economists.

Peter Bartelmus

Acknowledgements

I am grateful for comments and technical help from my former colleagues at the United Nations Statistics Division, Reena Shah and Viet Vu, and at the Wuppertal Institute for Climate, Environment and Energy, Hans Kretschmer, Michael Ritthoff, Tilman Santarius, Mary Walker and Jola Welfens. Mark de Haan and Mohan Munasinghe provided valuable advice on thorny questions. The book also benefited considerably from the discussion with students of Columbia University and the Bergische Universität Wuppertal. I also wish to express my appreciation of the patient and efficient help of the editorial staff at Springer Science and Business Media.

Permission to reproduce text and illustrative material by Berliner Wissenschafts-Verlag, Stefan Bringezu, Center for International Earth Science Network, Elsevier, Eolss Publishers, Globus Infografik GmbH, Intergovernmental Panel on Climate Change (Secretariat), Island Press, The Johns Hopkins University Press, Kunert AG, New Economics Foundation, Seppo Leinonen, Dennis Meadows, Oxford University Press, Inc, Royal Swedish Academy of Sciences, Springer Science and Business Media, Alexander Steinbrink, Stora Enso Kabel GmbH, Taylor & Francis Group, UNEP/GRID-Arendal, Vislab Wuppertal Institute for Climate, Environment and Energy, Göran Wall, John Wiley & Sons, The World Bank, and Klaus W. Zimmermann is gratefully acknowledged. There are instances where we have been unable to trace or contact the copyright holder. If notified, the publisher will be pleased to rectify any errors or omissions at the earliest opportunity.

Acronyms and Abbreviations

ACCA Association of Chartered Certified Accountants
ADB Asian Development Bank
AIDS Acquired immune deficiency syndrome
BOD Biological oxygen demand
BSE Bovine spongiform encephalopathy
COD Chemical oxygen demand
CAC Command and control
CARICOM Caribbean Community
CBA Cost-benefit analysis
CCICED China Council for International
Cooperation on Environment and Development
CFC Chlorofluorocarbons
CGE Computable general equilibrium
CSR Corporate social responsibility
DM Deutsche mark
DMC Domestic Material Consumption
DMI Direct Material Input
DPO Domestic Processed Output
DPSIR Driving forces, pressures, states, impacts, responses
DSR Driving-force, stress, response
E Exponent
ECF Environmentally adjusted net Capital Formation
EDP Environmentally adjusted net Domestic Product
EE Eco-efficiency
EEA European Environment Agency
EF Ecological Footprint
EKC Environmental Kuznets Curve
ELR Environmental Loading Ratio
em$ emergy-dollars
EMA Environmental management accounts/accounting
EMAN Environmental Management Accounting Network
EMAS Environmental Management and Audit Scheme
EMC Environmentally weighted Material Consumption

EMS Environmental management system
Eolss Encyclopedia of life support systems
EPA (US) Environmental Protection Agency
ESI Environmental Sustainability Index
ETR Ecological tax reform
EU European Union
EVA Environmentally adjusted net Value Added
EYR Emergy Yield Ratio
FAO Food and Agriculture Organization
FDES Framework for the Development of Environment Statistics
FR Further reading
FSDI Framework for Sustainable Development Indicators
FSDS Framework for developing and integrating
Social and Demographic Statistics
FSI Framework for Statistical Integration
GATS General Agreement on Trade in Services
GATT General Agreement on Tariffs and Trade
GDP Gross domestic product
Gen Genesis
GEO Global Environmental Outlook
GIS Geographic information system
GNP Gross national product
GOS Gross operating surplus
GPI Genuine Progress Indicator
gr. Greek
GREENSTAMP GREEned National STAtistical and Modelling Procedures
GRI Global Reporting Initiative
GWP Gross world product
HDI Human Development Index
HIV Human immunodeficiency virus
ICLEI International Council for Local Environmental Initiatives
IDRC International Development Research Centre
IDS International Development Strategies
IISD International Institute for Sustainable Development
ILO International Labour Organization
IMEI Index of Material and Energy Intensity
IMF International Monetary Fund
IPAT Impact = population x affluence x technology
IPCC Intergovernmental Panel on Climate Change
ISEE International Society for Ecological Economics
ISEW Index of Sustainable Economic Welfare
ISO International Organization for Standardization
IUCN International Union for Conservation of Nature and Natural Resources
LCA Life cycle analysis
LDC Least developed country

LTG Limits to growth
MDG Millennium Development Goals
MEA Multi-lateral environmental agreements
MEB Material and energy balances
MEW Measure of Economic Welfare
MFA Material flow accounts
MI Material input
MIT Massachusetts Institute of Technology
n/a not applicable
NAMEA National Accounting Matrix including Environmental Accounts
NAS (1) National Academy of Sciences
NAS (2) Net Additions to Stock
NCF Net capital formation
NDP Net domestic product
NEW Net Economic Welfare
NGO Non-governmental organization
NI National income
NIC Newly industrializing country
NIEO New International Economic Order
NLG Dutch gulden
NPISH Non-profit institutions serving households
OECD Organisation for Economic Co-operation and Development
p.a. per annum
para. paragraph
p.c. per capita
ppm parts per million
PIOT Physical input-output tables
PPP Polluter-pays principles
PRED Population, resources, environment and development
PSR Pressure, state, response
ROW Rest of the world
R&D Research and development
SAM Social Accounting Matrices
SAR Special Administration Region (China)
SDA Structural decomposition analysis
SDI Sustainable Development Index
sej solar energy joules
SEK Swedish krona
SEEA System for integrated Environmental and Economic Accounting
SEPA State Environmental Protection Agency (China)
SESAME System of Economic and Social Accounting Matrices and Extension
SIA Strategic impact assessment
SNA System of National Accounts
SSDS System of Social and Demographic Statistics
SUT Supply and use tables

TEC (WTO) Committee on Trade and Environment
TMO Total Material Output
TMR Total Material Requirement
TNC Trans-national corporation
TRIPS Trade-related aspects of intellectual property rights
UBA Umweltbundesamt [Federal Environmental Agency] (Germany)
UK United Kingdom (of Great Britain and Northern Ireland)
UNCED United Nations Conference on Environment and Development
UNCTAD United Nations Conference on Trade and Development
UNDP United Nations Development Programme
UNECE United Nations Economic Commission for Europe
UNEP United Nations Environment Programme
UNESCO United Nations Educational, Scientific and Cultural Organization
UNFCCC United Nations Framework Convention on Climate Change
UPP User-pays principle
US United States (of America)
USA United States of America
VA Value added
WCED World Commission on Environment and Development
WEO World Environment Organization
WHO World Health Organization
WMO World Meteorological Organization
WRI World Resources Institute
WSSD World Summit on Sustainable Development
WTO World Trade Organization
WWF World Wildlife Fund
WWI World Watch Institute
ZERI Zero Emissions Research Initiative

Contents

List of Boxes

List of Colour Plates

List of Figures

List of Tables

Part I
Questions, Questions, Questions

This introductory part raises the main questions, which the book seeks to answer. Chapter 1 identifies the planet's environmental problems and describes defensive action by the international community. The scattered evidence does not confirm predictions of environmental doom; it does reveal, though, human responsibility for environmental deterioration.

Chapter 2 answers the question about the role of economics as a matter of interaction between economic activities and the provision of environmental services; both also affect human welfare. Among different schools of environmental-economic analysis, two approaches represent a fundamental dichotomy between environmental (market-oriented) and ecological (market-sceptical) economists. Both schools want to maintain environmental services and human well-being but offer different concepts of these maintenance goals. *Economic sustainability* relies on produced and natural capital maintenance; *ecological sustainability* seeks to reduce the burden on the environment by a dematerialized economy. The term '*eco*–nomics' stands for both schools and sustainability concepts. Parts II, III and IV will look for ways of bridging, or at least clarifying, this dichotomy by quantitative assessment and analysis.

The picture gets more complicated when introducing further social, cultural and political goals into the sustainability discussion. The resulting popular paradigm of *sustainable development* is opaque and suffers from an implementation deficit. This is the reason why Chapter 3 dares to ask whether the paradigm has run its course. The final chapter of the book raises all these questions again. It will provide some answers without pretending to know them all.

Chapter 1
What on Earth is Wrong?

Environmental doom-and-gloom literature created awareness of environmental problems, as well as advocacy for environmental action. The international response produced declarations, action plans and conventions. Global conferences propagated the paradigm of 'sustainable development' but did not succeed in penetrating economic policy.

Vision, advocacy and action plans are important means of spreading the idea of sustainable development. They need to be questioned and modified if facts and figures do not support their predictions and strategies. Available indicators and reports do show symptoms of environmental *non*-sustainability of particular economic activities. They are inconclusive as to the overall effect on human welfare and the sustainability of economic growth and development. Extended economic analysis (Ch. 2) provides the framework for assessing sustainability and its benefits.

1.1 Paradise Lost

It began with innocence lost. Human awareness of good and evil was punished by extradition from paradisiacal harmony with nature. Together with the biblical call to 'subdue the earth' (Gen 1:28), this powerful metaphor dramatizes human aggression of nature by technology and unrestrained proliferation. Forty years ago, White (1967) set off a heated debate with his claim that Christian arrogance towards nature is responsible for the contemporary environmental crisis. Most religions embrace now the notion of stewardship of the environment by the current generation for future generations [FR 1.1].[1]

Far beyond the reach of Judaeo-Christian mythology, environmental destruction and catastrophe show the cost of human ingenuity in exploiting nature's resources.

---

[1] References to the further-reading section at the end of each chapter are shown in brackets as FR and section number.

P. Bartelmus, *Quantitative Eco-nomics*,

Overuse of natural resources contributed to the downfall of ancient cultures and empires like Mesopotamia, classic Maya and the Roman Empire. Land degradation, brought about by overirrigation and ensuing water logging and salinization, is the main reason for the breakdown of agricultural systems. Overpopulation, overtaxation, rebellion and war are socio-political factors in the collapse of ancient societies. In the eighteenth and nineteenth centuries the search for new sources of natural and human-made wealth drove the needs of warring and colonizing Europe [FR 1.2]. Among others, securing energy supply has been a motive for the current war in Iraq. Perhaps the most ominous development is that technological advances in harnessing nuclear power and genetic resources have now the power to endanger the survival of the planet.

Is apocalypse the inevitable consequence of heeding the biblical advice? Or are environmental concerns just another bug in our social systems geared towards the creation of ever-greater wealth? Can paradise be regained or at least some semblance of it re-established? When and where, and for whom? The answers range from predictions of environmental doomsday, calling for a new environmental ethics [FR 13.3], to faith in technological progress. Obviously, we need to examine these proclamations with hard facts and figures – hence the book's focus on quantification. The need to bring in economics may not be that obvious at first sight. A closer look at the environmental conundrum reveals, however, that economic activities can be both the cause of the problem and part of its solution.

The following section sets out, therefore, from an examination of early doomsday scenarios and international reactions and responses. Next, key indicators of the state of the environment are assessed as to their capacity of alerting to possibly disastrous transgressions of environmental thresholds. The purpose is to set the stage for examining (in Ch. 2) the ability of economics to deal with environmental limits in our quest for prosperity.

1.2 Environmental Doomsday and International Reaction[2]

Conspicuous pollution incidents in the 1960s and neo-Malthusian views of demographic and economic growth led to the appearance of environmental doomsday literature. Titles like *The Death of Tomorrow* (Loraine, 1972), *Silent Spring* (Carson, 1965), *Blueprint for Survival* (Goldsmith et al., 1972), or *Conservation for Survival* (Curry-Lindahl, 1972) are indicative of the environmental mood in the late 1960s and early 1970s. The use of a seemingly objective computerized global model gained widespread attention for the Club of Rome's *Limits to Growth* report (Meadows et al., 1972). The model predicted 'a rather sudden and uncontrollable

[2] Most of the first part of this section is (with some modifications) from Bartelmus (1994a, pp. 5–8; with permission by the copyright holder, Taylor & Francis).

decline in both population and industrial capacity' within the current century if growth trends remain unchanged. To avoid the disastrous consequences of transgressing these limits the authors called for 'a controlled, orderly transition from growth to global equilibrium'. Chapter 11 will critically review the assumptions and results of the model.

All these publications deserve credit for creating awareness of environmental concerns and alerting us to potentially disastrous trends of environmental deterioration. However, countries in the early stages of economic development could not accept zero-growth strategies with an exclusive focus on ecosystems. For them, improving the standards of living appeared to be more important than concern about wildlife or global pollution. In their view, only affluent countries could afford the luxury of diverting some of their wealth to environmental protection. Moreover, the high and wasteful consumption of the industrialized nations generated most of the stress on the resources of poor countries. Developing countries thus reacted with suspicion to proclamations of global solidarity for our planetary home. The only view rich and poor countries seemed to share at the time was the conviction that environmental conservation and economic development are in conflict.

The international community opened the dialogue on environment and development between developed and developing countries. A preparatory seminar for the global United Nations Conference on the Human Environment (Stockholm, 5–16 June 1972) concluded that environmental problems result not only from the development process itself but also from the very lack of development (United Nations Conference on the Human Environment, 1972). Poor countries have to cope with lack of clean water, inadequate housing and sanitation, malnutrition, disease and natural disasters. The metaphor 'pollution of poverty' illustrates this aspect of the environmental question. Consequently, environmental goals should provide a new dimension to the development concept. The Conference itself endorsed the principle of integrating environment and development. It also established a small, but rapidly expanding secretariat, the United Nations Environment Programme (UNEP) to implement and monitor an Action Plan for the Human Environment (United Nations, 1973).

Despite the call for integrating environment and development, integration did not take place. Issues of population growth and urbanization, economic development, desertification, pollution and resource exploitation continued to be the responsibility of specialized departments, while macroeconomic policies focused on maximizing economic growth. Relatively weak environmental agencies addressed environmental impacts, albeit without much influence on socio-economic decision-making by the central government.

'A widespread feeling of frustration and inadequacy in the international community about our own ability to address the vital global issues and deal effectively with them' (WCED, 1987) motivated, therefore, the United Nations to establish a World Commission on Environment and Development (WCED). Under the generic label of sustainable development, the WCED proposed a large variety of policy recommendations that should meet 'critical objectives' for such development. The objectives included:

- Reviving growth while changing the quality of growth
- Meeting essential needs for jobs, food, energy, water and sanitation
- Conserving and enhancing the resource base
- Reorienting technology and managing risk, and
- Merging environment and economics in decision-making.

The idea of effectively merging environmental protection into socio-economic planning and policies had been discussed extensively in the wake of the Stockholm Conference. The WCED advanced, however, a new approach for implementing the integration of environment and development. The idea was to move from dealing with environmental effects, after their occurrence, to focusing on the 'policy sources' of these effects for preventive action. This approach shifts the discussion from *environment* and development to *development* and environment. The purpose is to include environmental issues in mainstream policy rather than to change socio-economic policies from the periphery of the environmental movement.

In follow-up to the WCED recommendations, the 1992 United Nations Conference on Environment and Development (UNCED), the Earth Summit in Rio de Janeiro, attempted to translate the new paradigm of sustainable development into a globally adopted philosophy, an Earth Charter, and an international action programme (United Nations, 1994). Figure 1.1 provides a synopsis of the results of UNCED, comprising a watered-down Rio Declaration (as compared to a Charter) [FR 13.3], the action plan of Agenda 21, the adoption of two conventions on biodiversity and climate change, and a statement of forest principles. Immediate reactions to UNCED differed widely (Bartelmus, 1994a), ranging from

- Describing Agenda 21, as 'the most comprehensive, the most far-reaching and, if implemented, the most effective programme of international action ever sanctioned by the international community' (closing statement by the Conference's Secretary General, Maurice Strong) to
- Considering the Conference as 'a failure of historic proportions' (Greenpeace summary critique of UNCED results).

Five years after the Rio Summit, disillusion spread widely. The special session of the United Nations General Assembly, known as Rio + 5, achieved, in the words of its President Ismail Razali, an 'honest appraisal' of meagre progress (Osborn & Bigg, 1998). Most governments did not commit to implementing Agenda 21 and the Rio conventions. Contrary to the North's promises in Rio, 'new and additional' resources for the implementation of Agenda 21 had not come forth (with notable exceptions), and official development aid decreased in general. A renewed focus on economic growth, thinly veiled by sustainability rhetoric apparently prevailed.

It comes as no surprise, therefore, that the Plan of Implementation of the 2002 World Summit on Sustainable Development (WSSD) in Johannesburg (United Nations, 2003) presented mostly a perfunctory summary of Rio's Agenda 21. The declared objectives of the WSSD were 'to take stock' since Rio and foster implementation by means of work plans and new 'public-private partnerships'. It remains to be seen whether explicit targets (for sanitation, biodiversity, use of chemicals, and

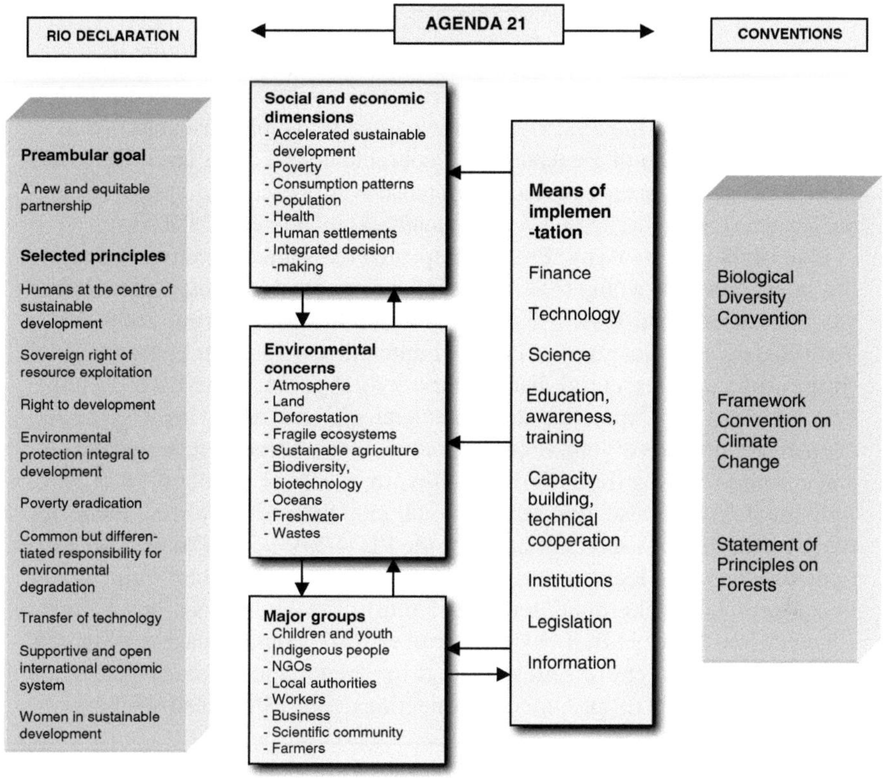

Fig. 1.1 Results of the Rio Earth Summit
Source: Bartelmus (1994a, fig. 6.1, p. 146; with permission by the copyright holder, Taylor & Francis).

harvesting of fish stocks), the inclusion of new topics (energy, transport, globalization), and some focus on regions, sustainable production and consumption, and poverty can overcome global lethargy.[3] At least the United Nations Framework Convention on Climate Change (UNFCCC) has now been translated into a concrete – some will say insufficient – commitment by governments through the ratification of the Convention's Kyoto Protocol (Box 1.1).

Judging from the flurry of publications on the sustainability of economic growth and development, there seems to be no consensus on the exact meaning and implications of these concepts. It is easier, therefore, to look first into the main symptoms of non-sustainability, which after all gave rise to the call for sustainable development. Chapters 2 and 3 will then explore possibilities of defining sustainability in more operational terms.

[3] Further information on the results and follow-up to the Summit can be found on www.un.org/esa/sustdev. For a more critical evaluation of the Summit outcomes, see WWI (2003).

Box 1.1 Framework convention and Kyoto Protocol on climate change

At the first Earth Summit in 1992, the international community adopted the *United Nations Framework Convention on Climate Change*. Its 'ultimate objective' is to 'achieve stabilization of greenhouse gas concentrations ... at a level that would prevent dangerous anthropogenic interference with the climate system' (http://unfccc.int/essential_background/convention/background/items/2853.php).

Five years later its Kyoto Protocol replaced the vague objective of dangerous interference by a target for industrialized countries to reduce greenhouse gas emissions by at least 5% below total 1990 levels during 2008–2012. With Russia's ratification the Protocol entered into force in February 2005 (http://unfccc.int/essential_background/kyoto_protocol/items/2830.php). The Protocol also specifies key 'mechanisms' for achieving this target: cooperative projects of joint implementation, clean development mechanism, and emission trading (http://unfccc.int/kyoto_mechanisms/items/1673.php). Individual greenhouse gas reduction targets for industrialized countries range from −8% of 1990 emissions for the EU (USA: −7%, Protocol not ratified) to +10% for Iceland during the 2008–2012 period (http://unfccc.int/kyoto_protocol/background/items/3145.php). The United Nations Climate Change Conference in Bali (3–14 December 2007) could not agree on targets for the post-Kyoto era; it settled instead for 'negotiations' to this end to be concluded by 2009 (http://unfccc.int/meetings/cop_13/items/4049.php).

1.3 Reaching the Limits?

The above-cited doomsday literature and subsequent international environmental conferences drew attention to the sorry state of the environment. Activist individuals and groups like Greenpeace, the World Watch Institute or the Club of Rome keep the environmental movement alive with unrelenting warnings about reaching the limit of the earth's carrying capacity. Box 1.2 presents typical proclamations about imminent environmental calamity.

Environmental indicators – like those in Plate 1.1 – have been put forth as evidence of environmental deterioration. Typically these indicators refer to three main categories of environmental impacts:

- Natural resource depletion – of forests, fish, soil/land, minerals, metals and water
- Degradation of ecosystems – involving loss of species, genetic resources and wilderness
- Pollution – either local (air, water, waste) or global (greenhouse gas emission and climate change, ozone depleting substances).

Add population growth and hunger, and you obtain what one 'skeptical environmentalist' calls 'the Litany of our ever-deteriorating environment' (Lomborg, 2001). The reactions by environmentalists to Lomborg's claim that we have mostly expe-

Box 1.2 Reaching the limits? Some warnings

- When the last tree is cut, the last river poisoned, and the last fish is dead we will discover that we cannot eat money (*Greenpeace*: www.greenpeace.org).
- Climate change and global warming are matters of life and death; increasing levels of air pollution threaten the survival of nature and the well-being of people around the world (*World Wide Fund for Nature*: http://www.panda.org/about_wwf/what_we_do/index.cfm).
- Our world is in a state of pervasive ecological decline; our current economies are toxic, destructive on a gargantuan scale, and grossly unfair (*WWI*, 2003).
- New insights have arisen, which not only confirm the impending disasters but also indicate that the limits to growth may well have been exceeded (*van Dieren*, 1995).
- Exponential growth has taken us from a relatively empty world to a relatively full world – full of people and our things, empty of natural life-support systems (*Daly*, 1996).
- What happens here on Earth could make the difference between a near-eternity filled with evermore complex and subtle forms of life and one filled with nothing but base matter (*Rees*, 2003).
- Humans are fundamentally, and to some extent irreversibly, changing the diversity of life on Earth (*Millennium Ecosystem Assessment*, 2005)

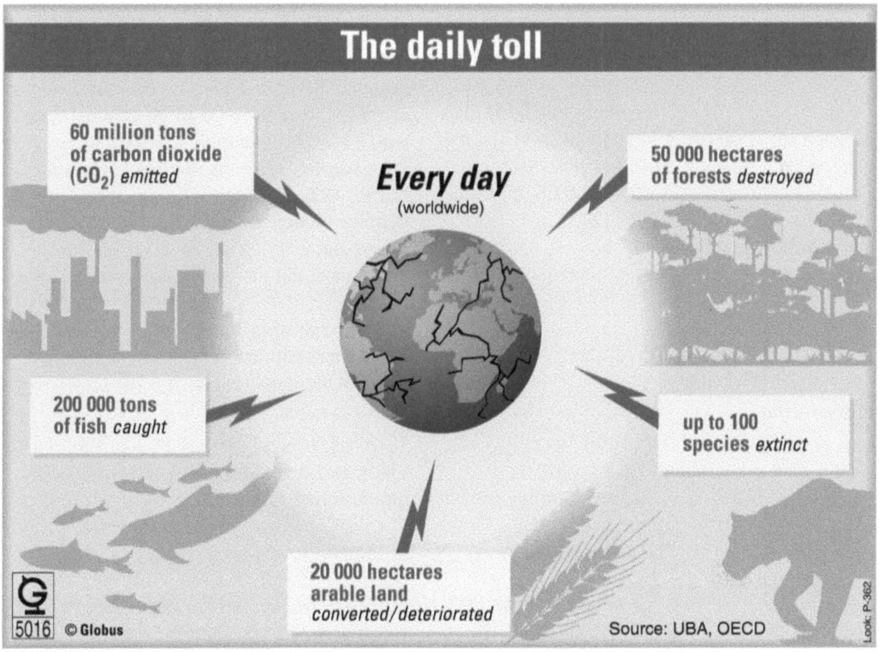

Plate 1.1 Environmental Indicators (*See Colour Plates*)
Source: Globus Infografic GmbH.

rienced an improvement rather than decline in these issues were harsh [FR 1.3]. Clearly, there is a need for assessing the validity of available data and their interpretations. How close are we indeed to life-threatening environmental limits? The World Resources Institute once proclaimed that in the field of environment one should 'let the facts speak for themselves' (WRI, 1992). But do they, and, if so, are they understandable?

Table 1.1 presents a more detailed (than Plate 1.1) but still selective list of environmental concerns and indicators. The indicators are typical for demonstrating the non-sustainability of global economic growth and development. They cover different aspects of the same concern, often in different units of measurement.

Table 1.1 Indicators of global non-sustainability

Environmental Concern/Indicator	Estimate [Reference]	Evaluation
Climate Change		
- CO_2 emissions (billion tons p.a.)	26.4 (2000–2005) [5] 47.3 (2100, no governmental control) [10] 110 (2100, worst cases) [5]	- Anthropocentric warming and sea level rise continue for centuries, even if greenhouse gas concentrations were stabilized [5, p. 11]. - Ecosystems, human health and economy are all sensitive to climate change; many regions are adversely affected, some effects are beneficial for some regions [3, p. 215].
- CO_2 concentration, increase (%)	26 (since pre-industrial times) [5] 32.5 [4]	- Climate change is an overriding challenge facing our global civilization [4a, p. 16].
- Global warming (° C)	0.74 (1906–2005) [5] **1.8–4.0 (by 2100)** [5] 2–2.5 (likely, by 2100) [2] 5–6 (likely by 2050) [12]	- Temperature increase of 0.6 °C is not a dramatic divergence from previous centuries; it will be far more expensive to cut CO_2 emissions radically than to pay the cost of adaptation to global warming [2, pp. 317, 318].
- Average sea level rise (cm)	17 (1900–2000) [5] 28–43 (1980/1999–2090/2099) [5]	
- Cost with 5–6 °C global warming (% of world GDP)	5–10 [12]	- Climate change could disrupt economic and social activity at a level similar to the great wars and the depression of the 1930s [12, p. ii]. - Climate change is unlikely to be catastrophic in the near term, but potentially highly damaging in the long run [10, p. 178].
Deforestation		
- Rate (million ha p.a.)	9.4 (1990–2000) [3] **13 (net loss: 7.3) (2000–2005)** [3a] 18 [Fig. 1.1]	- Tropical deforestation of 11.3 million ha vindicates fears about alarming rate of global forest loss [3, p. 91]. - Tropical countries lose more than 15 million ha a year (according to United Nations reports) [4, p. xx].
- Change in forest cover of total land area (%)	- 53.4 (original to 1996) [1] - 20–25 (original to 997) [2] - 0.85 (1950–1994) [2] - 0.9 (2000–2005) [6]	- Basically forests are not under threat: forest area has not changed since Second World War [2, p. 117].

(continued)

Table 1.1 (continued)

Environmental Concern/Indicator	Estimate [Reference]	Evaluation
		- In most countries the marketed values of ecosystems associated with timber and fuel wood are less than one third of the total economic value of forests [11, p. 9].
Species Loss		
- Total, inclusive unknown (million)	14 [3]; 2–80 [2]	- Global biodiversity is changing at an unprecedented rate; decline and extinction of species have emerged as major environmental issues [3, p. 121].
- Threatened (no., %)	3,679 (vertebrate species) [3] 16,118 (2006) [7] 10–30%[11]	- Losing 0.7% of species per 50 years over a limited time span is not a catastrophe but a problem [2, p. 257].
- Extinct (no.)	**68 (since 1970)** [3] 1033 (since 1600) [2] 40,000 (p.a.) [Fig. 1.1]	
Land Loss		
- Agricultural production p.c., increase (%)	52 (since 1961) [2]	- No clear indication that the rate of land degradation has decreased [3, p. 64].
- Loss p.a. (million ha)	7.2 (Fig. 1.1)	- During 1985-1995 the trend showed population growth racing ahead of food production in many parts of the world [3, p. 62].
- Degraded land per usable land area (%)	**23** [3] 17 (of all land) [2]	
- Starvation in developing countries (million people)	920 (1971) [2] 792 (1997) [2] 824 [9]	- There is no imminent agricultural crisis or any approaching scarcity of food [2, p. 109].
Overfishing		
- Annual fish catch (million tons)	18 (1950) [2] 19.3 (1950) [6a] 80 (since 1980s) [3a] 134 (2002) [6a] 72 (Fig. 1.1)	- Decade-long decline in the global fish harvest [4, p. xx]. - Exploitation of living marine resources and loss of habitats are now recognized as being at least as great a threat to oceans as marine pollution [3, p. 180].
- Fish stocks depleted, overexploited or recovering (% of total)	**27** [3] 70 [8]	- 10 million tons of catch foregone, the price of overfishing, is equivalent to 19 days of increased agricultural production [2, p. 108]. - The proportion of overexploited stocks increased from 10% (mid-1970s) to about 24% (2004, stable since the 1990s) [6a, section A.1].
Water Scarcity		
- Water availability p.c., p.a. (m^3)	8,549 (2005, actual renewable water resources) [1] 2,052 (1996, total accessible runoff) [2]	- More countries are facing water stress or scarcity [3. p. 157]. - Total use of water is less than 17% of accessible water and will require just 22% in 2015; basicaly we have enough water [2, pp. 149, 150].
- Population in countries or regions with water shortage (% of world population)	3.7 (chronic water scarcity) [2] **40** (serious water shortage) [3]	- Water withdrawal, to which the majority of the global population has access, amounts to 40–50% of the continental runoff [11, p. 106].

(continued)

Table 1.1 (continued)

Environmental Concern/Indicator	Estimate [Reference]	Evaluation
Ozone Layer Depletion		
- CFC 11, 12, 113 production (million tons)	0.04 (1950) [3] 0.06 (1970) [3] 1.04 (1988) [3] 0.14 (1995) [3]	- Ozone layer depletion has now reached record levels; return to per-1980 levels by mid-21st century [3, pp. 212, 213]. - Today we have pretty much done what we can [2, p. 274].
- Ozone layer decline(%)	3–6 (1998, below 1979) [2] **5–6** (current, in mid-latitudes) [3]	

References: [1]=WRI (2006); [2]=Lomborg (2001); [3]=UNEP (2002); [3a]=UNEP (2006); [4]=WWI (2003); [4a]=WWI (2000); [5]=IPCC (2007); [6]=FAO (2005b); [6a]=FAO (2005a); [7]=IUCN (2006); [8]=World Bank (2003); [9] United Nations (2006); [10]=Nordhaus and Boyer (2000); [11]=Millennium Ecosystem Assessment (2005); [12]=Stern (2007). In turn, many sources are based on primary data from national and international statistics. More commonly accepted data are shown in bold.

The last column of the table indicates the variety of sometimes contradictory conclusions about environmental impacts and the sustainability of economic activity. As a consequence, some of the so-called facts might indeed raise more questions than answers, for instance,

- What are the likely consequences of different degrees of global warming?
- Is the global forest cover decreasing or increasing? Where and when?
- How does human-induced species loss compare to natural losses, and what is the value of these losses?
- Are soil erosion and land degradation harbingers of increasing starvation?
- Is the depletion of fish stocks more of an economic or ecological (species and habitat loss) problem?
- Is local water scarcity mainly a management problem of facilitating access to available water resources?
- Did we solve the ozone-layer-depletion problem?

The list of questions could be easily extended. The reports shown as references in Table 1.1 are among many more that raise environmental issues and suggest how to tackle them [FR 1.3]. Other indicators might cover further environmental concerns, and differing criteria could be used to assess the sustainability of human activities at different regional levels. For years, politicians, researchers, the public media and the general public have been exposed to an information overload of hardly comparable numbers, tonnes, kilowatts, centimetres, ppm, cubic metres, or hectares.

There is a need to reduce this overload, but indicators and indices of sustainable development, environmental quality, quality of life or genuine social progress proliferate. The reason is that we still lack internationally agreed concepts and statistical standards, and in particular a quantifiable notion of sustainability. In many cases the purpose of advancing new indicators seems to be more to disqualify conventional economic indicators and policies than to support scientific measurement and rational decision-making (see Chs. 4 and 5).

For now, Table 1.1 identifies a broad set of commonly cited environmental impacts (in bold) as evidence for the non-sustainability of human activity on the planet. These impacts include, in particular:

- 1.4–5.8 ° C of global warming by the year 2100
- A net loss of 7.3 million ha per annum of forest cover
- A loss of 68 species since 1970
- Degradation of 23% of usable land area
- Overexploitation of 27% of fish stocks
- 40% of the world population facing serious water shortage
- 5–6% of ozone layer decline in the mid-latitudes of the earth.

It is far from clear whether these data indicate non-sustainability of economic performance and growth; nor do they give a conclusive picture of the comparative and overall severity of environmental problems. However, most environmentalists interpret these data as indicators of looming disaster. For instance, Daly (1996) sets out from his 'pre-analytic vision' of an expanding human subsystem of nature's overall system and sees a previously 'empty world' as 'full' with disastrous consequences (Plate 1.2).

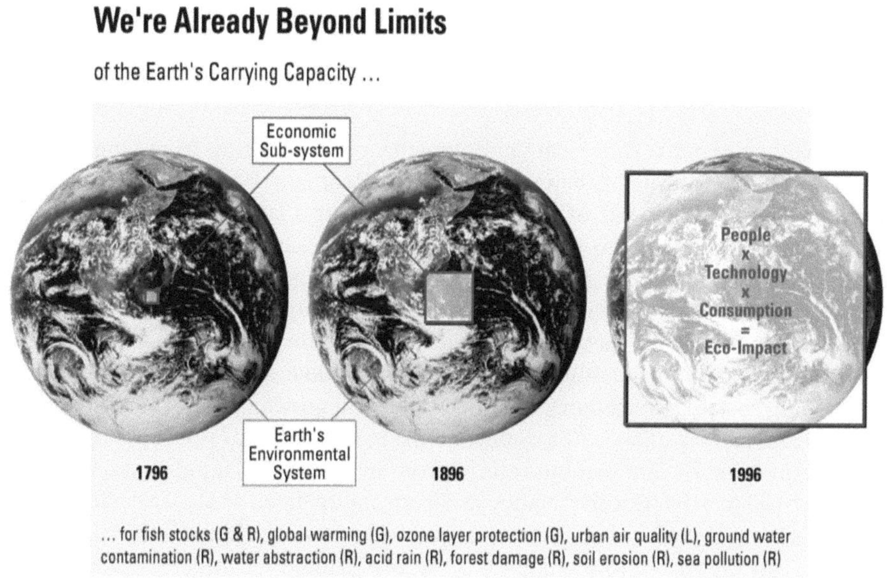

Plate 1.2 Full World? (*See Colour Plates*)
Source: Based on Daly (1996); copyright VisLab/Wuppertal Institute for Climate, Environment and Energy; with permission by the copyright holder.

The merit of this reasoning is that it brings into the open hidden convictions, which appear to motivate the normative and sometimes moralistic argumentation of environmentalists. Later chapters will examine the mixing of norms and science (Sections 3.3.3, 13.4.2). For now, we have to leave unanswered this section's question about reaching the limits. Part IV of the book will attempt to find some answers using the more systematic compilation of accounting indicators described in Parts II and III.

Further Reading

FR 1.1 Religion and Environment

Gore (1993), among others, attempted to refute White's (1967, reprinted in Gottlieb 1996) claim of Christian-faith motivated exploitation of nature, distinguishing biblical 'dominion', which calls for stewardship of the Earth, from 'domination'. Gardner (2003) expects the greening of religions to overcome the 'chasm between science and spirituality'. Gottlieb (1996) provides an overview of the responses by religions to the environmental problem. Daly (1996) maintains that religious insight leads us to sustainability and its associated principles of sufficiency, equity and efficiency. See Section 13.4.2 for a brief discussion of environmental ethics and sustainable development.

FR 1.2 Environmental Destruction and Collapse of Societies

Archaeological records show that environmental crisis resulting from human interaction with nature is not a new phenomenon (Redman, 1999). Hardesty (2001) describes the nature and use of such records, depicting a large variety of changes in life support systems due to climate fluctuations, population growth, agricultural failure and warfare. Deforestation, for example, is a significant contributing factor in the collapse of ancient Greece and Rome, according to Hughes and Thirgood (1982). Diamond (2005) describes selected cases of ancient and modern societal collapse; he also draws lessons for changing individual and governmental values for attaining sustainable resource use and population growth. Tainter (1988) considers the sustainability of societies as beyond the maintenance of life support systems and as a question of diminishing returns from increasingly complex societies. He also describes modern Europe's lucky escape from the fates of ancient civilizations due to innovation and colonization (see for an overview, Tainter 2001). Hughes (2000) stresses the need to include the – largely ignored – ecological process as a major theme in a new narrative of world history.

FR 1.3 Reports of the State of the Environment

Proliferating environmental and sustainable development indicators (see Section 4.2) brought about widely differing assessments in national and *international environmental reports*. UNEP coordinates a number of global reports under its Earthwatch programme (http://earthwatch.unep.net/world/index.php), including the Millennium Ecosystem Assessment (2005), the annual GEO Yearbooks (UNEP, 2006) and the biennial World Resources Series (UNDP et al., 2003). Most of the statistics presented in these reports are compiled by the specialized agencies of the United Nations, notably FAO, WHO, UNESCO and WMO; their databases can be found on their websites. Regional organizations such as the regional commissions of the United Nations, the OECD (for industrialized countries) and the European Environment Agency (EEA) also publish environmental reports and compile data through their statistical offices.

Other *non-governmental organizations* (NGOs) tend to be more advocatory in their views of the (dismal) state and trends of the environment or the world (see the respective publications and websites of the WWF, WWI or the Club of Rome). They (and to some extent also the above-mentioned intergovernmental organizations) have been accused, notably by the irreverent 'skeptical environmentalist' (Lomborg, 2001) of bias in painting an overly pessimistic picture. The ensuing heated debate is at least an eye-opener on widely differing conclusions, often from the same data, on the state of the world and the significance of environmental problems for human well-being. See for the critique of Lomborg the *Scientific American* (January 2002) and www.anti-lomborg.com, and for the author's rebuttal www.lomborg.com.

Review and Exploration

- What are the roles of religion and spirituality in assessing and tackling environmental questions? Can we overcome the chasm between 'rational' science and 'moralistic' religion?
- Are natural resource constraints a leading cause for war? Explore the motives and impacts of imperialism and colonization in this regard.
- Environmental protection: a luxury of the rich? Explain 'pollution of poverty'.
- Do you think our planet is at risk of being destroyed? What do statistics tell us? Select a topic of Lomborg's 'litany' and assess his data in the light of critique and rebuttal.
- Assess the key aspects of the state of the environment in your home country, province or town. What are the international and global implications? What should we do about it?
- Assess the progress made by the international community in tackling environmental problems and fostering sustainable development; check out the United Nations' mega conferences (from Stockholm, via Rio, to Johannesburg).

Chapter 2
What's Economics Got to Do with It?

Much of this book deals with the question of how to reduce the above-mentioned information overload while generating more relevant information for integrative long-term environmental and economic planning and decision-making. Economic theory and statistics have well-established techniques for compressing scattered data in a systematic fashion. This chapter reviews what economic thought has to offer for assessing the environment–economy interface. The result is quantifiable concepts and definitions of the sustainability of economic performance and growth. Chapter 3 extends this analysis to include social and institutional aspects of 'development'.

2.1 Economics Out of Sync?[1]

Bashing economics for wrong diagnoses and projections and misleading policy advice has tradition. The oil crises of the 1970s, social upheaval following structural adjustment in developing countries, the chaotic transition from centrally planned to market economies, and conspicuous environmental impacts of economic activity cast doubt on the predictive and advisory capability of conventional economic analysis. Galbraith (1986) launched an eloquent critique of the basic tenets of neoclassical economics, viz. optimal resource use under ideal conditions of atomistic markets. It is quite curious that Galbraith's attack on the fundamentals of widely taught and applied economics has brought about so little change in economic policy analysis. Galbraith himself pointed out that this is due to an alliance of 'mature' corporations and government; both hide their common goals of economic growth and power behind the screen of allegiance to – powerless – perfect competition. Much later, a growing discontent with the post-communist resurgence of neo-liberal *laissez-faire* economics castigated preference for formalistic rigour over real-world vision [FR 2.1].

[1] Parts of this section are from Bartelmus (1997b).

P. Bartelmus, *Quantitative Eco-nomics,*

In the field of environment, 'ecological' economists[2] Funtowicz and Ravetz (1991) claim that potentially irreversible environmental impacts and externalities make mainstream economics irrelevant. As one of the protagonists of ecological economics puts it: 'the planetary boat might sink if overloaded by people and their useful and wasteful things, however optimal the distribution of its load' (Daly, 1996). Conventional economics is thus seen to be in denial, as it clings to its formalistic axioms of rational behaviour in perfectly competitive markets. Experiments, testing the rationality axiom of neoclassical economics, came up indeed with cases of irrational behaviour. However, more comprehensive experimentation would be needed for drawing general conclusions about the possible demise of the profit and utility maximizing *homo oeconomicus* [FR 2.1].

Mainstream economists defend their basic rationality axiom. They suggest that action based on an ideal situation might contribute to achieving this situation, possibly by a 'sequence of policy reforms' (Dasgupta, 1994), or by using 'economics in a vacuum' to gain insight into complicated problems (Samuelson & Nordhaus, 1992). As long as nothing drastic happens, one can probably live with the 'semi-fiction' (Solow, 1992) of perfect markets. The admission of semi-fiction opens the door, however, to second-best solutions that might or might not take us closer to the elusive optimum of general equilibrium (Lipsey & Lancaster, 1956).

Yet drastic things do happen in the natural environment. Kapp (1950) and Mishan (1967) were among the first economists to warn us about environmental disruption from economic growth. Conventional economics typically dismissed, however, environmental phenomena as 'external' to market activities.[3] Mainstream economists thus tend to ignore evidence of numerous cases, where the inclusion of the 'social' costs of externalities generates total cost in excess of economic benefits (revenues) of production. A frequently cited example is the case of Kiribati, a Pacific Ocean island living off its phosphate deposits. Depletion of these deposits terminated all mining activities in 1982, and GDP dropped to less than half of its average level of the previous 4 years (OECD, 1985). Dismissing this issue as the concern of a small island will not do. Industrialized countries may have dumped a good deal of increasing environmental problems on financially starved developing countries. Importing natural resources, e.g. oil, fish or timber, and translocating unsafe and polluting industries is equivalent to exporting depletion and degradation to the Third World under the cloak of market liberalization.

If environmental phenomena were independent of economic activity, there would be no need to trouble economic analysis. Environmental and economic policies could each pursue their own agendas without risk of impairing each other's achievements. Environment and economy do interact, however. This, at least, can be derived even from the otherwise quite inconclusive assessments in Chapter 1. The World Commission on Environment and Development (WCED) came to a similar conclusion. According to the WCED (1987), policy failures in both environment and

[2] See Section 2.2.3 for the distinction between ecological and environmental economics.

[3] See Annex I, Section I.1 for the definition and categorization of externalities.

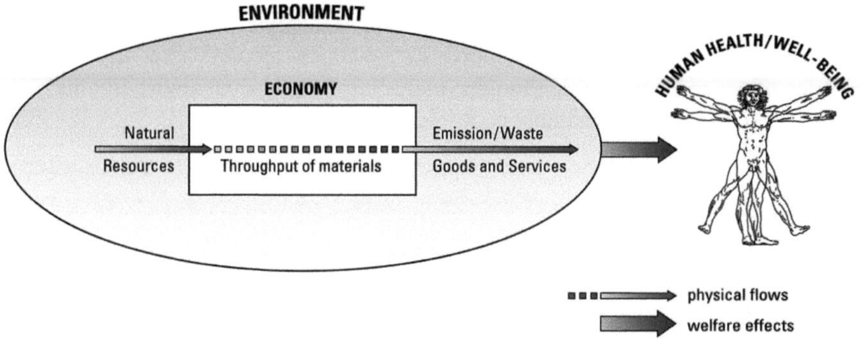

Fig. 2.1 Environment–economy interaction and effects
Source: Bartelmus (2001). Accounting for sustainability: Greening the national accounts, fig. 1, modified; Copyright Eolss Publishers; with permission by the copyright holder.

economic development are the result of neglecting economic and ecological 'interdependences' by compartmentalized line ministries and agencies.

Figure 2.1 is a stylized description of the environment–economy interaction. It describes this interaction in terms of the (re)source and sink (waste disposal), and welfare (human needs, health, life support) 'functions' of the environment (Hueting, 1980; de Groot, 1992; Ekins et al., 2003). The figure also indicates direct welfare effects from the consumption of produced and natural goods and services, and indirect effects on health and other amenities from environmental degradation.

Even if one questions the predictive and analytical power of economics, there can be no doubt that actual impacts and repercussions between economic activity and the environment reveal a new, or newly perceived, scarcity of environmental services. The only way to assess the scarcity of non-marketed goods and services and to compare it to that of market products is to draw non-market goods into the pricing system. Chapter 8 will show how green accounting can achieve this in a practical manner. Note however that – unconvinced – environmentalists reject any monetary valuation of the environment, giving rise to a serious and seemingly irreconcilable dispute between environmentalists and environmental economists (Section 2.2.3).

2.2 Schools of *Eco*–nomic Thought

Much of the discussion of the relevance of environmental phenomena for economic development and *vice versa* is still unsubstantiated. The questions left open in Chapter 1 resonate: is sustainable growth a bad oxymoron (Daly, 1991) or a *sine qua non* for development (Boutros-Ghali, 1995)? Are environmental externalities overwhelming economic analysis (Martinez-Alier, 1987), or can they be efficiently internalized in the budgets of households and enterprises? Various schools of 'green' economics tackle these questions. *Eco*–nomics, as used here, encompasses all shades of green in economic thought, including neoclassical environmental economics and more environmentalist ecological economics.

2.2.1 A Historical Overview

It is always useful to lend historical perspective to different lines of thought before ordering and comparing them as schools or domains of a broader discipline. Plate 2.1 provides a rough and necessarily incomplete indication of when and how economics and ecology attempted to bridge a profound gap between natural and social sciences. Considering economics as the art of managing scarce resources and ecology as the 'economics of nature' (Haeckel, 1898) gives a first indication of the potential relations between the two sciences. Haeckel (1866) is also credited with the first definition of ecology as the 'total science of the relationships of the organism with its surrounding outer world' (own translation). Referring to human organisms this definition gives us a generic definition of the *human* environment.

An early-eighteenth-century forestry and mining official from Saxony in Germany was probably the first to coin the notion of sustainability. In his *Sylvicultura Oeconomica*, von Carlowitz (1713) (Plate 2.2) called on humans to 'act with nature, and not against it'. Specifically, he postulated that the 'conservation and cultivation of timber should be conducted so as to provide a continuous, persistent and sustaining utilization' (own translation).

The eighteenth-century physiocrats, with their main protagonist François Quesnay, made, however, the first systematic and quantitative attempt at linking the power (gr. *kratos*) of nature (gr. *physis*) with the management of the national 'household'.[4] Quesnay himself called his famous *Tableau Economique* (Quesnay, 1759) a booklet of housekeeping (*livret de ménage*) (Kuczynski, 1971). The *Tableau*'s criss-crossing flows of money and product link landowners and 'productive' farmers with the 'sterile' class of industry and commerce. Even if the sterile class was to become the most productive one in the industrial revolution, the *Tableau* can be considered as a forerunner of environmental accounting. The *Tableau* also reflects ideas of sustainability in the relationships between society and economy and the process of their reproduction and maintenance.

Adam Smith (1776) dismissed the physiocratic 'Political Economy' as 'that system which … exists only in the speculation of a few men of great learning and ingenuity in France'. His derisive critique and the success of industrialization sent Quesnay's *Tableau*, and in fact environmental concerns in economics and accounting,

[4] The following description of the role of François Quesnay and Adam Smith in the analysis of environmental sustainability is taken from Bartelmus and Seifert (2003).

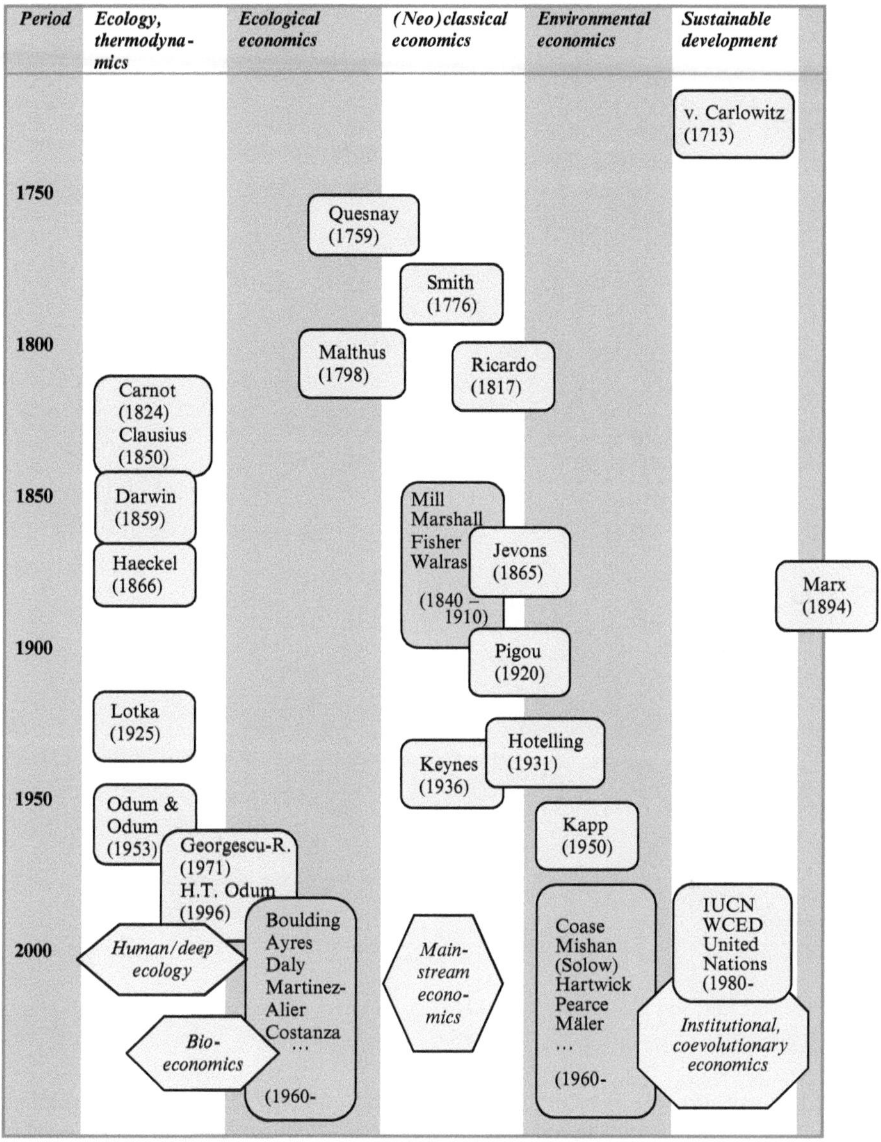

Plate 2.1 Historical perspective of *eco*–nomics (*See Colour Plates*)

into oblivion. In the wake of unprecedented economic growth in industrialized countries, classical and neoclassical economic theory (see brown box in the (neo)classical economics column of Plate 2.1) could not be bothered by dire warnings about population pressures on limited agricultural land (Malthus, 1798), diminishing returns from natural resources (Ricardo, 1817), or minor deviations

Plate 2.2 Hans Carl von Carlowitz (1645–1714) (*See Colour Plates*)

from perfect general equilibrium.[5] Perfect market equilibrium could be formalized with mathematical rigour in models, where disturbances from less-than-perfect markets or localized pollution incidents would be externalized. If need be such externalities could be easily internalized into mainstream economics by appropriate Pigovian taxation (Pigou, 1920). De facto they were ignored.

[5]One notable exception is Jevons's (1865) warning of running out of coal, a key natural resource at the time. Much later, Daly (1996) uses Mill's (1848) evaluation of the 'stationary state of capital and wealth' for arguing his own vision of sustainable development, based on a 'steady-state economy' (Section 2.4.2).

It took a looming environmental crisis and the visionary intellect of Kapp (1950) to make mainstream economists look beyond the microeconomics of the optimal use of an exhaustible natural resource (Hotelling, 1931). *Environmental economists* (brown box in the environmental economics column of Plate 2.1) made it their task to seriously study the full macro- and microeconomic cost and welfare implications of resource scarcity and environmental quality deterioration [FR 2.2].

At about the same time, and in opposition to the monetary analyses of environmental economists, a new branch of 'ecological economics' took its clues from the natural sciences. *Ecological economists* (brown box in the ecological economics column) studied the physical thresholds posed by the limited carrying capacity and resilience of ecological systems (Lotka, 1925; Odum & Odum, 1953); they also explored the effects of the dissipation (entropy) of energy (Carnot, 1824; Clausius, 1850) and matter (Georgescu-Roegen, 1971). Ecological economics covers a wide range of topics with spillovers into environmental economics and spin-offs like human ecology and bioeconomics. Ecological economics also claims to be the protagonist of 'sustainability science' [FR 2.2].

A broader approach to *sustainable development* emerged, in particular at the international level. The World Conservation Strategy (IUCN et al., 1980) together with the Third Development Decade Strategy (Box 3.2), the Brundtland Commission (WCED, 1987) and ensuing World Summits of the United Nations (1994, 2003) included, besides economic and environmental dimensions, social concerns of equity in the distribution of income, wealth and environmental impacts. Institutional economics and its particular, Darwin-inspired, co-evolutionary version describe the interrelationships between changes in natural and social systems (Section 3.3.1). The link of Marxist notions of greater equality in capital ownership and sustainable development is tenuous. However, recent takeovers of oil deposits by Latin American governments and the anti-globalization movement's calls for distributional equity and curtailment of corporate power (Ch. 14) do carry socialist ideas.

2.2.2 From Mainstream Economics to Deep Ecology

Depending on the particular world view of nature and human activity, different schools of thought address the environment–economy interface, either from an environmental or from an economic angle. Table 2.1 categorizes, or maybe forces, the different approaches into four basic schools, ranging from conventional (neoclassical) economics to 'deep' environmentalist thought. The first three rows refer to the underlying world visions in terms of the respective tenets, objectives and sustainability notions. Related strategies and assessment tools make the visions more visible; they are further explored in other parts of the book. This brief overview can of course not do full justice to the many variants of environmental–economic analysis. The further reading section [FR 2.2] provides some direction for gaining deeper insight.

Table 2.1 Schools of *eco*–nomic thought

	Conventional (neoclassical) economics	Environmental economics	Ecological economics	Deep (human) ecology
Basic tenets	Consumer sovereignty; frontier economics; utilitarian	Consumer sovereignty, limited by government intervention and environmental costing; utilitarian	Collective responsibility for protection of nature's assets; reformed utilitarian	Equality of species; symbiotic relationship with nature; non-utilitarian
Objectives	Profit, utility, welfare and economic growth maximization	Profit, utility, welfare and growth maximization, taking environmental social costs into account	Reduced or zero growth rates; qualitative development	Negative growth of economy and population
Sustainability concepts	Produced capital maintenance (very weak sustainability)	Produced and natural capital maintenance (weak sustainability)	Dematerialization of the economy (relatively strong sustainability)	Restoration and preservation of nature (strong sustainability)
Strategies and policy instruments	Economic efficiency; unfettered markets set environmental priorities	Eco-efficiency; environmental cost internalization by market instruments	Eco-efficiency and sufficiency; delinkage of growth and environmental impacts according to environmental norms and standards	Sufficiency and consistency; command and control; moral suasion
Assessment and monitoring	National accounts (GDP, capital formation, etc.)	Integrated environmental and economic accounts (environmentally adjusted economic indicators)	Material flow accounts (material input and output); indicators of sustainable welfare and development; indicators of human quality of life	Assessment of carrying capacity and resilience of ecosystems; ecological footprint

The basic tenets of the four schools range from advocacy to rejection of individual responsibility and capability of environmental problem-solving. Unlimited and somewhat limited consumer sovereignty characterizes the view of conventional and environmental economists. In contrast, the relatively new schools of ecological economics and related bioeconomics emphasize the vital value of natural systems for human survival. The complexity of these systems thwarts in their view any market-based evaluation by households and enterprises.

Human ecology, as far as it stresses 'the means of applying ecological principles to the management of the human population' (Odum, 1971), is close to ecological economics. 'Deep ecologists' go beyond the 'shallow' (Naess, 1976) anthropocentric view of nature's value for human health and well-being, insisting on the equality of all species. The ultimate step in deeply ecological thought is the near-religious appreciation of Earth as a living, self-regulating entity – the so-called Gaia hypothesis. From these eco-centric perspectives, governments as bearers of collective responsibility are obliged to defend nature against its ruthless exploitation by economic agents [FR 2.2].

The various world views translate into different individual and social objectives for production and consumption, and long-term economic growth. Again, there is a split between economists, seeking utility, profit and growth maximization, and environmentalists calling for a halt of economic growth or even negative growth. On one side, neoliberal economists justify the rejection of market intervention by suggesting more or less automatic improvement of environmental quality at some point of economic growth – the so-called Environmental Kuznets Curve (EKC) hypothesis. Transition to a service-oriented and hence dematerialized society supposedly explains this automaticity. Dematerialized post-industrial societies are also rich enough to afford environmental protection. Figure 2.2 shows the inverted U-curve, indicating that environmental impacts begin declining once economic growth reaches a certain level.[6] On the other side of the pro- and contra-growth discussion, ecological economists refute the EKC hypothesis and argue that the physical scale of economic growth has already violated vital environmental thresholds.

To achieve their objectives economists rely on the invisible hand of unrestrained markets for the efficient use of scarce human-made and natural resources. Where markets fail to mitigate environmental problems, environmental economists pursue an optimal balance between total (including social environmental) costs and economic benefits with the help of 'market instruments' of social cost internalization. Environmentalists, on the other hand, call for supplementing or even replacing eco-efficient production and consumption processes by alternatives that are in harmony with nature ('consistent') or voluntarily curbed ('sufficient'). Standard setting, rules and regulations, and education should bring about these new production and consumption patterns. Chapter 13 addresses these strategies and policies in some detail.

[6] The EKC hypothesis is named after Kuznets' (1955) similar assessment of correlation between the level and distribution of income. Section 11.1 assesses the results of testing the hypothesis and their implications for policymaking.

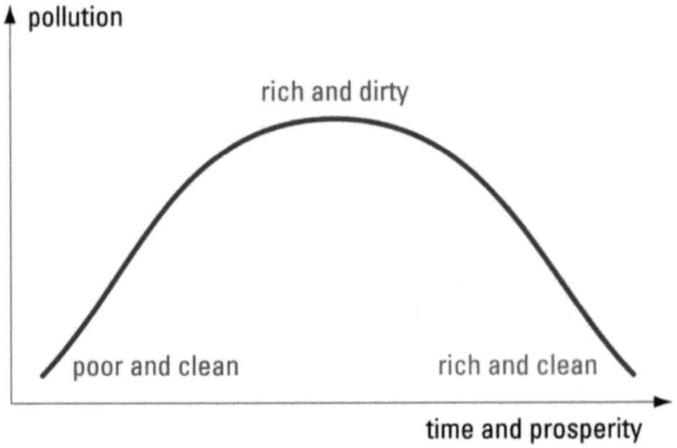

Fig. 2.2 EKC hypothesis
Copyright VisLab/Wuppertal Institute for Climate, Environment and Energy; with permission by the copyright holder.

All serious schools of *eco*–nomics make some attempt at operationalizing their philosophies, moving from theoretical notions to quantifiable variables. Economists show a preference for monetizing environmental problems, which they perceive as new scarcities in nature's services. Environmentalists, on the other hand, reject monetary valuation in favour of physical (non-economic) indicators that reveal the real pressure of economic activity on ecosystems.

Different concepts of the environmental sustainability of economic performance and growth thus emerged. Economics offers its fundamental concept of capital maintenance as one requisite for avoiding decline in future production, income and consumption. It is a logical step to extend this concept into newly scarce natural assets by incorporating 'natural capital' into economic theory and accounting. The ecological side offers a different view of sustainability. Considering economic growth as the culprit for most environmental impacts, rather than as a desirable goal, environmentalists focus on human threats to the carrying capacities of nature.

The distinction between economic and ecological sustainability polarizes environmental and ecological economics. Section 2.2.3 describes this polarization in some detail, as both sides offer important tools of analysis, measurement and policy. Note though that both approaches limit themselves to assessing the *environmental* sustainability of economic performance and growth. Chapter 3 broadens this view by introducing further social, political, cultural and institutional dimensions in a broader concept of sustainable development.

2.2.3 Ecological Versus Environmental Economics

Let us look more closely at the question of what *eco*–nomics can contribute to tackling simultaneously environmental and economic goals. To this end we zoom in on the core columns of Table 2.1. These two columns describe the main schools that address the environment–economy interface from an economic and ecological perspective. The two approaches differ distinctly, reflecting the above-mentioned polarization of environmentalist and economic views about the use and abuse of environmental functions.[7]

Environmental economics is about one part of mainstream economics where it did not work: the use of nature. If nature's services would be free, a science focusing on scarcity could ignore them. On the other hand, if these services were scarce but markets would ensure their efficient allocation to users, there would be no need to distinguish environmental economics from conventional economics. The fact is, nature's services have become scarce but markets either do not exist, as in most cases of waste/emission absorption or, at least in some cases, ignore the limited availability of natural assets. Existing markets then fail to ensure an efficient (optimal) use of nature's services.

Environmental economics is thus about correcting market failure in the provision and use of environmental services. It seeks to draw the environment into the exchange system of goods, services and money in markets. Environmental economists expect markets to give a monetary scarcity value to environmental services that had once been freely available. According to the widely advocated polluter/user-pays principles (Section 13.3.2), economic agents of households, enterprises, and governmental and non-governmental institutions should be held accountable for the environmental impacts and social costs they generate.

Ecological economists distrust the problem-solving ability of economics as they see economic activity as the main reason for environmental decline. In their view, the economic exchange system is a black box, eating up limited natural resources and disgorging the indigestible parts as waste back into nature. This appropriation of nature has now reached a level that violates vital life-support functions. Environmental impacts are therefore no longer an issue of optimal resource allocation but a matter of 'scale', calling for the reduction of the 'physical size of the economy relative to the ecosystem' (Daly, 1996). Consequently, ecological economists are less interested in marginal costing and pricing of environmental services. Rather, they want to prevent or minimize environmental impacts at the outset by reducing the amount of primary material flows from the environment into the economy.

A dichotomy of monetary valuation of environmental goods and services by individual preferences vs. collective *e*valuation according to physical (non-monetary) norms and standards is thus at the heart of the dispute between economists and environmentalists. The latter, and a good part of ecological economists, consider the environment a national heritage, bequeathed to us from the past generation and borrowed from the next. Moral attitudes and convictions (Doob, 1995), rather than

---

[7] The crude distinction between economists (including environmental ones) and environmentalists (including ecological economists, bioeconomists and industrial ecologists) is of course a simplification of various shades of green found in different branches of *eco*–nomics. It serves, however, to expose the main alternative concepts and measures of sustainability.

a willingness to pay (Jacobs, 1994), should express our appreciation of this heritage. Misleading economic valuation bears the risk of the environmental movement 'being colonized by the economy' (van Dieren, 1995).

Plate 2.3 illustrates the economic colonization of the environment by a strangling price tag of 60 billion (bio) Deutschemarks for annual environmental deterioration in Germany. The figure also depicts the alternative of weighting the burden on nature by the weight of a material requirement of 80 tons per head annually in most industrialized countries.[8] Calls for reducing this requirement by 'factors' of 2 or 10 (von Weizsäcker et al., 1997; Factor 10 Club, 1994) look in turn like attempts at colonizing the economy by environmental restrictions for economic activity.

The polarization of the two schools of *eco*–nomics thus spills over into measurement and assessment. Part II of the book describes physical indicators and material/energy balances, while Part III elaborates the environmental adjustment of the monetary national accounts. Operational concepts of economic and ecological sustainability focus on easier-to-measure causes of non-sustainability, rather than defining what should be sustained. It is a chicken and egg question whether statistical indicators of capital loss and material flows, or economic (capital, growth) theory and thermodynamics gave rise to two fundamental concepts of environmental sustainability (Bartelmus, 2003):

- *Economic sustainability*, which aims at keeping (produced as well as natural) capital intact so as to facilitate non-declining economic performance and growth

Plate 2.3 Getting physical or monetary?
Copyright VisLab/Wuppertal Institute for Climate, Environment and Energy; with permission by the copyright holder (*See Colour Plates*).

[8] See Sections 6.3.2 and 8.3 for the presentation and interpretation of these and further results of physical (material flow) and monetary (environmental) accounts.

- *Ecological sustainability*, which seeks to diminish pressure on the carrying capacities of natural systems by 'dematerializing' the economy.

Both concepts search for the conditions that maintain environmental source and sink functions. Their ultimate and common goal is to avoid a persistent decline in generic human well-being. Ecological sustainability specifies the physical requirements for preserving environmental functions. Environmentalists and ecological economists base the need for such preservation on the moral conviction that our natural heritage has to be passed on undiminished and if possible unchanged to future generations. Economic sustainability, on the other hand, assesses the value of the assets used up in providing environmental functions to the economy. This value represents a capital depreciation, which is the social and private cost allowance required for reinvestment in reproducible production factors.

The *raison d'être* for sustainability analysis is thus non-sustainability – either of life support and other amenities of the natural environment, or of economic activity depending on produced and natural capital services. Ecological sustainability is therefore less concerned with any kind of socio-economic progress. Economic sustainability, on the other hand, aims at netting out the value of natural and produced (and possibly also human and social) capital loss from its measures of economic performance and growth. Table 2.2 summarizes the main features of ecological and economic sustainability, discussed in detail in the following sections.

2.3 Economic Sustainability: Maintaining Capital and Welfare

Environmental source and sink functions and environmental assets that provide them obtain their economic value as nature's capital input into the production of economic output. Capital maintenance is the operational concept of economic sustainability as it aims at sustaining economic growth in terms of net output of the economy – net of capital depreciation cost. In turn, this cost allowance represents the funds necessary for replacing the real (non-financial) assets used up in production.

Some environmental economists consider environmental deterioration directly as a matter of broadly defined loss of economic welfare. However, non-declining welfare is a rather abstract concept of economic sustainability, especially at the national level of economic activity. It is shown in Table 2.2 because economic analysis generally adheres to utility and welfare maximization. This is the case in particular in microeconomics, which deals with individual 'rational', i.e. optimizing, behaviour. To attain environmental goals in an efficient, least-cost manner, macroeconomic policies use microeconomic rationality of utility and profit maximization. Policy instruments of environmental cost internalization make households and enterprises account not only for their own private costs but also the social costs of environmental impacts they generate. The expectation is that such accounting will restore or at least approach overall (Pareto-)optimality of national economic performance.

Table 2.2 Environmental sustainability: concepts and analysis

	Ecological sustainability (dematerialization)	Economic sustainability (capital maintenance)	Economic sustainability (non-declining welfare)
Rationale	Preserving nature	Sustaining economic performance and growth	Sustaining human welfare
Strategy	Decoupling economic growth from environmental pressure on carrying capacities	Maximizing economic efficiency and growth while keeping produced and natural capital intact	Maximizing income and utility from produced and non-produced environmental goods and services
Accounting tools	Material and Energy Balances (MEB) and Material Flow Accounts (MFA)	System for integrated Environmental and Economic Accounting (SEEA)	Welfare indices
Policy analysis (modelling)	Modelling trends of material flows; hybrid (physical–monetary) input–output analysis	Environmental cost internalization; in economic growth models with natural capital stock and use	Environmental damage costing in general equilibrium and optimal economic growth models
Strength of sustainability	Strong: reduction of material and substance flows to meet sustainability standards; maintenance of critical capital	Weak: overall capital maintenance, allowing for substitution between produced and natural capital and other production factors	Weak: typically assuming perfect substitution in production and consumption functions

Source: Based on Bartelmus (2003), table 1, p. 68; with permission by the copyright holder, Elsevier.

2.3.1 Environmental Macroeconomics: Assessing the Sustainability of Economic Growth

Neoclassical economists investigate the welfare relevance of environmental protection expenditures and environmental degradation in models of optimal (maximum) economic growth. The corresponding strategy is to achieve intergenerational equity in terms of non-declining welfare (Pezzey 1989) while taking environmental restrictions and welfare losses into account. For reasons of measurability, social welfare is usually replaced by per capita consumption or national income. Even a narrow focus on exhaustible natural resources produced widely differing results, depending on model features and their underlying assumptions. Beyond demonstrating a need for maintaining or widening the productive capacity of both produced and natural assets for long-term economic growth, the abstract notions of welfare maximization and optimal growth stand little chance of practical application (Section 12.3.2).

One way of skirting the difficulty of predicting the sustainability of future growth is to look at *past* economic performance. Indices of sustainable welfare estimate the damages and benefits generated in consumption and production for deduction from and addition to national income or consumption. Chapter 7 will show, however, that the indices suffer from conceptual and statistical deficiencies. Another way is to forego welfare measurement and to account more systematically for the social *costs* of economic production and consumption. Solow (1992) suggested to measure the depreciation of non-renewable natural resources and environmental assets as a 'practical step toward sustainability': 'maintaining the broad stock of society's capital intact' is indeed the rationale underlying the greening of the national accounts (Section 8.2.1).

In quantifiable terms, economic sustainability is thus simply the extension of the conventional economic notion of capital maintenance to natural assets. Extending this notion into the future,[9] the long-term maintenance of the total value of capital represents, however, a *weak* sustainability notion. The reason is that maintaining a total capital value implies possible substitution of non-produced natural assets by produced capital. Where 'complementarities' in production and consumption processes thwart substitution, weak sustainability criteria hide possible constraint for economic growth in real (constant price) terms and with current production and consumption patterns and technologies. Ecological economists call therefore for applying a 'stronger' sustainability criterion, i.e. the preservation of 'critical' (non-substitutable) natural capital categories (Section 2.4.2). Also, countries with significant population growth would be well advised to allow for further capital formation so as to ensure non-declining per capita output and income.

The identification, measurement and evaluation of complementarities in capital use are an unresolved issue. One can safely assume, though, that persistently negative or greatly reduced total net (accounting for produced and natural capital consumption) capital formation in the past would warrant changing established growth, investment and savings policies. Moreover, the narrow focus on environmental sustainability ignores other capital categories of a human, social and institutional nature. Strictly speaking, produced and natural capital maintenance may thus only improve rather than ensure economic (growth) sustainability. Again, we face here a measurement problem in the absence of regular accounting for human and social capital.

---

[9] Hartwick's (1977) well-known rule for reinvesting net returns from the use of exhaustible resources in reproducible capital calls for keeping the value of the total capital stock intact in order to achieve long-term constant consumption (Section 12.3.2). As discussed in Section 8.2.1, the descriptive (green accounting) concept of capital maintenance does not assume reinvestment, as desirable as it may be; it simply records the loss and degradation of produced and natural capital as additional production cost and as an indicator of potential non-sustainability.

2.3.2 *Environmental Microeconomics: Cost Internalization, Cost–Benefit Analysis, Optimal Use of Natural Resources*

Allocating the cost of mitigating or reducing environmental impacts to those who cause them is the objective of market solutions to achieving sustainability and optimality. Fiscal incentives and disincentives and other market instruments are an effective way of inducing economic agents to internalize these costs into their budgets. Cost internalization and incidence reflect consensus among economic agents of policymakers, households and corporations on the significance of environmental impacts. Markets negotiate such consensus, whereas rules and regulations obtain it more forcibly. Market solutions thus minimize interference with personal values and preferences while insinuating social concerns into individual decision-making. Chapter 13 discusses the ecological and economic efficiency of different strategies and instruments of environmental cost internalization.

Environmental costing and adding up the costs in indicators guiding macroeconomic policies points to the need for connecting macroeconomic sustainability concerns and microeconomic optimizing behaviour. However, much of this 'micro–macro link' (MML) remains murky when it comes to real-world analysis – not withstanding the abstract analysis of general equilibrium. Chapter 9 explores the relevance of the MML as a matter of harmonizing corporate and national responsibilities and accounting. However, given the aggregative nature of economic growth and development, the sustainability of these paradigms is first of all a macroeconomic concern. Microeconomics may make economic agents contribute to the sustainability of overall economic performance but the economy's sustainability is hardly a primary objective of individual behaviour.[10]

The following brief review of microeconomic approaches to dealing with environmental depletion and degradation costs helps sorting out practical concepts and methods for environmental measurement, valuation, aggregation and accounting.

2.3.2.1 Environmental Cost Internalization and General Equilibrium

Policy instruments of social cost internalization are to ensure the optimal use of environmental source and sink services. Under perfect market conditions 'rational' choices of economic agents bring about general (Walras) equilibrium and Pareto optimality. This is of course standard microeconomics. Most textbooks on environmental economics focus, therefore, on environmental cost internalization for devising environmental policy [FR 2.2]. Annex I.2 provides an illustration of how environmental cost internalization by means of the prototype Pigovian eco-tax on emission can maintain or at least approximate optimality.

[10] This is despite proclamations on the accountability or social responsibility of corporations, and calls for moderation in the consumptive behaviour of individuals (Section 13.4.1).

Popular computable general equilibrium (CGE) models proceed from the Walras model of market clearance of all outputs and inputs in an economy under perfect market conditions. CGE models attain their computability by aggregating individual economic activities into sectors and establishing behavioural consumption, production, investment and even utility functions for these sectors by means of econometric parameter estimates (Section 12.1).

CGE analysis links individual profit and utility maximizing behaviour to macroeconomic outcomes. In the environmental case, CGE models seek to compare the effects of different policy measures, such as environmental regulations or fiscal (dis)incentives, on short-term economic performance (Conrad, 1999). Existing market disequilibria (e.g. in the case of persisting unemployment), other non-environmental externalities, oligopolistic and monopolistic markets, the second-best conundrum,[11] assumptions about functional relationships in supply and demand, and lack of data render the results of such modelling questionable.

2.3.2.2 Natural Resource Economics

Environmental CGE models typically ignore natural resource depletion. The reason might be the separate development of natural resource economics as a special branch of microeconomics. Natural resource economics explores the optimal long-term exploitation of exhaustible natural resources [FR 2.2]. The approach is to predict costs and revenues over the expected lifetime of a resource and compare the – discounted – net return from resource extraction with alternative investments. Natural resource accounts pick up this analysis and apply the net present value to stocks and stock changes of non-marketed natural resources (Section 8.1.1).

The choice of rapid, slow, or non-exploitation depends crucially on the discount rate. The discount rate reflects the long-term profitabilities of different investments, including financial ones. Sustainability objectives could thus be introduced into investment decisions by stipulating a 'social' discount rate for the use of environmental assets. The social discount rate would usually be lower than the economic one to ensure the availability of natural assets for future generations. However, the rate is difficult to determine and may have opposing effects on the environment (Pearce et al., 1990; Pearce & Ulph, 1995).

Owing to their free-for-all and rival nature, previously abundant and self-renewing natural resources 'in the public domain', such as fish in the ocean or wood in tropical forests, are now threatened with overexploitation and destruction. This has become known – somewhat misleadingly – as the 'tragedy of the commons' (Hardin, 1968).

---

[11] Lipsey and Lancaster (1956–1957) advanced the general theorem of the second best. The theorem states that if one of the Pareto optimum conditions cannot be met, a second-best 'optimum situation can be achieved only by departing from all the other Paretian conditions'. As a consequence, situations in which some (but not all) Pareto conditions are met cannot be considered superior to others, which satisfy a lower number of Pareto conditions.

In fact, traditional communities managed common-property resources with great efficiency (World Bank, 1982; Upadhyay, 2004). The problem lies therefore more in the nature of open-access resources that lack clear definition and enforced, individual or common, ownership (see Annex I.2).

Figure 2.3 is a simplified illustration of why profit maximization in harvesting a renewable open-access resource might lead to its depletion. Total fish catch (effort) at maximum revenue level x_{my} is actually higher in the figure than profit-maximizing catch x_{mp}. The reason is open access to the fishing ground, which makes other fishermen come in as long as they find some profit. Lacking or ignoring knowledge about potential depletion of their fishing grounds, the fishermen continue exploiting the resource up to a total catch effort of x_{oa}, where their total cost equals revenue. At this level, the critical minimum stock, required for the regeneration of the resource, might have been surpassed. Catch beyond x_{cms} then triggers by definition the eventual destruction of the fish stock under the prevailing ecological conditions.

2.3.2.3 Cost–Benefit Analysis (CBA)

The evaluation of particular projects and programmes is not really a feature of microeconomic behaviour. Neither is it an approach of overall macroeconomic policy. The limitation of CBA to a particular project or programme has the advantage, however, to facilitate the measurement of economic and environmental project costs and benefits (Cooper, 1981; Dixon et al., 1994; Russel, 2001). On the other hand, CBA loses the capacity of CGE to capture the impacts and repercussion of environmental action on all sectors of the economy. Consequently, one cannot consider CBA as a tool of overall environmental or sustainability policy.

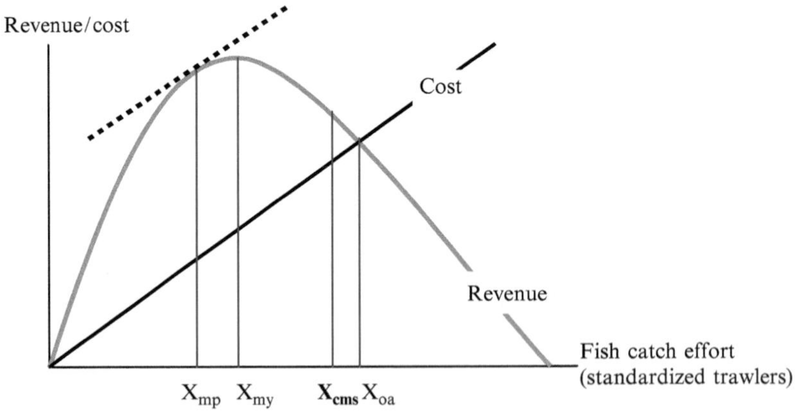

Fig. 2.3 Tragedy of the commons
Source: Turner, R. Kerry, David Pearce, and Ian Bateman (1993). Environmental Economics: An Elementary Introduction, p. 209; Copyright 1993 Kerry Turner, David Pearce and Ian Bateman, reprinted with permission of The Johns Hopkins University Press.

CBA does provide the rationale and tools for selecting projects with the highest net benefits where, as in the case of public goods and services, markets cannot make the selection. Basically CBA compares the total, current and future (discounted) costs and benefits of projects. For example, Annex I.2 (with Fig. I.1) shows how CBA can determine the optimal amount of environmental protection expenditure for reducing emission to a desirable level. Experience with environmental cost and benefit measurement improved knowledge about economic valuation in the field of environment. Environmental accounting employs therefore some of the CBA techniques (Section 8.1, Box 8.1).

Unfortunately, less-than-ideal real-world market conditions render the connection between short-term microeconomic optimality and long-term macroeconomic sustainability of economic growth exceedingly tenuous. However, the analytical rigour of environmental economics helps formulating clear concepts and indicators of environmental and economic interaction, required for sustainability measurement and valuation. Annex I illustrates, therefore, the core tenets of environmental economics. The annex discusses market and policy failure due to environmental externalities and describes the internalization of their social cost as a means of maintaining general equilibrium and optimality.

2.4 Ecological Sustainability: Dematerialization

2.4.1 Carrying Capacity, Ecosystem Resilience and Environmental Space

Ecological economists reject the treatment of environmental effects as a side effect of running a self-contained, unlimited growth machine. They see the economy as part of the natural world, which can only be exploited up to its physical limits or carrying capacities. The concept of carrying capacity of an ecosystem, country or the planet is a core tenet of ecological economics and of its operational definition of ecological sustainability.

Carrying capacity is a specific property of ecosystems. It is related to their limited resilience to any kind of shocks from the outside. Ecologists describe the range of resilience in terms of a 'homeostatic plateau', within which ecosystems maintain their equilibrium (Odum, 1971). Ecosystem resilience has therefore been considered as the axiomatic concept of sustainability (Perrings, 1995, 2006) [FR 3.3]. Measuring resilience requires the specification of thresholds, beyond which the ecosystem loses its adaptive capability and is pushed into a new, potentially disastrous equilibrium. Owing to the widely differing and complex nature of ecosystems and their vulnerabilities, resilience measurement at national and global levels remains elusive (e.g. Common & Stagl, 2005). If at all, the assessment of resilience can be usefully applied at local levels only, notably for sustaining agricultural production (see Section 3.2.3).

The focus has been, therefore, on the – limited – support, which ecosystems can give to the human population of a particular territory, i.e. its carrying capacity. Carrying capacity thus assesses more narrowly the biophysical limits to environmental impacts from economic activity within a region. Ecological economists stress not only the transgression of these limits but also the need for equity or 'fair share' in the limited availability of nature's services. The notion of equal 'environmental space' for everybody, or at least all countries, expresses this sentiment [FR 2.3]. Even these anthropocentric concepts face definition and measurement problems. Differing and mostly judgemental assumptions about standards of living, income distribution, technology, national and international policies, environmental conditions, the time frame of predictions, and individual preferences impair the meaning and validity of concrete carrying-capacity estimates (Cohen, 1995; see also section 5.2 with regard to the ecological footprint indicator).

Ecological economists are not deterred, though, from pointing to the threat of overloaded carrying capacity. The metaphor of the full world (Fig. 1.3) describes the economy as an 'open subsystem of the finite natural ecosystem', whose expansion has now reached global limits of carrying capacity (Daly, 1996). Consequently, our planet should be treated as a self-contained 'spaceship' rather than an ever-expanding production system (Boulding, 1966).

2.4.2 Material Throughput and Dematerialization

Another metaphor, social metabolism (Fischer-Kowalski & Haberl, 1998), explains threats to carrying capacities in more operational terms as the physical material and energy 'throughput' through the economy. Georgescu-Roegen (1979) deserves the credit for relating the metabolic flows of both energy *and* matter to the economy in extension of the first (conservation of energy) and second (its dissipation) thermodynamic laws. Chapter 6 shows how these physical laws can be related to the economy for a systematic accounting of material flows.

Following again Daly's (1996) line of thought, there is a maximal and a somewhat lower optimal level ('scale') of physical throughput. Sustainable development is achieved in a 'steady-state economy', where throughput does not increase at a level that does not impair the regenerative and absorptive capacities of the natural environment. In this situation, quantitative economic growth yields to qualitative development. The 'big question for environmental macroeconomics' (op. cit.) is indeed at what level of throughput sustainability turns into non-sustainability. Unfortunately, this question remains unanswered unless we accept Daly's affirmation, shared by most environmentalists, that 'the greenhouse effect, ozone layer depletion, and acid rain all constitute evidence that we have already gone beyond a prudent Plimsoll-line for the scale of the macroeconomy' (op. cit.). The Plimsoll line may mark the level at which the planetary boat begins to sink, but our first review of non-sustainability indicators (Section 1.3) did not yield any such line at national or global levels.

Austrian and German scholars attempted therefore to operationalize ecological sustainability by using material flows as indicators of the dematerialization of an economy. In order to determine how much dematerialization is needed for attaining ecological sustainability one would have to set clear dematerialization standards. Perhaps the best-known overall target for dematerialization is the Factor 4 goal (von Weizsäcker et al., 1997). It calls for increasing natural resource productivity by halving material input into the economy while allowing a doubling of wealth or output, within 50 years. At the global level, Factor 4 supposedly reflects the long-term ecological equilibrium of the planet and an anticipated, desirable or unavoidable, economic growth rate. The above-discussed concept of environmental space specifies planetary equilibrium as a state of the world, in which equal rights of access to environmental services are ensured without violation of the planet's carrying capacities.

Some of the Factor 4 protagonists seem now to dissociate themselves from the double-wealth and hence 'quantitative' growth part of the Factor. In their view, Factors 4 or 10[12] are more in the nature of general 'guard-rails' (Hinterberger et al., 2000) or a 'leitmotiv' (Bringezu, 2002) for policymaking. The purpose is to steer policymakers in the right direction of 'decoupling' the use of materials and hence environmental pressure from economic growth. This leaves the operational definition of ecological sustainability, i.e. the level of dematerialization needed for sustaining economic activity, to the political process; it also opens the door to normative advocacy, moving ecological economics even farther away from 'hard' scientific analysis into the realm of 'soft' ethical 'post-normal science' (Funtowicz & Ravetz, 1991).

However strict the adherence to Factor X targets, the purpose is to apply a relatively strong sustainability criterion.[13] Nevertheless, the Factor 4 advocates do put their faith on substitution and resource-saving possibilities, as indicated by the many technologies proposed for Factor 4 implementation (von Weizsäcker et al., 1997).

There is, however, some scepticism about the role of technology as the saviour from environmental collapse (Costanza et al., 1991; Daly, 1996; Fischer-Kowalski & Haberl, 1998). The argument is that the occurrence of critical capital prevents any substitution (even at the margin) of irreplaceable natural assets. Furthermore, 'rebound effects' may offset resource savings from improved production processes by increased (less expensive) consumption (Sachs, 1995). The specification of critical capital categories represents a much stronger sustainability notion, demanding the full preservation of these types of natural assets. Critical capital is usually

---

[12] To allow developing countries some catch-up in living standards the Factor 10 Club (1994) suggests applying Factor 10 to industrialized countries. Admittedly (by oral communication from staff of the Wuppertal Institute for Climate, Environment and Energy), these factors are crude estimates, anticipating an equal distribution of access to natural resources, of which 80% are currently used by 20% of the world population.

[13] The relativity stems from the rarely admitted fact that overall dematerialization still allows substitution among different types of materials.

linked to the resilience of ecosystems by setting 'safe minimum standards' for its preservation (Ekins et al., 2003). Lack of operational definition and standardized measurement has thwarted, however, the systematic identification and assessment of critical natural capital.

There has been progress in the assessment of ecological sustainability by means of material flow accounts. When it comes to describing *future* possibilities of dematerialization much of ecological economics seems to stay in the relatively anecdotal and metaphorical stage, though. Noteworthy new developments are hybrid models that introduce material flows in input–output analyses. Their objective is to present scenarios of economic growth within limits for flows of materials from, and pollutants into, the environment (Section 12.1.2).

To summarize: Economic sustainability, based on produced and natural capital consumption, reflects the goal of *preserving business*, not only for corporations but also the whole nation. Capital maintenance is a minimum condition for avoiding decline in economic activity and, in the long term, economic growth. Ecological sustainability, on the other hand, aims to *preserve nature*, whose assets are owed to future generations rather than owned by corporations, consumers or governments. Preserving nature is, from this point of view, not so much a matter of increasing or maintaining human welfare but of complying with carrying capacities of natural systems 'at any price' (or level of economic activity).[14]

Both preservation goals thus address the future use of nature's services. To this end, economic sustainability applies the preferences of the current generation, discounting uncertain needs and preferences of future generations. In contrast, advocates of ecological sustainability use their own insight to set risk-averse standards and regulations for current and future uses of the environment (Perrings, 1995; Rennings et al., 1999). Desirable norms for dematerialization such as the above-mentioned Factors 2 and 10, rather than individual preferences, reflect the norm setters' view of current and future human, environmental and economic needs.

Economic and ecological sustainability concepts coalesce where natural capital consumption impairs human welfare permanently because complementary assets of nature cannot be regenerated or replaced by reproducible capital or labour. The essential differences between the two sustainability concepts are assumptions about substitution and regeneration of nature's assets. Setting standards and regulations is the ecological economists' response to complementarity. Environmental economists contend, on the other hand, that overall (produced and natural) capital maintenance is in most cases the 'eco-efficient' and more democratic solution to the environmental problem. They put their faith, therefore, in the 'invisible hand' of the market rather than in 'strong-armed' regulation and control.

Chapter 13 will further discuss the policies and policy tools of implementing both sustainability concepts. Before exploring the tools of measuring economic and

[14] Of course individual environmentalist scholars have deviated from these principles by allowing for some growth as, for instance, in the Factor 4 stipulation.

ecological sustainability in Parts II and III, we have to clarify *what* we want to sustain – narrowly defined economic activity and growth, or the broader concept of socio-economic development? This will avoid using economic growth and development interchangeably – an unfortunate practice in many proclamations on sustainable development.

Further Reading

FR 2.1 Neo-liberalism and Environmentalist Critique

Downsizing government through privatization of public goods, deregulation, tax cuts and trade liberalization in the course of globalization (Ch. 14) has been the battle cry of *neoliberalism* (e.g. *The Economist* of 10 April 1999; Henderson, 2001). The EKC hypothesis (Sections 2.2.2 and 11.1) claims that unfettered markets and resulting economic growth are quite capable of dealing with the environmental effects of economic activity. Ecological economists disagree and counter with suggesting a regulatory framework for attaining sustainable development (Rennings et al., 1999; Sections 2.2.3 and 13.2). Heilbroner and Milberg (1995) and Kuttner (1997) are representatives of the 'literature of discontent' with neo-liberal economics.

More fundamentally, environmentalists attacked the behaviour of income-restrained, utility-maximizing *homo oeconomicus* as unworthy of *homo sapiens*. They point out that, at least in part, *homo sapiens* is a *homo politicus*, seeking justice, freedom and common (community) wealth (Faber, Petersen & Schiller, 2002; Söderbaum, 1999). Siebenhüner (2000) calls for personal growth into a *homo sustinens*, who lives in accordance with the requirements of sustainability. Instead of evoking an altruistic or political human being, *experimental economics* (Gintis, 2000) uses tests of human reactions to various challenges for providing evidence of biased (irrational) economic decisions. Simon (1982) describes 'local knowledge' constraints in human behaviour as a matter of 'bounded rationality'.

FR 2.2 Schools of Eco–nomics

Environmental economics is anchored in neoclassical economics in its attempt to remedy market failure through environmental cost internalization. Textbooks of environmental economics focus usually on the (market) instruments of cost internalization, and environmental CBA (Russel, 2001; Field & Field, 2002). Turner et al. (1993) provide a clear and easy ('elementary') introduction. In contrast, a three-volume handbook of environmental economics (Mäler & Vincent, 2003, 2005) comes not really handy as a guide through the subject, presenting a kaleidoscope of scholarly articles on welfare economics, modelling and market instruments.

Natural resource economics emerged originally as a special branch of economics, dealing (since Hotelling's 1931 pioneering publication) with the optimal use of exhaustible resources (Barnett & Morse, 1963; Field, 2001; Shogren, 2001). Possibly under the influence of the sustainability discussion, textbooks seem now to combine environmental and resource economics (Perman et al., 2003; Tietenberg, 2005), as does a voluminous reader (van den Bergh, 1999). There is much less literature on the macroeconomics of sustainability. Munasinghe's (2002) reader stands out as a more systematic review of environmental macroeconomics.

One of the best introductions to *ecological economics* is by Costanza et al. (1997a). The authors describe their domain clearly, but with an unabashed disavowal of mainstream economics. Costanza (1991) presents a wide range of topics and authors as the outcome of workshops, whose purpose was 'to produce a consensus on the … emerging field of ecological economics'. Daly (1996) is an eminently readable rejection of quantitative economic growth for environmental reasons. Ecological economists also seek to raise their field to scientific levels, describing it as 'sustainability science' (Costanza, 1991; Kates et al., 2001; Waggoner & Ausubel, 2002). Common and Stagl (2005) extend ecological economics by taking in environmental economics and sustainable development in an easy-to-read introduction. The *Ecological Economics* journal of the International Society for Ecological Economics (ISEE) is the platform for the writings of most of the above-mentioned (and many other) authors on the interaction of ecology, environment, economy and society.

The work of Austrian and German scholars on material flow analysis (see Ch. 6) gave rise to a new field of *industrial ecology* (Ayres & Ayres, 2002). Another spin-off of ecological economics is *bioeconomics* (European Association for Bioeconomic Studies, 1997). Bioeconomics focuses on the harmonious integration of human beings into the ecological cycles of nature, a view that is probably shared by most ecological economists.

Naess (1976) introduced *deep ecology* for a normative, eco-philosophical view of equal rights of all species. Sessions (1995) provides a reader on the deep ecology movement. The Gaia hypothesis goes beyond the egalitarian view of species, considering the whole Earth as a living organism (Lovelock, 1988, 1995). In comparison, *human ecology* is more anthropocentric, 'applying ecological principles to the management of the human population as part of that self-contained ecosystem, the biosphere' (Odum, 1971). Again, there is a Society for Human Ecology (www.societyforhumanecology.org) and a *Human Ecology Review* as the Society's mouthpiece.

FR 2.3 Fair Share of Environmental Space

Ecological economists presented evidence for the transgression of environmental limits in terms of limited photosynthetic capacity for food production. They claim that the human appropriation of 40% of the terrestrial net primary productivity (Vitousek et al., 1986) is a measure of a full world that leaves little or no room for human expansion (Daly, 1996). Weterings and Opschoor (1992) advanced the concept of environmental space as an indication of this limitation. Sachs et al. (1998), the Friends of the Earth (http://www.foe.co.uk/pdf/sustainable_development/tworld/summary.pdf), and the Fair Share Initiative (http://www.fairshareinternational.org/) advocate a fair share of this space for countries and their populations. The Ecological Footprint indicator (Section 5.2) claims to measure the compliance with, or exceeding of, a 'Fair Earthshare' of ecologically productive area per capita (Venetoulis et al., 2004).

Review and Exploration

- *Homo oeconomicus* vs. *homo politicus*: are utility and profit maximization criteria of rational economic decision-making?
- Market and policy failure: interpret the different cells in Table I.1.
- Compare the main schools of *eco*–nomics. What are their sustainability notions? Why and how do environmental and ecological economists differ in dealing with environmental impacts?
- Are carrying capacity and ecosystem resilience useful concepts of ecological sustainability?
- Is nature a substitutable or complementary production factor?
- What is the tragedy of the commons? Can we prevent it?
- Can technology prevent natural resource depletion and environmental degradation in the long run?
- Is Factor 4 a valid goal for ecological sustainability? Or is the EKC hypothesis more plausible (cf. Section 11.1)?

Chapter 3
Sustainable Development – Blueprint or Fig Leaf?

Chapter 2 asked what economics has to do with it, 'it' being the natural environment. The response was *eco*–nomics, defined as a new field of economics that reaches beyond the marketplace to deal with non-marketed scarce environmental services. There is no reason why economics should not reach even further into other areas for making rational choices, as long as it involves other scarce amenities. Governments supply some of these amenities as public goods such as security, public health, distribution of wealth, education, culture, and in fact environmental protection. All these goods and services meet societal goals like the satisfaction of human needs, a better quality of life beyond material standards of living, or, in the long term, sustainable development. The question is whether general proclamations on these goals hide potential trade-offs and deflect attention from actual implementation. Cornucopian rhetoric on social progress and development risks to remain just this – rhetoric. Has the paradigm of sustainable development run its course?

This chapter explores first the meaning and operationality of 'development' as a broad paradigm for attaining social goals beyond, or possibly in opposition to, economic growth. It then specifies those limitations that might thwart the simultaneous attainment of the different goals, rendering development non-sustainable. The translation of these limitations into measurable limits is one way of operationalizing sustainable development. Setting standards and norms for development constraints is judgemental, however; it does not provide a definite blueprint for implementation.

3.1 What is Development?[1]

3.1.1 Goals and Decades of Development

Socio-economic development is a process that improves the living conditions of people. Most also agree that the improvement of living conditions relates to physical

[1] This section contains material from Bartelmus (1994a, pp. 1–5; with permission by the copyright holder, Taylor & Francis).

needs and non-material aspirations. Popular calls for the increase of human welfare or the quality of life reflect this agreement.

Measuring progress towards these development goals requires a quantifiable definition. A first step towards quantification is breaking down the overall goal of human welfare into more specific objectives. Box 3.1 offers a list of such objectives, condensed from a variety of national and international policy proclamations. Any such listing is necessarily subjective in its selection and description. Further breakdown into secondary objectives is even more judgemental; but it is also more concrete since secondary objectives are more in the nature of means for attaining primary objectives. Instrumental secondary objectives may serve various primary objectives at the same time; they also differ considerably among individuals and cultures. Even the narrow focus on minimum requirements or 'basic human needs' for escaping underdevelopment and poverty was doomed: developing countries rejected this approach as unwarranted intervention in their development priorities and policies [FR 3.1].

For a more down-to-earth exploration of development one can look at the characteristics and conditions of poor countries, deemed to be in need of 'developing'. In most developing countries low levels of living and productivity prevail together

Box 3.1 Basic human objectives

Primary objectives

- Affection/love
- Recreation/entertainment
- Education
- Human freedoms (security)
- Shelter
- Aesthetic/cultural values
- Equity
- Health
- Physiological needs
- Future quality of life

Secondary objectives/means

- Food
- Clothing
- Mobility
- Drinking water
- Social services
- Housing
- Conservation of the environment
- Stability and justice
- Nation building
- Distribution of income and wealth
- Social security
- Working conditions
- Employment
- Time and leisure
- Education and training
- Security

Source: Bartelmus (1980).

with high levels of population growth, unemployment, international dependence and a predominantly agrarian economy. Based on these common factors, the United Nations agreed on international development goals and strategies for International Development Decades. However, these agreements had to be revised repeatedly because of their failure at the national level. As indicated in Box 3.2, this stark picture of development brought back, in the last Decade of the1990s, the first (1960s) Decade's call for economic growth. The absence of a widely accepted indicator of development, and hence the common use of per capita GDP as a proxy, could have been a factor in the return to a growth-oriented strategy.

Forty years of international development strategies were thus unceremoniously dumped. It comes therefore as a surprise that the United Nations Millennium Declaration (General Assembly resolution A/55/L.2, A/56/326) brought about the adoption of a new set of *Millennium Development Goals* (MDG). The time-bound (mostly for 2 decades) and quantifiable goals and targets are meant to monitor the Declaration; they are summarized in Box 3.3. As a minimum, the focus on targets and indicators demonstrates the need to move beyond generic declarations about socio-economic development. The MDG also indicate shifts in priorities reflecting new international concerns such as the AIDS epidemic and globalization.

Box 3.2 International Development Strategies (IDS) – a history of failure

The International Development Strategy of the *First United Nations Development Decade* of the 1960s called for economic growth in the belief that its fruits would trickle down to the low-income population strata. Since the trickle-down effect did not materialize, the *Second Development Decade* added the objective of social justice in the distribution of the results of economic growth. The strategy for the *Third Development Decade* of the 1980s recognized that inequities and imbalances in international economic relations prevented meeting the objectives of the Second Decade. The strategy included, therefore, the goal of establishing a New International Economic Order (NIEO), earlier adopted by the United Nations General Assembly. However, negotiations on the implementation of the NIEO broke down, and with it the Third Development Decade. Facing falling economic growth rates in developing countries and a deepening poverty gap within and between countries, the Strategy of the *Fourth Development Decade* for the 1990s called again for (accelerated) economic growth. It also considered economic growth as a prerequisite for 'priority aspects of development' that included the eradication of poverty and hunger, human resource development and the protection of the environment.

Source: United Nations General Assembly resolutions: http://www.un.org/documents/resga.htm

Box 3.3 United Nations Millennium Development Goals

- Eradicate extreme *poverty and hunger* (reduce by half the number of people suffering from both)
- Achieve universal primary *education*
- Promote *gender* equality and empower women (in education by 2015)
- Reduce *child mortality* (by two thirds)
- Improve *maternal health* (reduce mortality by three quarters)
- Combat HIV/AIDS, malaria and other *diseases* (halt and reverse spread)
- Ensure *environmental sustainability* (integrate sustainable development into country policies, improve access to safe drinking water by 50%, improve lives of 100 million slum dwellers by 2020)
- Develop a *global partnership* for development (open and non-discriminatory financial system, needs of least developed and vulnerable countries, debt relief, work for youth, affordable drugs, benefits of new technologies)

Source: http://www.un.org/millenniumgoals/

It remains to be seen if these goals will succeed in establishing development, rather than economic growth, as the primary policy goal of poor countries. In 2002, the Secretary General of the United Nations commissioned the director of the Earth Institute at Columbia University to develop an action plan for the implementation of the MDG. The final report (Sachs, 2005) identifies the grassroots needs of the poorest countries and suggests detailed measures such as insecticide-treated bed nets for malaria control – hardly a blueprint for development, but a welcome measure of help [FR 3.1].

3.1.2 Which Countries Are Developing?

As discussed in Chapter 4, sets of indicators cannot define and assess development unequivocally. This is the reason for the wide use of GDP for categorizing countries into developed and developing. However, the above listing of development goals calls attention to the possible fallacy of using economic output as a welfare or development indicator.

Table 3.1 gives a first impression of the suitability of GDP as a proxy for development and country ranking. The table compares GDP per capita (grouped into five categories from A to E) with the Human Development Index (HDI) (categories a to e). The HDI claims to capture development in terms of income, health and education levels.[2] A further category of least developed countries

[2] See Sections 5.2 and 5.3 for a review of the HDI and sustainable development indices.

Table 3.1 Country categories by level of growth and development[a]

A	B	C	D	E
$470–1,833	$1,860–3,680	$3,720–6,400	$6,550–15,560	$16,060+
*Afghanistan[b]	Albania c	Algeria b	Antigua and	Australia e
*Bangladesh b	*Angola a	Belize d	Barbuda d	Austria e
*Benin a	Armenia c	Bosnia and	Argentina e	Bahamas d
*Bhutan b	Azerbaijan c	Herzog. d	Barbados e	Bahrain d
*Burkina Faso a	Bolivia b	*Cape Verde c	Belarus d	Belgium e
*Burundi a	*Cambodia b	China c	Botswana b	Brunei Dar. e
Cameroon a	*Comoros b	Cuba d	Brazil d	Canada e
*Centr. Afr. R. a	*Djibouti a	Dominica d	Bulgaria d	Cyprus e
*Chad a	Ecuador c	El Salvador c	Chile d	Denmark e
Congo b	Egypt b	*Equatorial	Colombia d	France e
*Congo, D.R. a	*Gambia a	Guinea[c] b	Costa Rica d	Finland e
Côte d'Ivoire a	Ghana b	Fiji c	Croatia d	Germany e
*Eritrea a	*Guinea a	Gabon b	Czech R. e	Greece e
*Ethiopia a	Georgia c	Guatemala b	Grenada c	Hong Kong,
*Guinea-Bissau a	*Haiti a	Guyana c	Dominican R. c	China/SAR e
Kenya a	Honduras b	Iran, Islamic R. b	Estonia d	Iceland e
*Kiribati[b]	India b	Jamaica c	Hungary d	Ireland e
*Lao's People	Indonesia b	Jordan c	Kazakhstan c	Israel e
D.R. b	Kyrgyzstan c	Lebanon c	Korea, R. e	Italy e
*Liberia b	*Lesotho b	Macedonia d	Kuwait d	Japan e
*Madagascar a	*Mauritania a	*Maldives c	Latvia d	Luxembourg e
*Malawi a	Moldova, R. b	Panama d	Libyan A.J. d	Netherlands e
*Mali a	Morocco b	Paraguay c	Lithuania d	New Zealand e
Mongolia b	Nicaragua b	Peru c	Malaysia d	Norway e
*Mozambique a	Occup. Palestine	Philippines c	Malta e	Portugal e
*Myanmar b	Territory c	Romania c	*Mauritius d	Qatar d
*Nepal a	Papua New	*Samoa (West.) d	Mexico d	Seychelles d
*Niger a	Guinea b	St. Lucia c	Namibia b	Slovenia e
Nigeria a	Pakistan a	St. Vincent and	Oman c	Singapore e
*Rwanda a	*Solomon Isl. b	the Gren. c	Poland e	Spain e
*Sao Tomé and	Sri Lanka c	Suriname c	Russian Fed. d	St. Kitts and
Principe b	Syria A.R. b	Swaziland b	Saudi Arabia c	Nevis d
*Senegal a	*Sudan b	Thailand c	South Africa b	Sweden e
*Sierra Leone a	Uzbekistan c	Tunisia c	Slovakia d	Switzerland e
*Somalia[b]	*Vanuatu b	Turkey c	Trinidad and	United Arab
Tajikistan b	Viet Nam b	Turkmenistan c	Tobago d	Emirates d
*Tanzania a	Zimbabwe a	Ukraine c	Uruguay d	United Kingdom e
(*Timor Lesté[d])		Venezuela d		United States e
*Togo a				
*Tuvalu[b]				
*Uganda a				
*Yemen a				
*Zambia a				

Notes: [a]Country groupings: *A* to *E* by GDP per capita (in purchasing power parities), *a* to *e* by corresponding HDI rank; an asterisk marks the 'Least Developed Countries' (LDC). [b]Not included in UNDP (2003a). [c]Position corrected according to World Bank data. [d]Not yet a nation in 2003.
Source: UNDP (2003a); http://www.un.org/special-rep/ohrlls/ldc/ldc%20criteria.htm (for LDCs).

(LDCs) (marked by an asterisk) adds political flavour to the identification of the poorest countries: the LDCs are determined by the United Nations as those countries which have a low GDP per capita (less than US$ 900), weak human assets in terms of health, nutrition and education, and high economic vulnerability (instability of agriculture and exports, low diversification, economic smallness, and exposure to natural disasters).

Low levels of output and income clearly dominate the position of the LDCs in the A column. Some small islands, which enjoy income from tourism but are vulnerable to natural disasters, are exceptions. Less pronounced, but still discernible, is the correlation between the HDI and GDP per capita (a major ingredient of the HDI). The fluctuation is usually restricted to neighbouring columns. Namibia and South Africa are notable exceptions: despite relatively high income they rank much lower on the HDI owing to low life expectancy, caused in particular by the sub-Saharan AIDS epidemic. All in all, these country rankings do not refute the significance of economic growth as a means of meeting human needs and as a significant factor in contributing to the three HDI dimensions of development.

3.2 Towards an Operational Definition of Sustainable Development

3.2.1 Cornucopia from Sustainable Development?

Section 3.1 gives a first impression of the scope and facets of development in terms of human needs and wants. Ranking and comparing countries with regard to their success in meeting these needs required the selection of pertinent indicators and combining them in an overall index. The assumption is that such an index, through its underlying indicators, represents the main features of human or sustainable development, human quality of life, social progress, and ultimately well-being or even happiness. All these paradigmatic notions have in common an implicit promise of cornucopia in everything and for everyone. It is no surprise that policymakers and other advocates of social progress make ample use of these concepts: they sound nice enough to gain popular acclaim and are vague enough to prevent accountability for their implementation.

Box 3.4 shows the example of the USA where the euphoria of independence created belief in potential happiness for all. The more realistic Constitution appears to dampen these expectations with its reference to tranquillity, security and the promotion of welfare. However, the success of industrialization and a concomitant growing materialism of society reduced the grand notions of happiness and general welfare to the pursuit of prosperity as the dominant paradigm for about two centuries. More recently, the environmental movement cast doubt on the unfettered pursuit of material wealth, warning that nature's capacity to support this objective might soon be exhausted.

Box 3.4 The rights to happiness and welfare

We hold these truths to be self-evident, that all men are created equal, that they are endowed by their Creator with certain unalienable Rights, that among these are Life, Liberty and the pursuit of Happiness. (*Declaration of Independence*, July 4, 1776)

We the People of the United States, in Order to form a more perfect Union, establish Justice, insure domestic Tranquility, provide for the common defense, promote the general Welfare, and secure the Blessings of Liberty to ourselves and our Posterity, do ordain and establish this Constitution for the United States of America. *(US Constitution, Preamble*, September 17, 1787)

Reacting to these warnings, the United Nations convened the first global conference on the environment in 1972 and later established the World Commission on Environment and Development (WCED) to explore the interaction of environment and economic development (Section 1.2). The WCED's (1987) popular definition of sustainable development resurrects the broad human needs concept:

> Sustainable development is development that meets the needs of the present without compromising the ability of future generations to meet their own needs.

Owing greatly to the WCED's all-encompassing approach, the sustainable development paradigm has shown a perhaps surprising staying power, insinuating itself into the policy agendas of governments and international organizations. The two Earth Summits in Rio and Johannesburg embraced sustainable development as their *leitmotif.* The Constitution of the European Union makes the transition from the more restrictive sustainability of *growth*, stipulated in the 1992 Maastricht Treaty, to sustainable *development* in the 1997 Amsterdam Treaty.[3] Sustaining development is thus perceived as an obligation for all countries at any stage of economic growth and development.

But what does sustainable development really do for us? Are we to believe that this new paradigm will meet all or most of our needs and wants (cf. Box 3.1)? Will sustainable development bring well-being or even happiness to all? The Brundtland definition does not answer these questions. In fact it remains vague: it does not specify the categories of human needs, gives no clear time frame for analysis (future generations!), nor does it indicate how economic performance, social equity

[3] 1992 Maastricht Treaty: Declaration 20 on assessment of the environmental impacts of Community measures applies the 'principle of sustainable growth' (http://www.eurotreaties.com/maastrichtfinalact.pdf); 1997 Amsterdam Treaty: Title I Common Provisions stipulates as the objective of the EU 'to promote economic and social progress and a high level of employment and to achieve balanced and sustainable development' (http://europa.eu.int/eur-lex/en/treaties/dat/amsterdam.html).

and environmental functions combine to satisfy human needs now and in the future. Perhaps most disturbing is the absence of any reference to trade-offs among the different needs, for instance between the welfare of the current generation (and its poor) and the (largely unknown) needs of future generations. Rather, with sustainable development all is 'in harmony' and set to 'enhance both current and future potential to meet human needs and aspirations' (WCED, 1987).

Environmentalists indicate what to look for in development. Their argument is that the current malaise stems from our strife for economic growth and prosperity, despite the fact that wealth does not make us happy. Rather, simplicity and frugality will lead us towards the 'good life' [FR 3.2]. For some this implies the generation of 'qualitative development' (Daly & Farley, 2004, p. 6) in a 'post-growth society' (Hamilton, 2004).

The ultimate rewards of renouncing or at least curbing prosperity are of course a near-religious question. Theoretical investigations into the determinants of tastes and preferences stress the lure of attaining or maintaining desirable social status by more or less 'conspicuous' consumption [FR 3.2]. However, frowning upon a sense of identity and security through income, wealth and consumption might have little impact on general consumption behaviour. Consider pleasure from a continuous flow of stimulating novelties (Scitovsky, 1976) and the attraction of choice from a large variety of goods and services (Broda & Weinstein, 2006), and you have a strong argument for working with, rather than against, economic wealth and consumption possibilities (Bartelmus, 2000).

But what do the consumers themselves have to say? Proliferating happiness surveys in the USA report little change in the happiness situation of its citizens [FR 3.2]. This appears to confirm Duesenberry's (1949) relative income hypothesis. The hypothesis claims that relative standards of living (as compared to the neighbouring Joneses) count more for most people than an absolute increase in prosperity. At the country level, the hypothesis appears to be responsible for the so-called Easterlin (1974) paradox, which finds that above a certain level of income and economic growth 'national happiness' remains stationary. Max-Neef (1995) uses the Index of Sustainable Economic Welfare (see Section 5.2) to illustrate his 'threshold hypothesis' of actually declining, rather than stationary, welfare or quality of life in rich and growing economies.

Proclamations on meeting human needs for development or on attaining happiness through frugality remain vague and judgemental. Should we really trust happiness surveys and statements that 'almost everyone says no' to the question: are you 'happier now than 40 or 50 years ago?' (Hamilton, 2004). Maybe a socialist playwright's answer is more to the point: *'Mir löst sich ganz von selbst das Glücksproblem, nur wer im Wohlstand lebt, lebt angenehm'.*[4]

[4] 'On happiness this I can tell, it is your wealth that makes you living well' (Brecht, 'Threepenny Opera', the ballad of good living; own translation).

3.2.2 From Sustainability to Feasibility of Development[5]

Rather than focusing on unmeasurable happiness, a more realistic approach is to assess glaring symptoms and sources of unhappiness: these are obvious detractors from social progress contributing to the *non-sustainability* of any type of development. Table 3.2 lists the main limitations of the three general – economic, ecological and social – 'dimensions' of sustainable development. Beyond the impairment of economic growth by the loss of environmental source and sink functions, the table refers to further social (including cultural and political) effects of human activity. These effects include poverty and overconsumption, inequity in the current and future (intergenerational) distribution of income and wealth, deteriorating social cohesion from crime and social and cultural exclusion, and the loss of security from war, riots and terrorism. An additional dimension of institutional sustainability is sometimes added. It reflects laws, regulations, customs, and educational and other institutions; usually it is either ignored or subsumed in the social dimension as a means of facilitating cooperation and participation.

Table 3.2 translates the limitations of sustainable development into more operational limits. In the absence of a real or simulated market for mitigating non-sustainability symptoms, monetary valuation of damages and benefits of development reaches its limits. Setting desirable standards for maximally tolerable damage and minimum

Table 3.2 Non-sustainability in development: From limitations to limits

Sustainability categories	Limitations	Limits
Economic: Sustaining production, consumption and economic growth	- Produced capital consumption - Natural (economic) resource depletion - Degradation of environmental media as sinks for wastes and pollutants	- Productive capacities (produced capital) - Natural resource availability (stocks) - Absorptive capacities of natural systems
Ecological: Maintaining environmental functions	- Environmental degradation (pollution, degradation of ecosystems) - Resilience of natural systems to disturbance	- Environmental space (guard rails for material throughput) - Environmental (quality, resilience) standards - Carrying capacities of natural systems
Social: Attaining social goals	- Unemployment - Distributional inequity, poverty - Crime and corruption - Health and education needs - Security needs - Cultural identity loss - Social exclusion	- Basic human needs (minimum standards of living for current and future generations) - Maximum consumption levels for sufficiency - Social norms and conventions

[5] See Bartelmus (1994a) for a more detailed analysis. Some of the text of this section is from this source (with permission by the copyright holder, Taylor & Francis) and from Bartelmus (1997b).

satisfaction of human needs seems to be the only way to link the wide range of development concerns to economic performance. Table 3.2 identifies, therefore, in addition to economic capacity, sustainable development limits as:

- Ecological thresholds of ecosystem resilience (Section 2.4.1) to environmental impacts
- Limits to material throughput and consumption such as Factor X guard rails to prevent exceeding local and global carrying capacities (Section 2.4.1)
- Minimum and maximum standards of human needs satisfaction, which in turn affect carrying capacities of human populations
- Other social, cultural and political norms.

The introduction of standards and targets in development analysis shifts the focus of sustainability from capital maintenance in economic growth to compliance with minimum and maximum standards of living, natural resource exploitation, environmental degradation and other social norms. Violation of standards or non-achievement of targets indicates thus a development path that *should not* be pursued in the long run. In this sense, normative targets replace the relatively neutral sustainability criteria of dematerialization and capital maintenance.

Development programmes would thus have to operate within a normative framework defining a feasibility (or, more accurately, permissibility) space for these programmes. The determination of feasible programmes is a forward-looking approach, which requires the modelling of future scenarios of activities and impacts (Section 12.2). Assuming that such modelling can be carried out and focusing on the basic objective of human needs satisfaction and its environmental and social implications, an operational definition of sustainable (feasible) development can be put forward as

> the set of development programmes that meets the targets of human needs satisfaction without violating long-term natural resource capacities and standards of environmental quality and social equity. (Bartelmus, 1994a)

Compliance with social goals or norms may thus override individual (market) preferences for goods and services by social fiat – however democratically such fiat might have been achieved. Market valuation would be replaced by social evaluation, and sustainability by *feasibility* of human activity in terms of non-violation of social norms. To the extent that such standard setting affects market exchange a radical change in economic analysis would take place – from a focus on individual preferences to those of society, the government or self-proclaimed experts. The invisible hand of the market is overruled in this case by the visible one of the standard setter(s).

3.2.3 Local (Eco-) Development

Scepticism about governmental development priorities and policies motivated a shift of attention to local-level 'eco-development' (Box 3.5) [FR 3.3]. Many interactions between human activities and the environment are best observed,

Box 3.5 Features of eco-development

A non-governmental organization, the Centre International de Recherche sur l'Environnement et le Développement tested the eco-development concept for UNEP. It established the following features for its case studies:

- Basic needs satisfaction
- Satisfactory social ecosystem
- Rational natural resource use in solidarity with future generations
- Eco-techniques
- Horizontal, participatory authority
- Environmental education.

At the heart of eco-development are eco-techniques of biological pest control, aquaculture, non-conventional energy, eco-dwelling and traditional medicine.

Source: Sachs (1976, 1980).

evaluated and managed *in situ*. Those directly affected by environmental impacts should be in a better position to assess their responses than planners and policymakers in distant capitals (Bartelmus, 1994a). Rather than forcing economic and non-economic values and activities together in an overwhelming normative framework, the closeness of people in local communities might achieve a more spontaneous merger of local values, traditions and conditions through participatory, grassroots-democratic procedures.

The United Nations Environment Programme defined eco-development as

> Development at regional and local levels … consistent with the potentials of the area involved, with attention given to the adequate and rational use of the natural resources, and to applications of technological styles … and organizational forms that respect the natural ecosystems and local sociocultural patterns (UNEP, 1975).

Such development would apply particularly to agrarian societies of developing countries, whose social and economic systems are closely tied to the rhythm and productivity of nature. The ecological resilience concept (Section 2.4.1) can provide a useful understanding and managerial advice in this agrarian context.

After a widely publicized flurry of case studies, the term eco-development disappeared from the vocabulary of local development strategies. One reason might be the variety of particular environmental, social, cultural and political conditions that thwart the promotion of locally conditioned programmes and techniques as a general development strategy. This did not dissuade the 1992 Earth Summit to launch a participatory *local Agenda 21* movement [FR 3.3]. A survey of progress made since then identified about 6,500 communities involved in local Agenda 21 activities. Most communities focused quite narrowly on municipal water supply, due to lack of resources and governmental commitment.

3.3 Normative Economics for Sustainable Development?

'Whatever the definition, sustainable development is undoubtedly normative' (Faucheux, 2001). As discussed (Sections 1.3, 2.2.3), faith, pre-analytic vision and moral convictions underlie the environmentalist view of development. Putting economic activity into a frame of minimum and maximum constraints may tame economic growth, but at the same time mixes normative standards with factual assessment of economic performance – anathema to mainstream economists (e.g. Caldwell, 1982; Samuelson & Nordhaus, 1992; Beckerman, 1994).

'Institutional' economists and like-minded ecological ones counter that much of the positivist mainstream-liberal economics has become irrelevant for uncertain but urgent policy concerns because of its 'puzzle-solving ... ignorance of the wider methodological, social and ethical issues' (Funtowicz & Ravetz, 1991). 'Value pluralism' (Martinez-Alier, 2002) dissuades in their view the integration of environmental issues with conventional economic analysis and policy. Instead, a co-evolutionary approach to development offers linkage of social values with ecological-evolutionary ideas, without risking 'colonization' by economists.

3.3.1 Co-evolutionary Economics

Institutional economists address directly some of the concerns specified in Section 3.2.2 as the normative framework of sustainable development. They view economic performance as a function of exogenous factors such as technology, power structures and organizations, rather than as the endogenous workings of the economic exchange system. Co-evolutionary economics extends institutional economics into the environmental field, and beyond into overall societal change [FR 3.4]. Its basic tenet is that the institutional framework connects everything to everything else (Fig. 3.1). Such a sweeping statement is, however, hardly conducive to practical application.

The co-evolutionary explanation of social change argues convincingly about the relevance of institutional change and society's evolution for sustainable growth and development. At the same time its advocate admits that institutional reform for re-embedding humans into their natural and cultural environment would be difficult to effect by top-down governmental policy. The co-evolutionary approach should therefore focus on the local level. At this level, a 'coevolving patchwork quilt of discursive communities' could attain decentralization, participation and cultural diversity more easily and democratically than by central authorities (Norgaard, 1994). Such escape to local levels reflects the communitarian roots of the above-discussed eco-development concept [FR 3.3]. From this point of view, co-evolutionary economics does not provide the means for achieving – national – sustainable development. Rather, it looks more like a passive hope for a trickle-up of grassroots values for resetting society's development path.

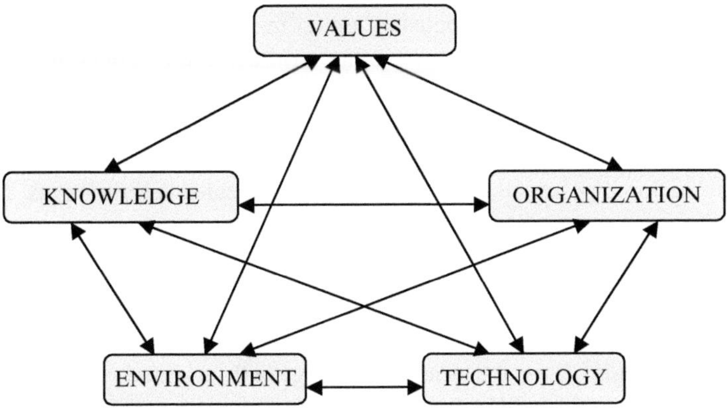

Fig. 3.1 Co-evolutionary process: Everything related to everything else?
Source: Norgaard (1994), fig. 3.1; with permission by the copyright holder, Taylor & Francis.

Ecological economists embrace the institutional approach (1) from an ethical-moralistic point of view, calling for a better life (Section 3.2.1) and (2) for setting standards guiding the co-evolution and sustainability of interdependent social, ecological and economic systems. 'Ecologically and economically sensitive safe minimum standards' that 'hold back' human activity (Perrings, 1995) would ensure the maintenance of carrying capacities and resilience of natural systems. However, carrying capacity and resilience are complex and even ambiguous concepts; they face corresponding measurement problems, notably at national and global levels (Section 2.4.1), and when extended to social systems.

3.3.2 Has the Paradigm Run Its Course?

Neither the generics of co-evolutionary economics nor the specifics of local-level case studies provide a blueprint for overall sustainable development. In practice, policymakers continue to see sustainable development as economic growth with some environmental protection. Even the WCED (1987) found it necessary to pacify growth-oriented governments by calling for the revival of economic growth. Undeniable successes of local pollution control appear to have turned the attention of industrialized countries to remote global concerns, at the expense of social problems and deteriorating environmental conditions in developing countries. Even if not openly admitted, EKC mentality seems to prevail: the ongoing transition to a non-material service society is expected to take care of the remaining environmental problems. It remains to be seen if the clamorous warnings of global warming reflect a genuine shift from economic to environmental priorities by society and policymakers (cf. Section 4.3)

In principle we have several options of addressing the dilemma of operationality vs. comprehensiveness in covering simultaneously all dimensions of sustainable development. They are

- Remaining operational by focusing on conventional economic growth and leaving environmental, social and other non-economic concerns to separate analyses and policies
- Implementing a normative notion of sustainability in terms of politically set constraints for economic activity
- Skirting national policymaking by focusing on local-level initiatives
- Finding a compromise by integrating only those non-economic concerns in economic analysis that can be defined and measured in consistency with operational economic concepts and indicators.

The first option ignores the interaction of economic growth with other environmental and social goals. It is the conventional growth-oriented view of economic development. We could call it the fig leaf approach as it may pay lip service to sustainable development but hides deferment of action or inaction behind the rhetoric of an opaque paradigm. At best, it assigns responsibility for social and environmental policies to comparatively weak (low-budget) departments.

The second option is the interventionist, normative strategy, favoured by institutional and ecological economists as the appropriate response to an environmental emergency. It sets out from the above-discussed definition of sustainable development as economic activity within a framework of social goals and constraints. Such constraints make the vision of sustainable development more visible, but remove it from objective analysis. At the same time, judgemental target setting reduces the chances of wide acceptance and impedes objective measurement of sustainability. Overlapping and interacting standards and norms make it indeed difficult to agree on a combined set of standards, even if there is consensus on individual targets.

This option leaves the implementation of sustainable development to the political process. Lacking a unifying theory or model, ecological economics has not solved so far the problem of merging positive science with normative prescription in a transparent and operational fashion. The generics of the co-evolutionary analysis is a point in case, as is the anecdotal treatment of environmental and social problems. One economist even considers sustainable development as either 'morally repugnant' (curbing economic growth in poor countries for the sake of unforeseeable preferences of future generations), or as 'logically redundant' (accommodated by welfare economics) (Beckerman, 1994, 1992).

The third option of focusing on community activity and local eco-development is escapist. It has merit in addressing local problems and creating awareness of social and environmental concerns among citizens. The hope is here for some upwelling of sustainability values and goals from the grassroots to civil society and government. Linking grassroots programmes with top-down policies could support the forming of alliances and the acceptance of sustainable development beyond the local level. It would require, however, some delegation of central authority to local institutions, which is unlikely to occur (Bartelmus, 1994a).

The last – compromise – option is this book's basic philosophy of integrative quantification of the environmental sustainability of economic performance and growth. The opaque concept of sustainable development seems to have done its duty of alerting us to interdependences among different dimensions of development. In principle, sustainable development calls for pursuing economic and non-economic societal concerns through combined policies. In practice, the paradigm has largely failed, though, to integrate and hence compare these concerns with the central societal goal of creating prosperity for all. On the other hand, there is still a lot of goodwill attached to the notion of sustainable development, notably with regard to environmental and social/distributional objectives. It also reminds policymakers to continue reviewing their goals and priorities and to communicate them transparently to their electorate. Calling for the demise of sustainable development could therefore be detrimental to the broad acceptance and pursuit of these goals.

All in all, the elusive and judgmental paradigm appears to be of limited analytical and practical value for dealing with non-economic impacts of economic activity. The following chapters try, therefore, to identify those environmental concerns that can be quantified in a systematic fashion, and whose significance can be compared to the outcomes of economic activity. For now, there is less doubt and better knowledge about measuring the interaction between economy and environment than about the interrelationships between the economy and other social, cultural, ethical or political concerns. Economic and ecological sustainability, as defined in Section 2.2.3, remain therefore a valid and vital concern for rational, integrative economic and environmental policies. Sustainability economics, dealing with the immediate environment-economy interface provides useful measurement and analytical tools for such policies. Sustainable development provides the décor and reminds us of other social concerns.

Further Reading

FR 3.1 Development and Human Needs

The failure of the International Development Strategies (IDS) (Box 3.2) casts doubt on the concept of development as a means by which poor countries would catch up with industrialized ones. Early surveys of development revealed uncertainty and conflicting views about the concept (Birou et al., 1977; Jolly, 1977; Todaro, 1977). Seers (1981) provides a critique of contemporary 'dependency theories' in which core dominant nations exploit peripheral dependent ones. A rejection of the Western development paradigm comes from the environmental and anti-globalization corner: the argument is that the global environmental crisis and security concerns in a limitless global economy undermine any solidarity between rich and poor countries (Sachs, 2000; see also Ch. 14). The UN Millennium Development Goals and Project are the latest attempt at advancing international development principles and recommendations (http://www.un.org/milleniumgoals/, http://www.unmillenniumproject.org).

The 1972 United Nations Conference on the Human Environment brought out doubts about narrow growth-oriented development strategies and led to the search for a new development paradigm (Section 1.2). As a first response, an international symposium in Cocoyoc (UNEP & UNCTAD, 1974) advanced the idea of first meeting the *basic human needs* or 'inner limits' with due consideration for global environmental risks or 'outer limits'. The 1976 World Employment Conference publicized the basic-needs approach to development but failed in its call for including it in the IDS (ILO, 1977).

Developing countries considered the strong support of the basic needs strategy by industrialized nations as a tactical move to divert attention from the implementation of a New International Economic Order. In addition they felt that an international strategy of meeting the basic needs of the poor represented an intrusion into sovereign national policies (Bartelmus, 1994a). Despite this rejection of the basic-needs approach, more recent views of 'new economics' and sustainable development aim to resurrect the human needs concept for operationalizing welfare (Max-Neef et al., 1990; Ekins et al., 1992) and for defining sustainable development (WCED, 1987).

FR 3.2 Consumerism, Happiness and the Good Life

Veblen's (1899) critique of 'conspicuous consumption', Duesenberry's (1949) 'relative income hypothesis' and its updated versions of 'new consumerism' (Schor, 1998) or 'luxury fever' (Frank, 1999) criticize human strife for ever-increasing consumption as gaining status rather than real well-being or happiness. Ecological economists picked up this argument, initially to defend their strategy of dematerialization as offsetting 'rebound effects' in demand from natural resource savings (Section 13.4.1). They also argue that moderation ('sufficiency') in consumption provides a *good life* of simplicity, contentment and less guilt about social inequities (Durning, 1992; St. James, 1996; Sachs et al., 1998; Segal, 1999).

Happiness web sites such as www.happyplace.net and www.sohp.com, where 'sohp' stands for the Secret Society of Happy People, have proliferated. The National Opinion Research Center of the University of Chicago (www.norc.uchicago.edu) and the general social survey of the University of Berkeley (http://sda.berkeley.edu:7502/D3/GSS2000/Doc/gss2.htm) conduct recurrent happiness surveys in the USA. A World Database on Happiness seeks to bring together research on individual and national happiness (www.eur.nl/fsw/research/happiness). Anielski (2007) advances 'genuine wealth' as a vision (and measure) of an economy of well-being and happiness.

FR 3.3 Eco-development and Related Local Initiatives

Communitarian thought (Sandel, 1982; Wolfe, 1989; Etzioni, 1993) shares anti-liberalism and a focus on community values and institutions with co-evolutionary analysis [FR 3.4]. The different focus on human rather than human-ecological communities prevented, however, a full integration of environmental concerns into the communitarian movement. More recently, ecological resilience has been advanced as a sustainability concept, notably for agrarian societies (Section 2.4.1). The Resilience Alliance offers numerous studies of 'resilience in social-ecological systems [as] a basis for sustainability' (http://www.resalliance.org/1.php). Reacting to central (top-down) planning and policy failures, *eco-development* programmes (Sachs, 1976, 1980) sought to empower local communities for tackling poverty and environmental impacts. The reason is to prevent the plundering of the fruits of local development by top-down intervention in local economies. Martinez-Alier (2002) describes the 'environmentalism of the poor' in defence against biopiracy, environmental impacts of local natural resource exploitation and ecological footprints of cities on their hinterlands.

The 1992 Rio Summit included a call for launching a *local Agenda 21* by local authorities (in its shortest chapter and without reference to the eco-development programmes of the 1970/1980s: United Nations 1994, ch. 28). As before by UNEP, the United Nations Secretariat charged another organization, the International Council for Local Environmental Initiatives (ICLEI) to foster the implementation of local Agenda 21 programmes and keep track of progress made (http://www.iclei.org/ICLEI/la21.htm).

FR 3.4 Institutional and Co-Evolutionary Aspects of Ecological Economics

Ecology textbooks (e.g. Odum's, 1971 classic) describe general ecosystem dynamics, the co-evolution of biotic communities and their resistance to perturbations. *Institutional economists* criticized neoclassical economics as removing economics from the social, political and legal institutions to which it is inextricably connected (e.g. Commons, 1934). Ecological economists co-opted the institutional and co-evolutionary approaches for sustainable development analysis (Opschoor & van der Straaten, 1993; Gowdy, 1994; Rennings et al., 1999); Söderbaum, 1999; Faucheux, 2001). Norgaard (1994) attempted to lift the co-evolutionary analysis from its ecological roots into an analysis of overall societal change; he argues, in particular, that social values and technology are interlocked in a path of social change, which generated environmental degradation and inequity. Kallis (2005) describes the co-evolutionary approach as a 'meta-theoretic guide, a mode of enquiry, not a specific theory'.

Review and Exploration

- What are, in your opinion, the high-priority goals of development?
- Why did the International Development Strategies fail? Do you give the Millennium Development Goals a better chance?

- Is meeting (basic) human needs a useful approach to poverty alleviation in developing countries?
- Compare your country's ranking with other nations in Table 3.1. Do you agree?
- Does wealth make us happy? Or does sufficiency/frugality?
- Is zero-growth an option? What are its consequences for human welfare?
- Does the Easterlin paradox hold? If so, should we discourage striving for status according to the relative income hypothesis? How?
- Compare different definitions of sustainable development as to their practicality.
- Is mixing positivist and normative analyses for assessing and promoting sustainable development a good idea?
- Do you agree with Daly (1991) that sustainable growth is a 'bad oxymoron'?
- Does co-evolutionary analysis help implement sustainable development?
- What is the contribution of local development programmes to national and global sustainability strategies?
- Describe a community's local Agenda 21 effort and assess its achievements.
- What's not measurable is not manageable! What's not countable does not count! Do these statements reflect the book's focus on quantification?

Part II
Assessing the Physical Base of the Economy

Part I raised questions about the role of economics in tackling environmental impacts. It found general principles of *eco*–nomics for defining and assessing the sustainability of economic growth and development, including

- The interactions and repercussions between economy and environment as the principal cause of non-sustainability
- A persistent dichotomy in defining economic sustainability as produced and natural capital maintenance in monetary terms vs. ecological sustainability as dematerialization in physical (non-monetary) units
- The need to specify social, ecological, institutional and other limits to economic activity for a more operational but normative concept of sustainable development
- A focus on environmental (economic and ecological) sustainability of economic performance and growth in measurement and policy analysis so as to avoid normative evaluations of an opaque development paradigm.

Parts II and III respond to the physical-monetary dichotomy, scrutinizing physical and monetary data as to their capability of capturing the economic, environmental, and perhaps even social sustainability dimensions. Part II aims to measure the biophysical (ecological) sustainability of economic performance and growth. Part III assesses economic sustainability by monetary indicators and accounts.

Figure 4.1 shows the scope of, and interfaces between, the principal economic, environmental and social data systems. The further reading section reviews the main international statistical systems presented in the figure [FR 4.1 and 4.2]. Chapter 4 discusses environment statistics and indicators of sustainable development. Chapter 5 reviews the challenges of aggregating these indicators into compound indices. Chapter 6 describes the more systematic approach of physical accounting in terms of energy flows and material throughput through the economy.

Chapter 4
Statistics and Indicators

The purpose of environment statistics is to assess the biophysical world ignored by conventional human-centred social and economic statistics. Interaction between environment and economy requires linking all categories of statistics in a common framework or system. A proposed Framework for Statistical Integration (FSI) (at the centre of Fig. 4.1) suggests organizing stock and flow data from the main statistical systems under common information categories.

Compiling all these data at the same time would generate an information overload for decision-makers. The purpose of selecting indicators that are representative of sustainability concerns is to reduce this load to a manageable level. The preference of policymakers, public media and the general public for 'nutshell' information produced short lists of key or core indicators, and what appears to be now *the* surrogate for the environment – global warming. The question is: do these indicators (or indicator) really tell the sustainability story?

4.1 Statistical Frameworks

4.1.1 A Framework for Environment Statistics

In 1972 the United Nations Conference on the Human Environment called for the development of environmental data and indicators in an 'Earthwatch' component of its Action Plan (United Nations, 1973). The newly established United Nations Environment Programme (UNEP) implemented Earthwatch as a global environmental monitoring system. Remote sensing, even with some ground truthing, cannot provide, however, the necessary data for monitoring the environmental situation at national and local levels, nor can it assess the economic activities responsible for environmental deterioration.

To this end, the Conference of European Statisticians (1973) was among the first to call for developing a *system* of environment statistics – parallel to those existing in the economic and socio-demographic fields, the Nobel-Prize winning System of National Accounts (SNA) and the System of Social and Demographic Statistics

P. Bartelmus, *Quantitative Eco-nomics*,

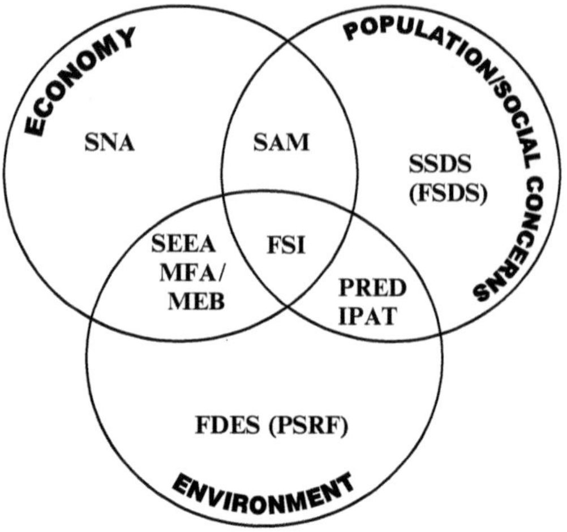

Acronyms:
FDES Framework for the Development of Environment Statistics
FSI Framework for Statistical Integration
FSDS Framework for developing and integrating Social and Demographic Statistics
IPAT Identity: Impact (I) = Population (P) x Affluence (GDP/P) x Technology (I/GDP)
MEB Material and Energy Balances
MFA Material Flow Accounts
PRED Population, Resources, Environment and Development (databank)
PSRF Pressure-State-Response Framework
SAM Social Accounting Matrix
SEEA System for integrated Environmental and Economic Accounting
SNA System of National Accounts
SSDS System of Social and Demographic Statistics

Fig. 4.1 Overlap and interaction in international statistical systems
Source: Bartelmus (1997a), Measuring sustainability: Data linkages and integration, fig. 2; copyright Wiley & Sons Ltd.; with permission by the copyright holder.

(SSDS). The idea was to cover in this manner the whole of the real world in all its physical, economic and social dimensions.

It became soon apparent that the system approach works fine when there is a generally accepted underlying theory and a common measuring rod for the different statistical variables. This is the case for the SNA, which connects statistical variables through additive and subtractive relations and balances, using market prices for the valuation of economic stocks and flows. The SNA also draws heavily on economic theory in depicting the circular exchange system of labour, products and money flows. However, as for the SSDS [FR 4.1], there is no such theory and common *numéraire* for the large variety of environmental impacts. The only option to organize environmental data was therefore to develop a looser organizational *framework*. After surveying existing frameworks and modules of environmental statistics, the author and his team at the United Nations Statistical Office advanced a Framework for the Development of Environment Statistics (FDES) (United Nations, 1984).

The FDES expands and modifies an ecology-based statistical system (Friend & Rapport, 1979) to better reflect the interaction of economic activities with the natural environment. The result is a two-way table (Table 4.1, part A), which relates major components or media of the natural and human-made environment to information categories. The information categories represent a sequence of activities, their environmental impacts and effects on human health and well-being, and responses to these impacts. A further information category of stocks and inventories facilitates the linkage of the FDES with environmental accounting systems (Section 7.2).

The contents of the framework are statistical topics, which are the quantifiable aspects of environmental and related socio-economic concerns (part B of Table 4.1). The FDES and its statistical topics served to develop detailed methodologies for a large list of statistical variables (United Nations, 1988, 1991). Part C of the table gives an example of selected variables for the topic of

Table 4.1 From framework to statistics: Format and use of the FDES

A. FDES				
	Information categories			
Components of the environment	A. Social and economic activities/ events	B. Environmental impacts of activities and events	C. Responses to environmental impacts	D. Inventories, stocks and background
1. Flora 2. Fauna 3. Atmosphere 4. Water (a) Freshwater (b) Marine water 5. Land/soil (a) Surface (b) Subsurface 6. Human settlements				
B. FDES: Statistical topics				
	- Use of natural resources and related activities - Settlements growth and change - Emissions, waste loadings and applications of biochemicals - Natural events	- Resource depletion and increase - **Environmental quality** - Human health and welfare	- Resource management and rehabilitation - Human settlement policies and programmes - Pollution monitoring and control - Private sector responses	- Biological resources - Cyclical and non- renewable resources - Energy stocks - Ecosystems inventory - Stocks of shelter and infrastructure - Background conditions

(continued)

Table 4.1 (continued)

C. VARIABLES OF FRESHWATER QUALITY			
	Variables	Classifications	Observations
B. 2.2.1 Inland water (quality)	a. Physical/chemical properties (μg/l, %, pH)	- Water body	- Incl. turbidity, salinity, acidity, conductivity
	b. Concentration of chemical contaminants (ppm, μg/l)	- Water body - Chemical compounds	- Stress on aquatic ecosystems and human health
	c. Nutrient indicators, e.g. chlorophyll a (μg/l)	- Water body	- Indicators of eutrophication
	d. Concentration of organic matter, e.g. BOD_5 (mg/l)	- Water body	- Level of dissolved oxygen
	e. Concentration of pathogens (μg/l, no./l)	-Water body - Type of pathogen	- Potability of water, e.g. faecal coliform count
	f. Areas with water-borne disease vectors (km^2)	- Water body	- e.g. bilharzia, onchocerciasis
	g. Water quality index (index value)	- Purpose - Water body	- Including aquatic habitat, drinking water index, recreation

Source: United Nations (1984, 1991).

freshwater quality. Table 4.2 illustrates how the variables can be presented in the questionnaire of a statistical survey.

Table 4.1 also shows that the FDES reaches beyond the 'pure' environmental field into socio-economic activities, health and welfare effects, and natural resource use and management. The FDES has therefore been used, sometimes under the pressure-state-response (PSR) label, to organize other integrative data sets such as those of environmental and sustainable development indicators (see Section 4.2.2).[1]

Presenting environmental data in a coherent framework serves several purposes, including the

- Determination of quantifiable aspects of environmental concerns
- Identification of statistical variables for the description of these concerns

[1] The OECD (1993) later relabelled the FDES information categories as pressure, state and response in its popular PSR framework for environmental indicators (Fig. 4.2). As the FDES is more systematic in translating statistical topics into statistical variables and includes both stock and flow variables, it is discussed here as the prototype framework.

Table 4.2 United Nations 2004 questionnaire on environment statistics, river quality

Section: WATER
Country:____________Contact person:__________Tel:______
Contact institution:__________E-mail:__________ Fax:______
Table W6A: Water Ouality of Selected Rivers
Name of River A:__________Sampling frequency: Minimum__________ /year
Name of Measuring Station:__________Maximum /______year
Distance to mouth or downstream frontier Sampling depth:__________m

Priority	Category	Unit	1990[a]	1995[a]	1996	...	2002
!	Annual average flow	m³/s					
!	Biochem. oxygen demand 5	mg O_2/l					
!	Dissolved oxygen	mg O_2/l					
	Chem. oxygen demand	mg O_2/l					
	Total dissolved solids	mg/l					
	Total phosphorus	mg P/l					
	Total nitrogen	mg N/l					
	Faecal coliform	MPN/100 ml[b]					
	Other, specify…………	………..					

Notes:[a] If data are not available for the years stated, please provide the data you might have for other years and add a footnote for the years to which the data apply.
[b] MPN/100 ml: Most Probable Number per 100 ml.
Source: http://unstats.un.org/unsd/environment/q2004water.pdf.

- Development of geographic and substantive classifications such as watersheds and environmental impact categories
- Assessment of the requirements, sources and availability of data
- Structuring of statistical databases, compendia and state-of-the-environment reports
- Facilitating data analysis through synthesis and aggregation into indicators or indices
- Improvement of international comparisons.

A number of countries and international organizations based their methods and surveys on the FDES or PSR framework. Figure 4.2 shows the PSR framework used by the OECD for the compilation of environmental indicators (OECD, 2003). In general the use of statistical frameworks does not extend to the structuring of databases and statistical compendia – possibly to hide data gaps. Statistical compendia thus present their data typically under selected environmental topics without attempting to carry over the framework's linkage of topics and statistics. This applies to Eurostat publications on environmental and related (transport, energy, environmental expenditures) indicators,[2] as well as to its publications for other regions such as *The CARICOM Environment in Figures 2002* (Box 4.1). Still, several member states of the Asian Development Bank (ADB) did organize

[2] Eurostat seems now to have discontinued its environment statistics publication, replacing it by sets of environmental and related indicators (http://epp.eurostat.cec.eu.int/portal/page?_pageid = 0,1136239,0_45571456&_dad = portal&_schema = PORTAL).

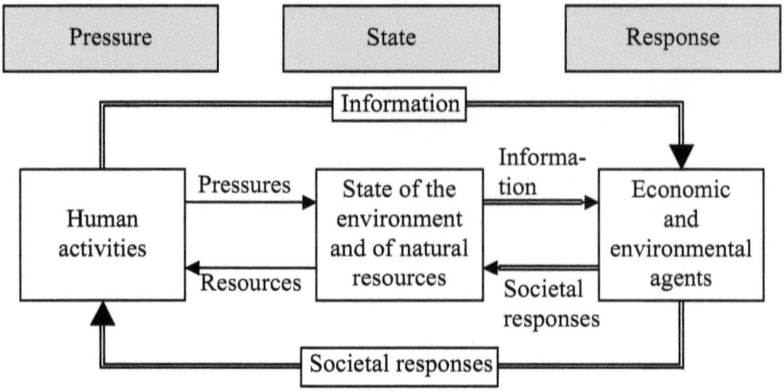

Fig. 4.2 Pressure-state-response framework (PSRF)
Source: OECD (1993, fig. 1a, simplified).

Box 4.1 Environmental topics in the CARICOM compendium

1–3 Background information
Geography, policy issues, socio-economic characteristics
4 Environmental health
Water and sanitation, environmental diseases
5 Tourism
6 Freshwater
Water abstraction, supply, use, treatment and quality
7 Coastal and marine resources
Water quality, vulnerable areas and protection, fisheries and aquaculture
8 Land use and agriculture
Land use, biochemicals
9 Forests
10 Biodiversity
Threatened species, protected areas
11 Minerals, energy and transport
12 Air
Greenhouse gas emissions, ozone-depleting substances, other, air quality
13 Waste
14 Natural and environmental disasters
Source: CARICOM (2003).

their environment statistics in the FDES format, as recommended for 'newcomers to this field' (ADB, 2002).

Following the lead of most international organizations, countries seem generally content to use the different frameworks for an initial check of data needs and availability, and to present their data for the traditional environmental media of air, water, land and, sometimes, human settlements. This allows flexibility, but introduces a certain degree of arbitrariness in selecting and presenting environment statistics.

The remaining question is what are the actual needs for and uses of a comprehensive environment statistics publication? One of the first assessments of an environment statistics compendium in Finland[3] came up with the surprising result of high schools as the main user. Most decision-makers apparently continued to rely on their own databases, tailored to their specific needs such as water data for a water department or hydrological institute. Of course, such specialization impairs data integration across institutions and environmental and socio-economic fields.

Policymakers tend to ignore the extensive and difficult-to-read measurement of environmental issues in large statistical compendia. The reason might be their preference for more selective and aggregate indicators related to pressing policy concerns. Moreover, statistical compendia rarely specify concrete use and applications, beyond generics like the objective of 'sustained monitoring and evaluation of the state of the environment and sustainable development' (CARICOM, 2003). Perhaps the most important use of a common framework for environmental topics and statistical variables is to foster better communication between data producers and users.

4.1.2 Integrating Economic, Environmental and Social/Demographic Statistics[4]

Chapter 2 identified interactions and repercussions between the environment and socio-economic activities as the cause of potential non-sustainability of economic activity. Obviously, environmental statistics cannot assess these interactions on their own, but need to be linked to the other statistical fields.

Figure 4.3 illustrates the numerous interrelationships among the stocks and flows of the three basic areas of economic, environmental and demographic (and social) statistics. The figure thus elaborates on Fig. 2.1, the environment-economy interface, in terms of statistical topics and variables. The (highlighted) sequence of flows of pollutants from production and consumption (flows 6 and 7) illustrates this interaction. Emissions are partially controlled by environmental protection (8), or escape control and accumulate in environmental media (14); subsequent 'consumption' of pollutants by humans is shown as part of 'other' population activities (18). A further extension of this sequence refers to health and welfare effects from pollution (19) and natural disasters (15).

---

[3] Oral communication by a representative of Statistics Finland at a United Nations expert group meeting.

[4] This section is based on Bartelmus (1987).

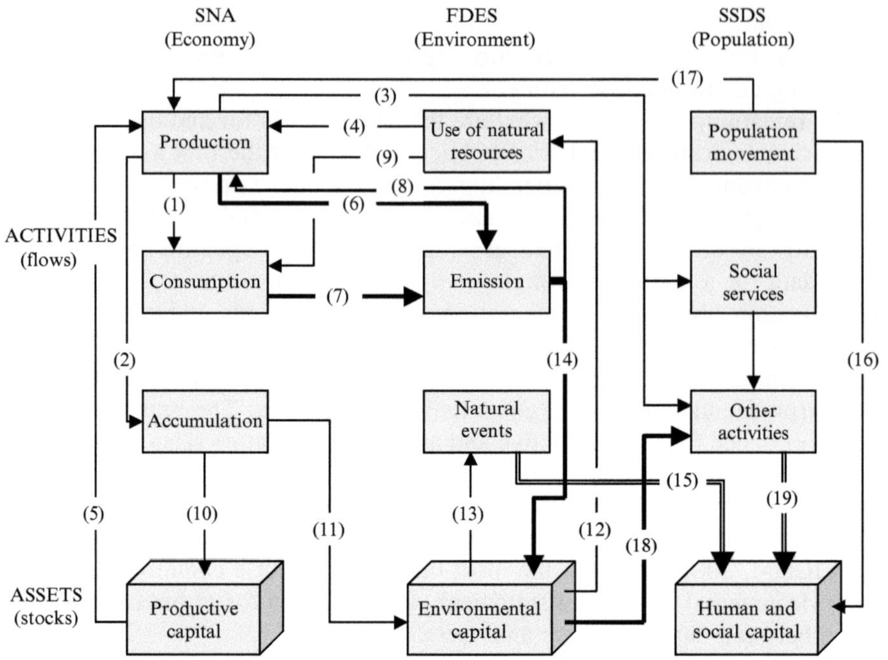

Fig. 4.3 Real world and statistical systems
(1) Goods and services for private and public consumption; (2) Capital goods; (3) Supply of social services and use of goods in 'other activities'; (4) Use of natural resources in production; (5) Fixed capital consumption; (6) Emission of waste and pollutants from production; (7) Emission from consumption; (8) Pollution control, environmental protection; (9) Consumption of natural resources (subsistence, physiological); (10) Capital formation; (11) Construction of shelter and infrastructure; (12) Depletion of natural resources; (13) Destruction of human settlements and natural resources by natural disasters; (14) Ambient concentrations in the human environment; (15) Loss of life and limb from natural disasters; (16) Net population growth; (17) Labour; (18) Human consumption of pollutants; (19) Health and welfare effects.
Source: Bartelmus (1987, fig. 1, modified).

Real-world complexities and interactions call for better and transparent linkage of the statistical systems across their conventional boundaries. The FDES offers some linkage through its information categories of economic activities, environmental impacts and social responses. The flexible structure of the FDES appeals as a tool for the coordination and connection of all basic statistical areas. Table 4.3 thus applies FDES criteria to the SNA and SSDS in order to outline an overall Framework for Statistical Integration (FSI) (or put more modestly, for statistical coordination). Contrary to a systems approach, the framework does not try to relate variables through strict functional or accounting relationships. It should help, though, developing common concepts, definitions, classifications and tabulations.

Table 4.3 Framework for statistical integration (FSI)

Information category / Statistical system (subject area)	Assets	Activities	Impacts on assets	Responses to impacts
SNA (economy)	- Financial assets - Net tangible assets	- Production - Consumption - Accumulation - Distribution - Rest of the world	- Saving - Net capital formation - Other volume changes	- Macroeconomic policies
FDES (environment)	- Stocks/reserves of natural resources and environmental assets - Human settlements	- Use of natural resources - Emissions - Natural events - Construction and use of shelter and infrastructure	- Resource depletion or increase - Ambient concentrations of substances - Ecological impacts - Welfare effects	- Environmental policies and programmes - Environmental management
SSDS (population)	- Population	- Population movements - Other social activities	- Population increase or decrease - Changes in public health - Other changes in human capital	- Population policy - Employment policy - Provision of social services

Source: Bartelmus (1987, table 1, modified).

The FSI largely maintains the original subject areas or topics of SNA, FDES and SSDS. On the other hand, it organizes these subjects around the FDES information categories. These categories simply reflect the common interest of all statistical systems in describing the state and changes of the real world, as well as the major activities responsible for these changes. Marking the first column of the framework as opening assets and adding a last column of closing assets could display accounting relationships. This is the approach taken in greening the national accounts (Section 7.2). Given ever-changing social concerns, the framework should also facilitate the evaluation of established but under- or overused statistical series, and reveal new data needs.

The category of social response is unusual in traditional statistics. Much of the dissatisfaction with conventional statistics stems from ignoring policy responses and their evaluation. This is one reason for the interest of policymakers in more flexible 'indicators' that relate to policy objectives and can be readily adapted to changing concerns and priorities.

4.2 From Statistics to Indicators 'for' Sustainable Development

4.2.1 Indicator Selection: Reducing Information Overload

The main drawback of the FDES is the generation of nearly 500 statistical variables in its follow-up methodological publications (United Nations, 1988, 1991). International organizations advanced, therefore, shorter lists of 'core', 'key' or 'headline' indicators of both the environment [FR 4.3] and sustainable development [FR 4.4].

Selecting key statistics as indicators for environmental assessment and policy analysis blurs the distinction between environmental statistics and broader indicators. Most definitions stress, indeed, the capability of an indicator to represent a social concern beyond the immediate meaning of the underlying statistic(s) (Box 4.2).

Indicator selection and definition are first steps towards aggregation of data for assessing the state of the environment or sustainable development. The inherent subjectivity in choosing indicators for multidimensional sustainable development, and a call by the Rio Summit to nonetheless develop indicators for the paradigm (United Nations, 1994, ch. 40) are the reasons for a flurry of different indicator proposals. International organizations, governments, NGOs and experts in the field proposed

Box 4.2 Indicator definition

The social indicator movement of the 1970s is probably the best-known attempt to reflect the standard of living by selected non-monetary statistics. It brought about a large and confusing variety of definitions and terms for indicators, statistics and indices (e.g. Gallopín, 1997). Most definitions refer to the broader 'representativeness' of a selected statistic or combination of statistics. This suggests the generic definition of an indicator as:

> simple average of a statistical variable or ratio of variables that provides an image beyond the immediate attribute or observation of the variable or ratio itself.

Besides selection of the statistics, the interpretation of the overall image introduces a further subjective element into indicator use. A good example is average life expectancy that is generally taken as a measure of population health.

In contrast to an average of statistical variables, an *index* is a combination of indicators. The index is usually calculated as a weighted or unweighted (equally weighted) indicator average; other more complex aggregation methods also apply (see Section 5.1).

widely differing indicator sets, undeterred by the largely negative experience of the social indicator movement in the 1970s [FR 4.3]. Indicator lists of varying length and contents reflect the concerns or interests of their authors [FR 4.4]. Typically these concerns refer to some or all of the following topics:

- Population (growth, migration, refugees)
- Human needs (health, food, housing, education, equity, security, etc.)
- Renewable and non-renewable natural resources
- Environmental quality (air, water, land)
- Ecosystems (acidification, eutrophication, biodiversity)
- Economic activities (and their impacts, including emissions, natural resource use, production and consumption patterns, technologies)
- Natural and man-made disasters
- Global environmental problems (climate change, ozone layer depletion)
- Globalization
- Institutions.

Clearly, these topics overlap. Determining the desired scope and coverage, minimizing overlap and choosing the best indicators for different topics requires a more systematic selection process – than ad hoc choices by interested parties. Urging the use of good criteria for indicator selection and definition, such as those of Box 4.3, may help improve indicator quality and validity. However, admonition will not do: what we need is a clear procedure, which identifies

Box 4.3 Principles and criteria for sustainable development indicators

Bellagio principles (http://www.iisd.org/measure/principles/bp.asp):

- Guiding vision and goals, holistic perspective, essential elements of sustainable development
- Adequate scope (temporal and regional)
- Practical focus (categories and framework, limited issues and indicators, standardization, targets and thresholds)
- Openness, effective communication, broad participation
- Ongoing assessment (iterative and adaptive indicator development) and institutional capacity

to which one could add OECD (2003) criteria:

- Representativeness of indicators
- Comparability for international comparison
- Analytical soundness and measurability

quantifiable topics of broad concerns and relates the topics to the appropriate data system. This is indeed the approach of the above-described FDES and similar indicator frameworks.

4.2.2 A Framework for Sustainable Development Indicators

As shown in Section 4.1.2, the FDES is capable of presenting different statistical fields in terms of stock and flow categories. The FDES also facilitates linking these variables across the different fields and categories through its action-impact-reaction structure. In principle the – expanded – FDES could thus facilitate the transparent selection and definition of a reasonable number of sustainable development indicators.

The Earth Summit's Agenda 21 (United Nations, 1994; see also Fig. 1.1) reflects international agreement on the scope and coverage of sustainable development. For developing a *Framework for Sustainable Development Indicators (FSDI)* (Bartelmus, 1994b),[5] Table 4.4 groups the Agenda 21 programmes under the economic, social, environmental and institutional dimensions of sustainable development. Cross-classification with the FDES information categories obtains a framework, which combines the concerns of potential data users (reflected in Agenda 21) with those of the data producers (presented as FDES-type statistical topics). Most indicator proposals applied, at least initially, some version of FSDI (mostly under the PSR label), but without resort to the statistical database [FR 4.4].

In the environmental field, the contents of the FSDI consist mostly of FDES statistical topics. For sustainable development indicators, new topics stem from other statistical fields for the economic, social and institutional dimensions of sustainable development.

The impacts/effects column shows the physical impact of economic activity on the state of the environment and on humans as welfare effects of these impacts; these are the symptoms of environmental non-sustainability of socio-economic development. The activities/events category refers to the causes (driving forces and pressures) of impacts and effects from production and consumption, population dynamics, natural resource use, emission of pollutants and waste, and natural and man-made disasters. The social response to impacts and effects can be carried out through natural resource management, pollution control, macro-policies of sustainable development, private sector adaptation and institutional change. Inventories/stocks describe the economic and environmental capacities of supporting sustainable growth and development in the long term; they are a key element of environmental sustainability and accounting.

[5] The original proposal was for a framework for indicators of sustainable development. The relabelling as Framework for Sustainable Development Indicators' is more in line with distinguishing between indicators 'of' and 'for' sustainable development (see Section 4.3).

Table 4.4 Framework for Sustainable Development Indicators (FSDI)[a]

	FDES information categories			
Agenda 21 clusters	Socio-economic activities, events	Impacts and effects	Responses to impacts	Inventories, stocks, background conditions
Economic issues 2. Cooperation, 4. Consumption, 33. Finance, 16/34. Technology, 8. Decision-making	- Economic growth - Trade - Production and consumption patterns	- Sustainability of economic performance and growth	- Private sector responses - Sustainability policies and programmes - Fiscal instruments - Environmentally sound technology	- Economic situation - Produced capital stock
Social and Demographic Issues 3. Poverty, 5. Demographics, 36. Education, training, 6. Human health	- Population growth and change	- Distribution of income and wealth - Human health and contamination	- Private sector response - Social policy and programmes	- Demographic and social conditions - Human capital stock
Environmental Issues 9. Air/climate, 10/12–14. Land/soil, 17/18. Water, 11/15. Other natural resources, 19–22. Waste, 7. Human settlements and natural disasters	- Emission into air, water - Application of biochemicals - Waste - Use of natural resources (fish, land, water, other)	- Quality of air, land/soil, water - Change in stock/depletion (fish, water, minerals, etc.) - Impacts of disasters - Human health, contamination	- Pollution monitoring and control - Resource management and rehabilitation - Private sector response	- Natural resource stocks (agriculture, fishery, hydro-systems, fauna, flora, minerals, lithosphere, ecosystems) - Weather, climate
Institutional Support 35. Science, 34. Capacity-building, 23/32. Roles of groups, 38-39. Institutional, legal arrangements, 40. Information for decision making			- Private sector response - Environmental law and legislation - Environmental data, information	- Institutional capacities

Note: [a]Based on Bartelmus (1994b), table 3.

The importance of frameworks in tracing generic concerns down to statistics becomes evident when indicators need to be defined rigorously and transparently in terms of their underlying statistics. Unfortunately, data users mostly ignore this aspect when negotiating for indicator lists that serve different policy agendas. Note that in comparison to the core FDES topics the statistical topics of socio-economic and institutional sustainability dimensions are quite undeveloped in the FSDI. This may have contributed to the later abandonment of the FSDI by data users. Typically, data users are less concerned or familiar with the nitty-gritty statistical work.

Table 4.5 shows – in the FSDI format and for the example of freshwater – different indicators advanced by the original FSDI, the United Nations and the European Environment Agency (EEA). Some relabelling and break-ups of the FSDI columns do not really alter the original framework.[6] Other organizations also use the general pressure-state-response idea for their own environmental and sustainability concerns. However, applying similar information categories to differing or differently clustered environmental and socio-economic concerns still generates different indicator sets [FR 4.3, 4.4].

Deviations from the FSDI and the DSR framework reflect an unwillingness by national and international data users to be bound by the – non-binding – recommendations of Agenda 21 and the resulting large number of over 100 indicators.[7] The OECD thus limited its 'core' environmental indicators to 40–50 indicators and reduced these further to 10–13 'key' indicators as 'signals to policymakers' (OECD 2003). Similarly, the EEA uses 12 indicators in its summary of the *Environmental Signals 2002*[8] report.

The same motivation seems to be behind the abandonment of the DSR framework by the United Nations in a more recent publication: on the one hand, policymakers did not want to be bothered by a cumbersome data framework, which, 'although suitable in environmental context, was not as appropriate for the social, economic,

[6] The DPSIR framework of the EEA distinguishes explicitly between a state category ('impact' in the FSDI/FDES) and an impact category ('effects' in the FSDI/FDES); the framework also extends the activities/events category by introducing 'drivers' (of economic sectors) and presenting activities and events as 'pressures' (of natural resource use and emissions). The DSR framework of the United Nations simply renames the FISD categories of activities/events as 'driving force' and impacts/effects as 'state'. Note also that the omission of a stock category shifts the availability of natural resources such as groundwater or mineral reserves to the state category in the DSR framework, and to the response (reservoir stocks) categories in the EEA's DPSIR framework (indicated by arrows in Table 4.5).

[7] An initial 'starter set' of FSDI indicators (Bartelmus, 1994b) came up with 107 indicators; later, the DSR framework generated 130 indicators (United Nations, 1996).

[8] http://reports.eea.europa.eu/environmental_assessment_report_2002_9/en (summary); discontinued in the EEA 2004 Signals which present the full set of 30 indicators (http://reports.eea.europa.eu/signals-2004/en/ENSignals2004web.pdf).

Table 4.5 FSDI and related frameworks: Freshwater indicators

Frameworks	Activities/events	Impacts/effects	Responses	Inventories/ stocks
FSDI (statistical topics)[a]	- Fisheries - Water use - Emissions into inland waters	- Fish stock changes - Water resource changes - Water quality	- Resource management and rehabilitation - Pollution monitoring and control	- Fish stocks - Hydrological systems
DSR (indicators)[b]	*Driving force*: - Annual withdrawal of ground and surface water as per cent of total available water -Domestic consumption of water per capita	*State*: - BOD in water bodies - Concentration of faecal coliform [- Groundwater reserves][d] →	*Response*: - Wastewater treatment coverage - Density of hydrological networks	
DPSIR (indicators)[c]	*Drivers and pressures*: Drivers: - Emissions of nitrates and phosphates from urban wastewater treatment Pressures: - Emissions of organic matter and hazardous substances - Mean water allocation for irrigation - Water exploitation index - Water use by sectors and in urban areas	*State and Impact*: State: - Concentration of ammonium, BOD, nitrates, phosphates, hazardous substances, nutrients, organic matter in rivers - Bathing water quality - Drinking water quality - Biological quality of lakes - Hazardous substances, phosphates in lakes (eutrophication) - Nitrates, pesticides in groundwater Impact: - National river classification schemes - Non-indigenous species in rivers and lakes - Saltwater intrusion - Water exploitation index	*Responses*: [- Overall reservoir → stocks][d] - Urban waste water treatment (effectiveness) - Water prices - Water use efficiency	

Notes: [a] Table 4.4.
[b] United Nations (1996); DSR is the acronym for Driving force, State, Response.
[c] European Environment Agency (http://themes.eea.eu.int/Specific_media/water/indicators); DPSIR stands for Driving forces, Pressures, States, Impacts and Responses.
[d] Arrows indicate a misplacement of stock variables in the respective frameworks.

and institutional dimensions of sustainable development' (United Nations, 2001b). On the other hand, discarding a framework that might reveal large data gaps, allowed ignoring missing issues and data, and facilitated agreement on a short 'core set' of 58 indicators for selected policy 'themes'.

The sometimes-heated discussion of theme and indicator selection reveals another dichotomy between data users and official (governmental) data producers. Impatient data users are eager to obtain rough-and-ready information, even at the cost of less clarity and accuracy, whereas statisticians may question the validity of crude estimates. This dichotomy carries over into the assessment of the sustainability of economic growth and development by means of ad hoc compilations of indices (Ch. 5) and more systematic environmental accounting (Chs. 6 to 8).

4.2.3 Indicator Use: Alert, Action or Evaluation?

Policymakers are usually unable to specify their data needs beyond generics such as to 'provide solid bases for decision-making' (United Nations 1994, ch. 40), 'reporting on the state of sustainable development', 'fulfillment of governmental goals and targets' (United Nations 2001b), or 'to support and illustrate country environmental performance' (OECD 2003). A 'short list' of 14 'structural indicators' is to measure progress towards the somewhat conceited goal of the European Union 'to become the most competitive and dynamic knowledge-based economy in the world capable of sustainable economic growth with more and better jobs and greater social cohesion.'[9]

Three general purposes of indicator use can be distinguished:

- Early warning about hazardous impacts of economic activity
- Assisting in policy formulation
- Evaluation of policy performance.

Implicit or explicit extrapolation of trends of environmental and social impacts of economic growth can alert us to risks of environmental degradation, exhaustion of natural resources and social problems. More sophisticated modelling of impacts and repercussions between environment and economic growth can provide more accurate prediction, if based on realistic assumptions and valid data (Ch. 11).

---

[9] 2000 Lisbon European Council Presidency Conclusions (http://www.europarl.europa.eu/summits/lis1_en.htm). The structural indicators can be found on Eurostat's web site: http://epp.eurostat.ec.europa.eu/portal/page?_pageid = 1133,47800773,1133_47803568&_dad = portal&_schema = PORTAL.

Policy formulation and evaluation require the specification of goals, targets or benchmarks, for which policy instruments need to be specified and against which progress or failure can be assessed. The political process of selecting themes and sustainable development indicators by the United Nations Commission on Sustainable Development did not succeed in specifying such targets. Rather, the weak assumption is that the indicators 'implicitly reflect the goals of sustainable development' (United Nations, 2001b). The most the United Nations could do was listing goals, targets and standards from international conventions and conferences for the different themes in an annex, but without direct link to the proposed indicators.

The Millennium Development Goals (MDG) indicator programme is a collaborative effort of the United Nations Statistics Division, the International Monetary Fund, the World Bank and OECD. It went further, using the goals and their time-bound targets (Box 3.3) to develop and compile 48 indicators for each goal and target. Table 4.6 illustrates this approach for the access-to-water-and-sanitation target. Simple extrapolation of the 1990 and 2000 indicators to 2015 indicates that for attaining the target of halving non-access by 2015 greater strides need to be made, especially in rural areas. Such oversimplified analysis also reveals the limitations of indicator use, due to lack of data: a linear extension of a decade's first- and last-year data is not a valid prediction of what could happen 15 years into the future. Data availability is, of course, better in rich countries. For assessing progress towards sustainable development, the OECD presents for its member states full time series of indicators and confronts them with various national and international standards and targets (OECD, 2003).

An interesting variation of policy evaluation guides China's search for indicators of 'sustainable and harmonious development'. A focus on the performance of local government officials reflects the continuing influence of the hierarchical structure of the Communist Party (Box 4.4).

The indicators in the above-mentioned examples show progress or regress in the particular areas they represent. They do not show the relative significance of any specific area or target. The reason is incomparability of the indicators used for different areas. Indicators may indeed alert us to negative trends and urge action where particular limits are at risk of transgression. However, they cannot set priorities for

Table 4.6 Trends towards meeting MDG targets for access to water and sanitation

	Sustainable access to improved water sources (% of population)			Access to improved sanitation (% of population)		
	1990	2000	2015	1990	2000	2015
Urban	94	95	96.5 [97.5]	81	85	91 [92.5]
Rural	64	71	81.5 [85.5]	28	40	58 [70]

Note: 2015: linear extrapolation; target values in brackets.
Source: http://unstats.un.org/unsd/mi/mi_worldregn.asp.

Box 4.4 Evaluation indicators for local government officials in China

China's focus on rapid economic growth largely ignored environmental trade-offs. A recent indicator project (CCICED, 2005; Li et al., 2007) aims to 'change the bias' towards economic growth by refocusing governmental policy on a 'scientific approach to development and harmonious society'. Measurement of the performance of local officials by new indicators is seen as 'a conductor's "baton" that manipulates government works'.

Two categories of the proposed indicators either provide 'scores', which can be added up for performance evaluation, or 'veto' (prohibit) further activity because its effects are measured as a violation of environmental or social limits. The claim, based on case studies, is that the indicator system 'is able to ... stimulate the local government to pay more attention to social development as well as ecological and environmental protection, and to give more respect to social justice and life'.

This assessment seems now to be overly optimistic: the recent 'quashing' of two reports on 'green GDP accounting' [FR 8.2] and of data on deaths from pollution (by the World Bank) 'appeared to suggest reluctance at the top of China's government to acknowledge the seriousness of environmental degradation' (M. Landsberg, *Los Angeles Times*, 24 July 2007). Other experts blame the resistance of local officials to any attempt at evaluating their environmental performance.

action according to the importance of different concerns. Stakeholder groups might pick up indicators for prodding government into action, but would of course advance their own priorities and agendas.

Scattered indicator use and proliferation of indicator 'menus' serving different policy agendas revived the idea of a common framework. It remains to be seen if the United Nations Commission on Sustainable Development, which rejected such a framework (Section 4.2.2), is now ready to reverse this decision after doubts about the relevance of its indicator work. An expert group addressed this critique and called for a 'capital-based' framework, which would combine the capital maintenance concept of sustainability accounting with the policy agenda of the MDG (Pintér, Hardi & Bartelmus, 2006).

4.3 Global Warming: The Indicator 'of' (Non)Sustainable Development?

Assessing the overall sustainability or non-sustainability of economic growth and development requires aggregation. Simple listings of indicators 'for' sustainable development cannot capture composite notions of social progress. This might explain why national policymakers tend to ignore long and complex indicator sets.

Understandably, they prefer to respond to a 'nutshell' indicator *of* the environment or sustainable development that caters to the social concern *en vogue*.

Environmentalists have drawn attention to what they consider the greatest threat to human survival: global warming. In this they found broad support, owing to public media campaigns such as former Vice-President Al Gore's 'Inconvenient Truth' or the 'Live Earth' concerts [FR 4.5]. Even corporations flaunt their concern about climate change and cash in on lucrative tradables of greenhouse gas (GHG) emissions.[10]

Estimates of the impacts of global warming vary widely (cf. Table 1.1), however. There is still uncertainty about the degree of warming itself, and more so about its effects on natural systems, human health, and human capability of dealing with these effects. The latest report of the Intergovernmental Panel on Climate Change (IPCC, 2007) puts global warming since pre-industrial times at about 0.8 °C and predicts a temperature increase between 1.8 and 4.0 °C by the end of the century (Box 4.5). Plate 4.1 dramatizes the impact and distribution of global warming by the third and last decade of the century [FR 4.5].

Eco–nomics plays an important role in keeping particular environmental concerns such as climate change in perspective, especially with regard to other environmental and economic goals. The Stern (2007) 'review of the economics of climate change' might have succeeded in doing this by monetizing the different, mostly non-comparable environmental effects of global warming. However, the review shows some bias in its valuations that makes the results questionable.

Box 4.5 IPCC (2007) report on climate change – key results

- Greenhouse gas (GHG) concentrations increased 'markedly' due to fossil fuel use and land-use change since 1750.
- Global warming is, with 90% probability, the net effect of human activities.
- Total temperature has increased since 1850 by 0.76 °C.
- 'Best estimates' indicate a global temperature increase within the 21st century of 1.8–4.0 °C (lowest and highest scenario).
- Effects of global warming in the 21st century:
  - Snow cover and sea ice is 'likely' to decrease
  - The intensity of tropical cyclones is 'likely' to increase
  - Precipitation is 'very likely' to increase in high latitudes and 'likely' to decrease in the subtropics
- Even with GHG stabilization, global warming and sea level rise are expected to continue for centuries.

[10] *The Economist* of 9 September 2006, 'The heat is on, a survey of climate change'. The convenience of a surrogate indicator for environmental impacts has made CO_2 emissions, the main GHG, also a favourite of index calculations (Section 5.2) and modelling (see Part IV).

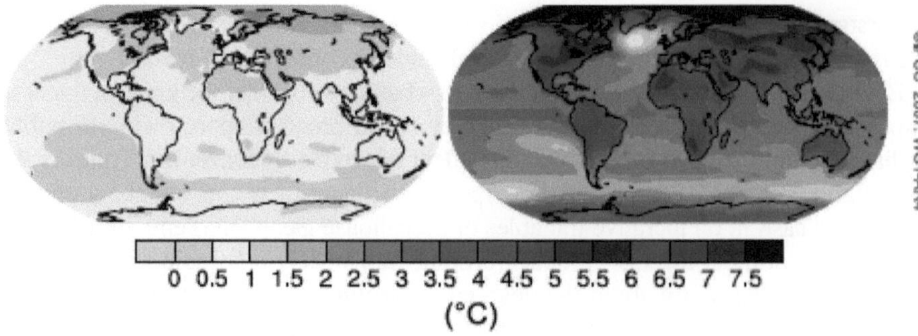

Plate 4.1 Projected surface temperature increase in the 21st century[a]
Note: [a]"Best estimates" for the high-impact scenario, compared to 1980–1999.
Source: IPCC (2007) – Climate Change 2007: The Physical Science Basis, Summary for Policymakers. Intergovernmental Panel on Climate Change (*See Colour Plates*).

The review uses cost-benefit analysis for evaluating the stabilization of climate change at a desirable level. Damage of non-action is measured as a welfare loss 'equivalent to a reduction in consumption per head of between 5-20%' ('now and in the future'). The net benefit is determined by comparing this welfare loss, deemed also to be 5–20% of world GDP, to the relatively low annual action cost of 1% of world GDP. A number of methodological flaws impair the estimates, including:

- The combination of different welfare valuations for health and environmental damage, and their incompatibility with the market values of GDP and consumption (see Section 8.1.3)
- The normative (ethical) choice of low social discount rates for international and intergenerational equity (cf. Section 2.3.2)
- The use of particularly pessimistic model assumptions, albeit with indications of risks and uncertainties.

The current media hype surrounding climate change risks ignoring or downgrading other environmental and sociol costs, e.g. of pollution or poverty. It is no surprise that in most countries proclaimed policies of sustainable development continue to focus on economic growth, catering to sustainability with some measures of energy saving and CO_2 emission control. Box 4.6 illustrates this reductionist view. As discussed in Section 3.3.2, one reason for this view is a persisting EKC mentality in economic policy; the expectation is that the transition to a dematerialized service economy will solve most other environmental problems.

The reductionist view overlooks, however, that

- Rich countries achieved some of their environmental successes by depleting the natural resources of developing countries and, in some cases, by translocating dirty production processes – in other words, by importing sustainability.
- Services and information technology still require large amounts of energy and material inputs, and infrastructure.

Box 4.6 A reductionist view

SUSTAINABLE DEVELOPMENT

⇓

SUSTAINABLE ECONOMIC GROWTH

⇓

ENVIRONMENTALLY
SUSTAINABLE GROWTH

⇓

[Climate change]

⇓

Economic growth with CO_2 control

- Risks of new (notably genetics and nanotechnology) and old (nuclear energy) technologies loom large.
- Rich countries mostly ignore 'pollution of poverty', i.e. poverty itself and environmental impacts in poor regions of the world (natural disasters, water shortage, soil degradation, deforestation, urban and indoor air pollution, and epidemic diseases).

One cannot dismiss, of course, the considerable ramifications and risks of an undeniable human-made global warming trend. But the potentially disastrous effects remain risks. Selecting one particularly ominous environmental problem and diverting funds from other social, economic and environmental concerns can only be justified when there is no doubt about an imminent disaster that dwarfs all other problems. The further reading section refers to some doubt, though, about exceedingly high damage cost of global warming and relatively low cost of tackling the damage without delay [FR 4.5]. Much of this book is therefore about comparison and evaluation, based on a *comprehensive* measurement of the environment-economy interaction rather than cost-benefit analysis of particular issues. The next chapter examines whether popular compound indices are up to the task of an overall assessment of interrelated environmental, social and economic concerns.

Further Reading

FR 4.1 Basic Statistical Systems

Figure 4.1 presents the main statistical systems and frameworks of the United Nations, recommended for worldwide application. Ward (2004) gives an overview of the history of all statistics developed and promoted by the United Nations.

The *System of National Accounts* (SNA) (United Nations et al., 1993) evolved out of interest in financing the Second World War mobilization and post-war recovery through economic growth. The initial focus on national 'income' accounting soon expanded to record all economic activity related to production, consumption, investment and foreign trade. A more concise introduction to the voluminous publication facilitates access to the complex accounting system for data users and 'first-time accountants' (United Nations, 2004). Another handbook of national accounting discusses use in policymaking and modelling, including green accounting (United Nations, 2002b).

In analogy to the SNA, the *System of Social and Demographic Statistics* (SSDS) (United Nations 1975) presented stocks and flows of individuals and social groups and their economic and social activities in an accounting system of life sequences, time budgets and cost-benefit distributions. Lacking a common *numéraire* and unifying theory, the United Nations Statistical Office abandoned the system approach for a *Framework for developing and integrating Social and Demographic Statistics* (FSDS) (United Nations, 1979); the result is a framework for the development of social indicators [FR 4.3].

In the field of environment, the same reasoning about the lack of a *numéraire* and theory brought about the *Framework for the Development of Environment Statistics* (FDES) (United Nations, 1984). The framework represents a combination of four common approaches (United Nations, 1982): the environmental media, stress-response, accounting, and ecological approaches. The latter represent a particular field of statistical analysis, referred to, often synonymously, as ecological statistics or statistical ecology (see, e.g. the journal *Environmental and Ecological Statistics*).

FR 4.2 Cross-disciplinary Statistical Systems

Figure 4.1 also displays cross-disciplinary statistics as interfaces in the Venn diagram. Besides the environmental-economic accounts of the *MFA* and *SEEA* (discussed in Chs. 6 to 8), *Social Accounting Matrices* (SAM) record the interactions between social groups and the economy. They expand the national accounts for the measurement of income distribution and labour market activities (United Nations et al., 1993, ch. XX). Data systems of environment-population interaction are least developed. The Population Division of the United Nations (2005) developed a *Population, Resources, Environment and Development* (PRED) databank, which seeks to capture some of the relations between these areas. Ehrlich and Holdren (1971) advanced in the early 1970s the *I = PAT identity* (see Ch. 13, Introduction). Harrison and Pearce (2000) used the equation as a framework for an atlas on population and environment; Waggoner and Ausubel (2002) applied IPAT for a systematic approach to 'sustainability science' (cf. FR 2.2).

FR 4.3 Social and Environmental Indicators

Social and environmental indicators were developed independently. The *social indicator* movement of the 1970s aimed at measuring the human quality of life as an alternative to economic (monetary) indicators (Drewnowsky, 1970, 1974; OECD, 1973, 1976). However, the quality of life and social indicator movements fizzled out when no agreement on the concept and its measurement could be reached (Hankiss, 1983). At the global level, only a 'minimum list' of social indicators for the follow-up of United Nations conferences on children, population and development, social development, and women survived (http://unstats.un.org/unsd/demographic/products/socind/default.htm). It remains to be seen if new attempts at a revival of quality of life measurement (Fergany, 1994, Henderson et al., 2000) will succeed in establishing these measures in recurrent (official) statistics.

Many national environmental agencies compile now *environmental indicators* as part of, or separate from, state of the environment reports. At the international level, the OECD compiles regularly 'core', 'key' and 'sectoral' environmental indicators (OECD, 2003). The European Environment Agency publishes 'environmental issues' and 'environmental headline' indicators, and 'environmental signals' reports (http://reports.eea.europa.eu/signals-2004/en).

FR 4.4 Sustainable Development Indicators

The pressure-state-response framework and its derivatives are now widely accepted tools for identifying, defining and organizing sustainable development indicators. The resulting indicators still differ, however, because the frameworks encompass different aspects of sustainable development, including particular 'themes' (Adriaanse, 1993; United Nations, 2001b; Eurostat: http://epp.eurostat.cec.eu.int/portal/page?_pageid = 1998,47433161,1998_47437052&_dad = portal&_schema = PORTAL), 'issues' (Kerr, 1997 for Environment Canada; OECD, 2003), 'syndromes' (Lüdeke & Petschel-Held, 1997), '(sub)systems' (Bossel, 1999) or 'policy fields' (Guinomet et al., 1997 for the European Union). Moldan et al. (1997) give an overview of approaches to developing indicators of sustainable development.

Time will show whether the 'core set' of indicators of sustainable development of the United Nations (2001b) or its current attempt at revision (http://www.un.org/esa/sustdev/natlinfo/indicators/isd.htm) will become the standard tool of assessing sustainable development. At present, the more practical, but limited (as far as sustainability is concerned) Millennium Development Goal indicators of the United Nations (http://milleniumindicators.un.org/unsd/mi/mi_goals.asp) appear to be more popular on the international stage. The International Institute for Sustainable Development (IISD) hosts a web site, which permits entries by indicator developers into a 'Compendium of Sustainable Development Indicators

Initiatives'; in March 2004, the Compendium included about 600 initiatives by individuals, governments, NGOs and international organizations (http://www.iisd.org/publications/pub.aspx?id = 607).

FR 4.5 Climate Change Assessment

The fourth assessment report by the *Intergovernmental Panel on Climate Change* (IPCC 2007; http://www.ipcc.ch/) provides the most authoritative assessment and prediction of global warming and its effects. The Climate Analysis and Indicators Project of the World Resources Institute presents climate indicators for countries and economic sectors in support of the United Nations Climate Convention (http://climate.wri.org/cait-project-93.html). Section 6.2 (Fig. 6.1) describes the greenhouse effect as a change in the global energy balance.

Mainstream economists expressed doubts about previous findings of the IPCC, stressing uncertainties in predicting the impacts of global warming (Beckerman, 1992; Nordhaus, 1998). Nordhaus and Boyer (2000) argue, with an optimal growth model, that setting limits to greenhouse gas emissions in the Kyoto Protocol lacks an assessment of implementation costs and benefits and achieves little in mitigating potentially high long-term damage. As discussed in the text, the *Stern* (2007) *Review* (http://www.hm-treasury.gov.uk/independent_reviews/stern_review_economics_climate_change/stern_review_report.cfm) does estimate the costs and benefits of tackling climate change. A Wikipedia web site provides an overview of first (positive and negative) reactions to the Review (http://en.wikipedia.org/wiki/Stern_Review).

Sounds and sights of Al Gore's 'An inconvenient truth' can be found on the movie trailer(http://www.apple.com/trailers/paramount_classics/aninconvenienttruth/trailer/). For world coverage of the Live Earth concerts see http://liveearth.msn.com/.

Review and Exploration

- A picture is said to be worth a thousand words. A statistical table may be worth a thousand pictures?
- Explain the difference between a statistical framework and system.
- Why do we need cross-disciplinary data frameworks? Describe the flows of natural resources in Fig. 4.3.
- What is the purpose of an indicator, as compared to a statistical variable? How can they help decision-making?
- Do the different indicator lists assess sustainability in economic growth and development? If so, how?
- Is there a communication gap between data users and producers? See also Sections 7.1 and 8.4.
- Does global warming adequately represent environmental and sustainability problems? What do cost and damage estimates tell us?

Chapter 5
Aggregation: From Indicators to Indices

Chapter 3 raised the question of quantifiability of the broad notion of development and its sustainability. The tentative conclusion was that the all-encompassing paradigm's promise of wealth and well-being for everyone appears to be rather empty, in the absence of verifiable results. The presentation of indicators *for* sustainable development in Chapter 4 neither confirms nor repeals this conjecture: assorted indicators fail to establish their contributions to sustainable development in a comparative manner; they cannot assess, therefore, overall progress towards implementing the paradigm's goals. For this, we need an index *of* sustainable development, which is built up comprehensively and consistently from the basic data. This chapter reviews critically different aggregation methods and the resulting indices as to their ability of conveying a coherent picture of sustainability. The flaws of these approaches direct us to the more systematic physical accounts and balances of Chapter 6.

5.1 Aggregation Methods

Building a compound index from indicators or statistical variables requires weighting the component variables according to their contribution to the overall index goal. The weighting problem is critical to both, ad hoc index calculations and to more systematic accounting (dealt with in Chs. 6–8). A brief review of different aggregation methods [FR 5.1] helps evaluate the numerous attempts at assessing environmental quality and sustainable development. Aggregation methods can be roughly categorized as judgemental, mathematical, scientific and empirical.

5.1.1 Judgemental Methods of Indicator Evaluation

Judgemental methods range from relatively informal, qualitative evaluations of different indicators to explicit procedures of reaching consensus in such evaluations.

P. Bartelmus, *Quantitative Eco-nomics*,

Facial icons added to indicators of the European Environmental Agency (EEA) (Table 5.1) are an example of a qualitative evaluation of a relatively short list of environmental indicators. The *Environmental Signals* report[1] describes these icons as

☺ Positive trend, moving towards target
😐 Some positive development, but either insufficient to reach target or mixed trends within the indicators
☹ Unfavourable trend.

Even then, the reader will be hard-pressed to give an overall evaluation of the European environmental state and its potential trend. *Traffic lights*, ranging from red alert, via yellow wait-and-see, to green o.k., are a similar, more advocatory presentation.

A first step from personal indicator evaluation to aggregative assessment is the simultaneous presentation of indicators in a geographical context, i.e. by *overlaying* indicators in *maps*. Such overlaying implies a possible correlation of one or more

Table 5.1 EEA indicator assessment

Environmental issue	Indicator	Assessment
	Tackling climate change	
Emissions of greenhouse gases	Trend in emissions and distance to 2008–2012 Kyoto target	😐
	Nature and biodiversity – protecting a unique resource	
Forest resources	Annual tree fellings	☺
Land resources	Land take and fragmentation of large habitats	☹
Emissions of acidifying substances	Trend in emissions and distance to 2010 EU target	☺
	Environment and health	
Emissions of ozone precursors	Trends in emissions and distance to 2010 EU target	😐
Urban air quality	Exceedance of ozone, fine particles, SO_2, NO_2	😐
Freshwater pollution	Concentration of phosphate and nitrate in rivers	😐
	Sustainable use of natural resources and management of wastes	
Material consumption	Total material requirement (vs. GDP)	😐
Fish stocks	Spawning stock biomass of the North Sea cod stock	☹
Urban waste generation	Trends in levels of municipal waste collected	☹
Water use	Water exploitation index	😐
Land take by development	Trends in built-up area, population and road network density	☹

Source:
http://reports.eea.eu.int/environmental_assessment_report_2002_9sum/en/signals2002_summary_en.pdf.

[1] http://reports.eea.europa.eu/environmental_assessment_report_2002_9/en.

indicators with regional characteristics. Plate 5.1 puts the above-described environmental surrogate of global warming into the world geography. The first part of the figure depicts a possible link of global warming to the colder areas of the Northern hemisphere. The figure also shows the limitations of such presentation, leaving a general impression of beneficial milder climates in the North. The lower part of the picture indicates, however, reduced precipitation and drought in subtropical countries as a further potential effect of global warming. Combining the two maps (and other mapped effects such as natural disasters) is not a solution since it would overload the graphical presentation. Judgemental selection, implicit equal weighting of indicators, and limited presentational capacity are the drawbacks of overlay mapping.

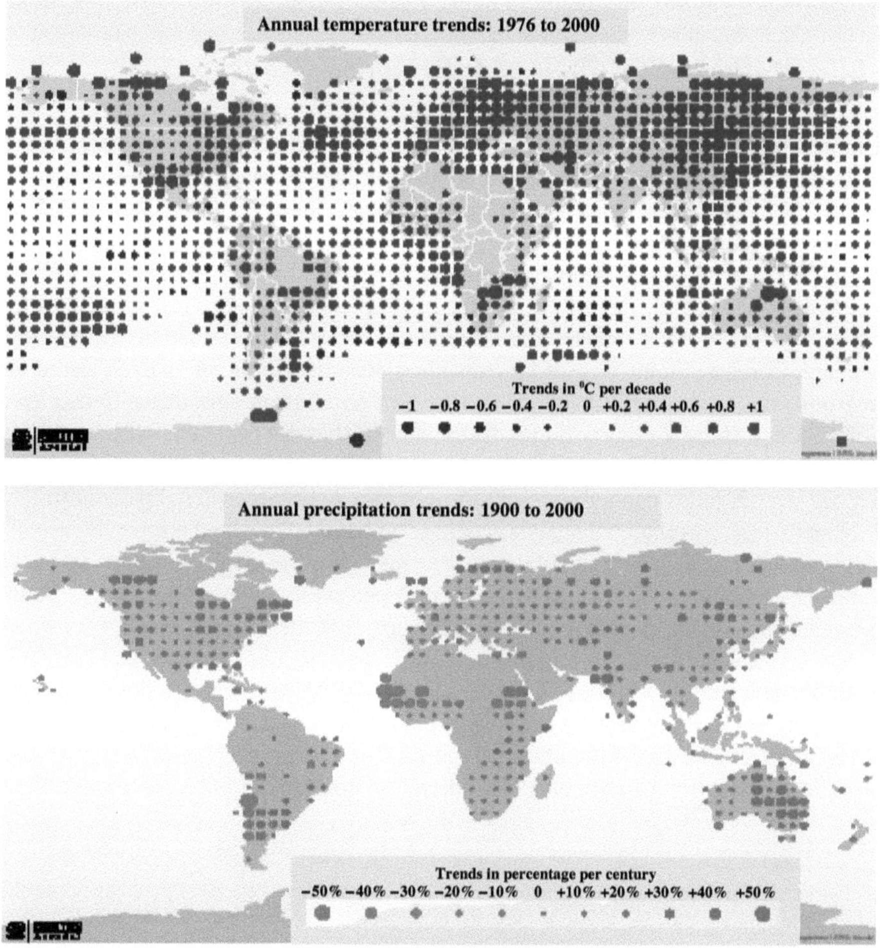

Plate 5.1 Overlay mapping: global warming and precipitation effects
Source: UNEP/GRID-Arendal (2005), Vital Climate Change Graphics. (*See Colour Plates*)

The *Delphi method* is a procedure for tapping into a broader range of expert knowledge. The idea is to generate evaluative consensus through several rounds of feedback from an expert panel. Jesinghaus (1997) suggested using this method for the 'expertocratic' weighting of Eurostat's environmental pressure indicators. However, to date, Eurostat has been hesitant to take up this suggestion.[2] The reason might be that even the use of expert knowledge remains subjective, depending on the selection of experts, their particular areas of expertise, methodological preferences and depth of knowledge, as well as the technical tools applied (such as questionnaire design).

Fuzzy logic combines qualitative evaluations of quantitative results in a more controlled manner. The method uses a range of results, rather than preset cut-off points, when determining combined indicator categories. Ultimately, even fuzzy logic has to set 'acceptability' standards.

5.1.2 Mathematical Tools of Information Reduction

A more objective method of reducing information overload is *factor* or *component analysis*. The approach is to reduce the number of intercorrelated variables to a lower number of independent variables that represent sufficiently indicator variance. If all indicators related to sustainable development could be reduced in this manner to one factor, this factor would indeed constitute an index of different levels of sustainable development. However, more than one factor is usually needed to reflect most of the indicator variance. The contributions of different factors to sustainability or development are, therefore, in most cases far from clear. Results of a factor-analytic test of the Environmental Sustainability Index (ESI) seem to confirm this (Yale Center etc., 1997–2006, ESI-2002 report). So far, factor analysis has not been applied successfully in deriving an index of sustainable development. This might indeed be an indication that one overall index cannot represent the complex multidimensional paradigm.

Nonetheless, correlational analysis could reveal redundancy and erroneous (usually equal) weighting in index calculations. For instance, the crude comparison of the Human Development Index (HDI) with GDP per capita (see Section 3.1.2) points to a relatively strong correlation – an indication that the HDI does not offer much to explain development beyond its economic growth component? On the other hand, low correlation of the ESI and GDP ($R^2 = 0.23$)[3] indicates independence of the ESI from GDP, suggesting original explanatory power of the index.

Despite its value-free computational analysis, judgment seeps into factor analysis through the selection of variables (or indicators) considered to be representative of

[2] The proposed European System of Environmental Pressure Indices (http://esl.jrc.it/envind/theory/handb_.htm), advanced in 1999, is yet to be adopted by Eurostat and/or the European Commission.

[3] Yale Center etc. (1997–2006), ESI-2005 report, summary for policymakers (http://sedac.ciesin.columbia.edu/es/esi/downloads.html#data).

an overall concern such as sustainable development. Also, the interpretation of the resulting factors as particular contributors to this concern remains to some extent subjective. Still, factor-analytic interpretations of long indicator lists can shed light on the most pertinent driving forces behind sustainable development, notably of economic growth and technological change.[4]

5.1.3 Scientific Criteria of Aggregation

The main drawback of the (presumably objective) scientific method of aggregation is its restriction to specific themes such as global warming. The weights for the underlying statistical variables are theme potentials that can be determined by physical or chemical analyses. Table 5.2 shows – for Dutch policy themes – the different potentials that can be applied as weights for variables to generate *theme equivalents* for each variable (Adriaanse, 1993). Global warming, acidity or eutrophication equivalents add up to respective theme indicators ('indices' in our terminology). Subjective elements may affect these indices through the selection and definition of the policy themes and their representative (emission) variables.

The table presents different emissions, measured in their particular theme equivalent units such as global warming contribution by different greenhouse gases. It also shows national policy targets for 1995 and 2000 in the same units of measurement for assessing progress made towards these targets. The indicators provide useful management information for particular policy themes. However, the different equivalent units do not permit inter-theme comparison and hence an overall evaluation of the success of environmental policy. To this end, the author applied distance-to-target weights for calculating two total environmental pressure indices (last two rows of Table 5.2). Supposedly these weights reflect the 'seriousness of environmental problems'. Still, the target setting is political or even subjective, notably for sustainability levels.

There is one attempt to overcome the theme restriction of the scientific method by using bio-productive areas as a common equivalency measure for all environmental impacts. The Ecological Footprint of nations converts flows of natural resources and residuals into bio-productive area 'required to maintain these flows' (Redefining Progress, 1994–2004). The 'scientific' validity of converting environmental source and sink functions into area equivalents suffers, however, from the numerous assumptions necessary to find appropriate land equivalency factors, e.g. for CO_2 absorption (cf. Section 5.3).

[4] Section 10.2.3 describes an alternative method of identifying the driving forces behind environmental impacts, decomposing (rather than reducing) the total variability in terms of variables, whose multiplicative product represents the total impact indicator.

Table 5.2 Dutch policy theme potentials for calculating theme equivalents and indicators

Environmental themes	Theme variables (and potentials)	Indicators (in theme equivalents) 1980	1991	Dutch policy objectives (indicator targets) 1995	2000
Global warming (climate change)	CO_2 (1), CH_4 (12), N_2O (290), CFC-11 (3500), CFC-12 (7300), CFC-113 (4200), CFC-114,115 (6900), Halons (5800)	286	239	205	195
Ozone layer depletion	CFC-11,12,13,114 (1), CFC-113 (0.8), CFC-115 (0.6), Halon-1211 (3), Halon-1301 (10)	20000	8721	54	0
Acidification	SO_2 (1), NO_2 (0.7), NH_3 (2)	6700	4100	4000[a]	2400
Eutrophication	P (1), N (0.1)	302	273		95
Toxic substances	Pesticides, radioactive and other substances (risk factors)	251	222	196	139
Solid waste disposal	Selected waste streams (weight)	15.3	14.1		5
Disturbance of local environments	Percent of people affected by noise and odour[b]	46	57		41
Total environmental pressure (indices)	Distance to (2000) targets	1335	1195		
	Distance to sustainability levels[c]	7999	6686		

Notes: [a] 1994.
[b] Excluding overlap (simultaneous exposure to noise and odour).
[c] Climate: 10, acidification: 400, eutrophication: 86, toxic substances: 12, waste disposal: 3, disturbance: 9 (preliminary, unofficial).
Source: Adriaanse (1993).

5.1.4 Empirical (Commensurable) Weighting

One of the theme indicators of Table 5.2, waste disposal, is in fact not a scientific index, but the result of direct measurement. Observable common units of measurement such as tons (of emissions or waste) or dollar values (of economic inputs and outputs) weight the importance of statistical variables implicitly by weight or economic value. The common *numéraire* of these variables permits direct summation (or subtraction) either for ad hoc calculations such as the Index of Sustainable Economic Welfare (Section 7.1.1) or for physical and monetary accounting aggregates, discussed in Chs. 6 and 8.

5.1.5 Indicator Averages

Perhaps the most common method of index construction is the averaging of indicators. The simple *unweighted mean* is a special case of the mathematical method of index building. Such a mean actually gives equal weight to the underlying variables. Where different units of measurement are used, the original variables need to be converted (standardized) into a common scale (e.g. from 0 to 1) for calculating the mean – an operation that distorts equal weighting to some extent. The reason is that the value of a variable is forced into a scale that shows relative distance, e.g. of individual country values, from their global mean (as for the ESI) or from maximum or minimum values (as for the HDI). As a consequence, differences from averages or target values replace the 'natural' (original) variation of indicator values, blurring their relative importance.

Weighted averages might be preferable to unweighted ones, provided suitable weights for the indicators underlying an index can be found. For environmental pressure indicators, damage and toxicity levels were suggested as weights (Section 6.3.3). In general, difficulties of linking actual or potential damage to particular indicators thwarted the calculation of weighted indices of environmental quality and sustainable development.

5.2 Indices of Environmental Sustainability and Sustainable Development

Table 5.3 shows the concepts and methods of four popular indices, which seek to measure the sustainability of economic development. They represent distinctly different notions of sustainability, applying correspondingly different calculation methods. With the exception of the ecological footprint, which is an indicator of environmental pressure, the indices attempt to capture most or all of the social, economic and environmental dimensions of sustainable development [FR 5.2].

5.2.1 Sustainable Development Index (SDI) and Human Development Index (HDI)

The SDI is a simple average of indicators for sustainable development. It shares, therefore, the indicator selection problem for a broadly and vaguely defined topic, discussed in Section 4.2.1. At the same time, the index attempts to improve on the United Nations (2001b) indicator themes by introducing 'politics and human rights' as an additional sustainable development dimension. On the other hand, it ignores largely environmental degradation, except for damage from CO_2 emission. Also, the indicators chosen differ considerably form the United Nations list without explaining why and how this particular choice is a better reflection of sustainable development.

The SDI shows 'progress … toward sustainable development' (Nováček & Mederly, 2002) as upward change in country ranking. Table 5.4 displays the SDI ranks and those of the HDI, which focuses on social and economic concerns ignoring environmental ones. Nonetheless, HDI and SDI provide similar rankings, owing to the relative weight of social and economic concerns in these indices. Exceptions are, notably, the oil-rich nations of Kuwait and the United Arab Emirates, where CO_2 emission and natural resource consumption seem to have weighed into the (higher score) of the SDI.

Table 5.3 Indices of sustainability: Concepts and methods

	Ecological footprint (EF)	Environmental sustainability index (ESI)	Sustainable development index (SDI)	Well-being Index (WI)
Definition	Biologically productive area, required for natural resource consumption and waste absorption by country or region, using prevailing technology	Average of indicators related to environmental sustainability	Arithmetic mean of indicators for significant aspects of sustainable development	Average of human and ecosystem well-being indices, whose underlying indicators are also averaged
Indicators	6 main categories of ecologically productive areas	20 indicators, which combine 68 variables	14 indicators (2 for each problem area), combining 58 variables	36 variables in 5 areas of human well-being and 51 variables in 5 areas of ecosystem well-being
Weighting	Area equivalents of world average bioproductivity for natural resource use and waste absorption	Equal weighting of standardized indicator scores	Equal weighting of standardized variables	Equal weighting of standardized variables (with reduced resource use 'when offsetting a poor state of the environment')
Sustainability concept	Ecological sustainability: inverse of the carrying capacity of a country's natural systems	Potential for environmental sustainability: ability to produce high levels of performance with regard to environmental and social conditions	Overall (development) sustainability: high index values in political, social, demographic, economic and environmental areas	Overall (development) sustainability: high levels (80+), of human and ecosystem well-being

References: EF: Wackernagel and Rees (1996); Redefining Progress (1994–2004).
ESI: Yale Center etc. (1997–2006), 2002 report
SDI: Nováček and Mederly (2002, ch. 4)
WI: Prescott-Allen (2001); http://www.sustainability.ca/index.cfm?body=SourceView.cfm&ID=422.

Table 5.4 Indices of sustainable development: Comparison of results

Country[a]	SDI	WI[c]	HDI	ESI	EF rank	EF ha p.c.
Norway	**1**	3	**1**	2	128	8.2
Finland	2	2	12	**1**	123	7.0
Canada	3	5	4	4	129	8.6
Germany	10	8	17	48	108	4.3
Belgium	12	25	6	118	116	5.1[b]
Australia	13	13	3	15	124	7.1
Japan	15	17	8	76	104	3.9
USA	19	18	7	43	**131**	**9.6**
United Kingdom	21	24	11	89	111	4.7
Costa Rica	27	34	34	8	71	1.9
Poland	28	39	28	85	97	3.4
Israel	29	58	20	61	105	4.0
Cuba	34	57	39	56	56	1.5
Jamaica	42	49	59	116	72	2.2
Brazil	43	65	54	19	77	2.4
Mexico	46	114	40	90	82	2.6
Korea, Republic	48	42	24	126	79	2.4
Venezuela	54	51	51	46	78	2.4
South Africa	55	103	87	75	98	3.5
Russia	57	46	43	70	109	4.3
China	63	119	70	122	49	1.4
Malaysia	66	71	45	66	88	3.0
Turkey	69	96	65	60	73	2.2
Kuwait	72	87	33	**131**	126	8.0
Botswana	79	26	94	12	86	2.7
Philippines	80	84	62	113	33	1.1
Iran	82	105	74	102	67	1.9
Saudi Arabia	84	120	57	129	106	4.1
Egypt	88	62	88	72	36	1.2
United Arab Em.	92	128	37	130	130	9.0
India	96	127	93	112	13	0.8
Bangladesh	100	98	100	84	**1**	**0.5**
Malawi	107	61	121	80	7	0.6
Nigeria	108	100	110	124	32	1.1
Pakistan	112	123	104	108	10	0.7
Sierra Leone	122	110	**131**	125	18	0.9
Kenya	124	108	108	87	30	1.1
Ethiopia	128	83	125	109	9	0.7
Angola	**131**	91	122	107	12	0.8

Notes: [a] Selected countries from a list of 131 nations covered by *all* indices; SDI in consecutive ranking order.
[b] Belgium and Luxembourg. [c] Sweden: 1, Uganda: 131.
References: SDI: Nováček and Mederly (2002), table 4.
WI: http://www.sustainability.ca/Docs/wonrank.pdf?CFID = 21400368&CFTOKEN = 23201840
HDI: http://hdr.undp.org/statistics/data/indic/indic_12_1_1.html
ESI: Yale Center etc. (1997–2006), ESI-2002, table 1
EF: Venetoulis et al. (2004), table 2.

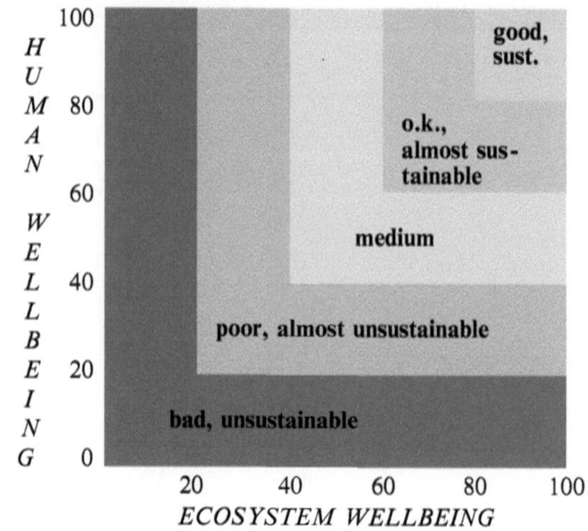

Fig. 5.1 Well-being Index: Barometer of sustainability
Source: Prescott-Allen (2001), The well-being of nations, fig. 1.5; Copyright by the author, with permission by the copyright holder.

5.2.2 *Well-Being Index (WI)*

Prescott-Allen (2001) advanced the WI as a 'barometer of sustainability'. The International Union for Conservation of Nature and Natural Resources (IUCN) and the International Development Research Centre (IDRC) later adopted the index as a 'synthesis of assessment approaches' (Sustainability Now, 2006). The index follows basically the SDI method, expanding and averaging the HDI areas and indicators. The WI includes more or less the same areas as the SDI with added data on air quality and greenhouse gas emissions. Sustainable development is 'equated' with the 'good life', which in turn represents high levels of 'human and ecosystem well-being', the two components of the WI. Colour codes for equal percentiles of the indices of human and ecosystem well-being visualize the qualitative evaluation of sustainable development performance – from bad (unsustainable) to good (sustainable) (Fig. 5.1). The sustainability target is thus met at a WI level of 80 + (out of 100).

The rankings of the WI and SDI are quite similar for rich countries but differ considerably, but erratically for poor economies in Table 5.4. The overall conclusion is that 'no country is … even close to sustainability' (Sustainability Now, 2006), with the first eight countries achieving a WI-level in the 60s only.

5.2.3 *Environmental Sustainability Index (ESI)*

The ESI extends ecological sustainability beyond stress on ecosystems from emission, resource use and population growth: The index covers also environmental quality and its effects on human health; the human, social and institutional capacity to potentially tackle these impacts and effects; and social response through policy, technology (eco-efficiency) and participation in international programmes and conventions. This extension looks like an attempt at covering all the above-described FSDI information categories of environmental stress, impact, response and conditions in one index. However, the focus of the index is on finding 'the most effective metric for gauging the prospects for long-term environmental sustainability' (Yale Center etc., 1997–2006, 2002 report). The resulting sustainability concept reflects therefore more a *potential* for reaching long-term sustainability than an assessment of the actual state of sustainability.

The potential for implementing sustainability depends not only on policy capabilities but also on current environmental and economic conditions, as indeed reflected in the index. This might explain the low, but still discernible correlation of the ESI with national income per capita; it also explains the occurrence of outliers like Belgium, the Republic of Korea and Kuwait (Fig. 5.2). In Belgium and Korea, high emissions of pollutants and stress on ecosystems are the culprit. Kuwait shows the worst performance of all countries in every category, except for a relatively good performance in the field of human health care – affordable by a rich country.

5.2.4 Ecological Footprint (EF)

The index refers to ecological sustainability as defined above and described in terms of reduced pressure on a nation's carrying capacity (Sections 2.2.3, 2.4.1). In fact, it can be interpreted as the inverse of carrying capacity. Redefining Progress (1994–2004) defines the EF as the 'amount of biologically productive land and water area required to produce the resources consumed and to assimilate the wastes generated, using prevailing technology'. The index thus relates per capita natural resource use, wastes and CO_2 emissions of nations to their biocapacity in terms of area equivalents. Results indicate that since the 1970s 'humanity's collective Ecological Footprint breached the sustainability mark', exceeding the world's biocapacity of about 27,500 million acres. By the year 2000 this ecological deficit reached nearly one acre per person (Venetoulis, Chazan & Gaudet, 2004). According to this measure, we have used nature beyond its regenerative capacity.

Table 5.4 shows that the EF runs mostly counter to the other indices whose economic dimension weighs heavily in their values. The USA obtains the lowest ranking with an EF of nearly 10 ha per capita, followed by most industrialized and oil-rich countries. Poor countries rank high as a result of low levels of economic activity. Plate 5.2 visualizes for selected countries the ecological deficit or surplus, i.e. the extent to which these countries have to import, or could export, environmental sustainability in terms of environmental source and sink functions. Among these countries, only Indonesia shows a surplus of biocapacity. For sub-national regions, and in particular cities, footprints on their 'hinterland' can be over 300 times the city area.[5]

[5] For instance the EF of Santa Monica, California, is 331 times the city's area (http://smgov.net/epd/scpr/ResourceConservation/RC6_EcologicalFootprint.htm).

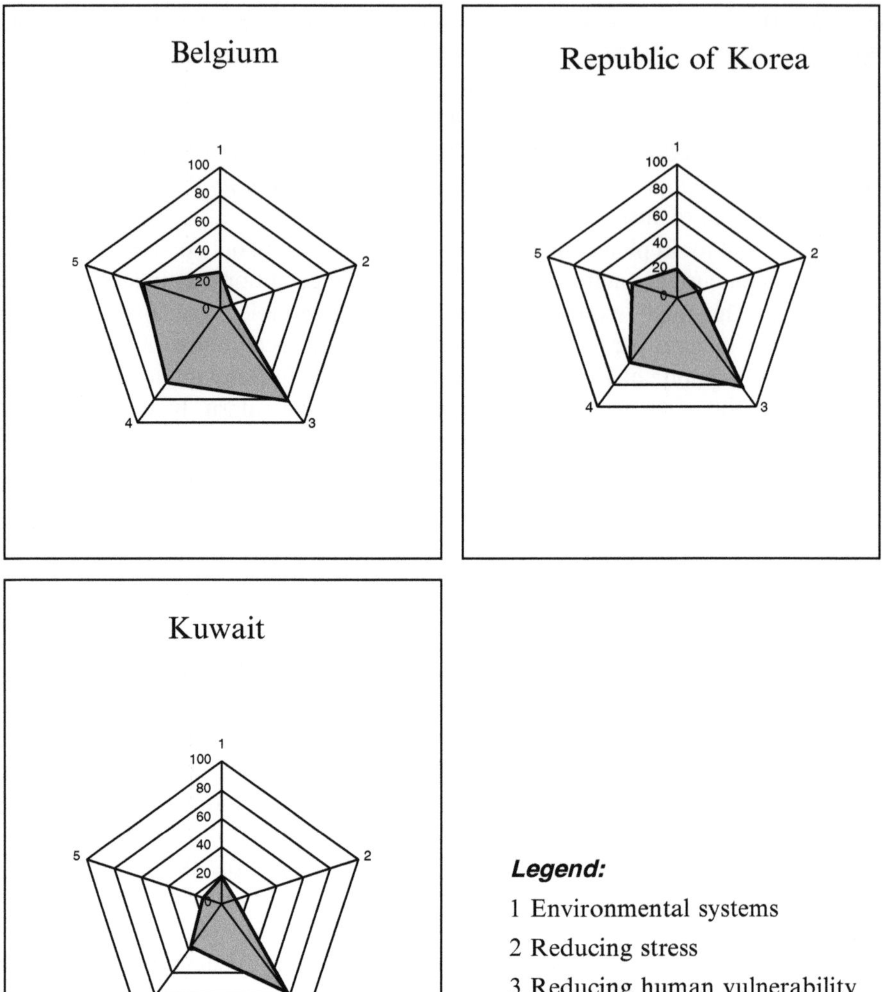

Fig. 5.2 Environmental sustainability index – country profiles
Source: Yale Center etc. (1997–2006), 2002 report, annex 5.

There can be no doubt about the existence of ultimate limits to expanding human activity on a finite 'spaceship Earth' (Boulding, 1966). The question is if indices like the EF can indeed determine when the limits were, or will be, transgressed. As to the number of people the earth can support, assessments are inconclusive.

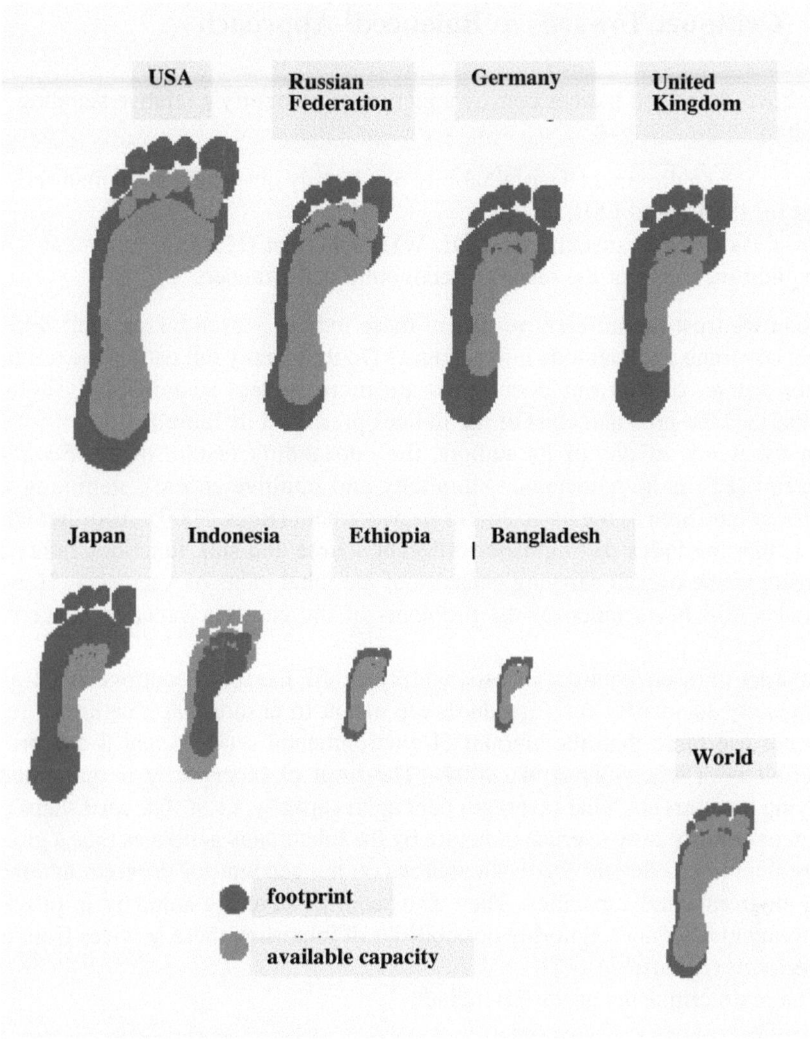

Plate 5.2 Ecological footprint
Source: Copyright VisLab/Wuppertal Institute for Climate, Environment and Energy; with permission by the copyright holder. (*See Colour Plates*)

Expecting a world population of 9.1 billion in 2050, according to United Nations forecasts, Cohen (1995) estimates that 'under the right conditions' the earth could support 15 billion people, but 'under other conditions' its carrying capacity could be less than one billion. Regional carrying capacities can of course vary according to their particular environmental conditions, required standards of living and possibilities of importing environmental services from other regions.

5.3 Critique: Towards a 'Balanced' Approach

The above-described indices convey a picture of a country's relative standing with regard to

- Either the environmental sustainability of – largely unspecified – human activities (as for the EF and ESI), or
- Progress towards sustainable (SDI, WI) or human (HDI) development – with explicit inclusion of the results of economic performance.

But can we trust the different results of these indices – even taking their different scope, coverage and methods into account? Do they really tell us that we are better off, or worse, or that our economies are more or less sustainable? Table 5.5 summarizes the pros and cons of the indices presented in Table 5.3.

In the words of one of its authors, the outstanding feature of the *Ecological Footprint* (EF) is its 'conceptual simplicity and intuitive appeal', stemming from the use of land area as the basic unit of measurement (Rees, 2000). A further advantage is that the index distinguishes different source and sink functions rather than referring to the overall carrying capacity of the inhabitants of a territory. However the index still bears much of the problems of the carrying capacity concept (cf. Section 2.4.1).

In order to assess ecological sustainability the EF has to set controversial biological capacity standards. Such standards are meant to ensure that a nation or region does not use more than the amount of environmental services that it can provide 'indefinitely' (Venetoulis et al., 2004). The limit of biocapacity is quantified by applying the current global (average) per capita capacity, i.e. a 'fair earthshare', to a region or country. Any overuse of nature by the inhabitants generates thus a globally unsustainable eco-deficit. Such calculations do not account for different national or local environmental capacities. They also seem to advocate autarchy in providing environmental services, ignoring possibilities of importing these services from other countries or regions.

The main criticisms of the EF include

- Ignoring technological abilities and substitution in alternative, environmentally sound production and consumption patterns
- Limited coverage of environmental impacts, omitting those that are difficult or impossible to convert into land area units such as toxic contaminants or depletion of subsoil resources
- Area equivalents that refer to environmental pressures rather than the ecological and health damage generated by these pressures
- Ignoring import of sustainability from other countries or regions, and
- The related autarchy objective in using national environmental services up to global average biocapacity per person.

Obviously the setting of biocapacity limits is crucial to the use of the EF for assessing the (non)sustainability of economic activity. Is it possible that – in view of the above critique – the EF might have set out from the pre-analytic vision and conclusions about

Table 5.5 Evaluation of indices of environmental and socio-economic sustainability

EF	ESI	SDI	WI
Plus	**Plus**	**Plus**	**Plus**
- Simplicity of index calculation and clarity of message - Definition of sustainability as compliance with biophysical limits - Application at different regional levels - International comparison of environmental performance: country ranking	- Extended coverage beyond stress: environmental quality and social response - Clear objective: assessing the likelihood of attaining sustainability - Transparent methodology and data evaluation - International comparison: country ranking and profiling	- Coverage and extension of the basic sustainable development dimensions, including human rights and freedom - Use of readily and regularly available data	- Coverage of sustainable development dimensions - Separate and combined assessment of human and ecosystem well-being - Extensive database
Minus	**Minus**	**Minus**	**Minus**
- Controversial setting of biocapacity limits - Methodological problems: conversion of environmental impacts into area equivalents, scope and coverage - Normative implications: autarchy in, and equal distribution of, environmental services - No link to economic growth and development	- Vague 'potential' sustainability concept - Equal weighting of indicators and variables - No direct link to economic growth or development	- Lack of sustainability and development concepts - Selection of indicators according to data availability rather than contribution to sustainable development - Equal weighting and intercorrelation of indicators	- Tenuous connection between well-being, good life and sustainable development - Indicator selection and availability - Intercorrelation and equal weighting may distort contributions to well-being and country ranking

a full world that has already transgressed its planetary limits (Section 1.3)? This at least seems to be the view of van Kooten and Bulte (2000), who consider the EF as non-analytic that is 'less a scientific measure than one designed to raise public awareness and influence politics.'

One virtue of the *Environmental Sustainability Index* (ESI) is its transparency with regard to (sustainability) objective, methodology and evaluation of data quality. The index also covers, beyond environmental stress, environmental impacts and health effects, and policies and programmes in response to impacts and effects. For the aggregation of these concerns the index has to resort to equal weighting, in the absence of a 'firm base for applying differential weighting' (Yale Center etc., 1997–2006, ESI-2002 report). However, the forward-looking sustainability concept, while a priori convincing, remains vague for the measurement of the 'relative likelihood' of achievement within the time frame of 'several generations'. Since the ESI does not establish a direct connection with economic performance (except for some correlational analysis) it remains essentially an environmental (outlook) measure. Reservations about differential weighting of indicators are abandoned in the Environmental Performance Index, which appears to replace the ESI at least for now [FR 5.2].

The *Sustainable Development Index* (SDI) claims to be a measure of sustainable development, averaging a large number of indicators *for* the paradigm. The index covers socio-economic and political dimensions that are not addressed by the environmental indices of the EF and ESI. However, as with most physical indicator combinations, the index suffers from problems of indicator selection, equal weighting and the absence of a clear, measurable sustainability concept. As to the latter, the SDI simply claims 'to show progress of individual countries toward sustainable development' (Nováček & Mederly, 2002). An equally generic pronouncement that 'sustainable development is not achievable if people have to live in totalitarian state, without the privilege of freedom' is not helpful in elaborating and quantifying the political dimension of sustainability. The question is indeed whether adding more indicators, as compared to the HDI's three-indicator average, improves the assessment of a country's development ranking.

The *Well-being Index* (WI) claims that a greater number of indicators does improve the validity of the index. The reason given is that a lower number of indicators (as for the HDI) makes an index 'more susceptible to distortion by missing data' (Prescott-Allen, 2001, p. 18). However, owing to its similarity to the SDI, the WI carries the same flaws. Indicator correlation and equal weighting increase in fact the risk of distorting the significance of different well-being components by a larger number of indicators. This might also explain the considerable difference of WI from SDI rankings for most mid-level countries.

In conclusion, neither short or long indicator sets nor their combination in ad hoc index calculations succeed in capturing a clearly defined sustainability concept. The following chapter explores the ability of physical material balances to assess ecological sustainability by means of a more systematic physical accounting approach.

Further Reading

FR 5.1 Aggregation Methods

In 2000 the United Nations conducted a review of different methods of aggregating its core set of 57 sustainable development indicators into a compound index (http://www.un.org/esa/sustdev/csd/csd9-aisd-bp.pdf); to date, however, there has been no follow up to the review, such as an attempt at index calculation.

Qualitative and hence *judgemental methods* of evaluating the results of a range of indicators include (1) the Delphi method of forming group judgement (see for a concise overview: http://www.iit.edu/~it/delphi.html), (2) multi-criteria analysis for appraising (scoring and weighting) options according to preset objectives (see, e.g. the U.K. government manual: http://www.odpm.gov.uk/stellent/groups/odpm_about/documents/page/odpm_about_608524.hcsp), and (3) fuzzy logic applied to indicator weighting (applied to environmental and sustainability indicators: Tulbure, 2001).

Principal component and factor analysis are well-known *mathematical multivariate tools* for reducing intercorrelated indicators into lower dimensions, and possibly a one-dimensional index. Darlington gives an accessible overview of this method (http://comp9.psych.cornell.edu/Darlington/factor.htm).

FR 5.2 Indices of Environmental Sustainability and Sustainable Development

Moldan et al. (1997) present examples of indices and indicators that get the 'big picture' of sustainable development and 'pieces' thereof. A compendium of indicator initiatives (http://www.iisd.org/measure/compendium/) is a source of national and international indicator and index programmes and projects. The tables of Section 5.2 indicate the sources for the description and evaluation of popular sustainability indices. The following provides background and references to these attempts.

The 'fair sharing' (Section 2.4.1, FR 2.3) of environmental space is the motive and rationale for *Ecological Footprint* (EF) compilations (http://www.rprogress.org/publications/footprintnations2004.pdf) and material flow indicators (Spangenberg et al., 1999). EF estimates for sub-national regions can be found in www.RegionalProgress.Org. A special section of *Ecological Economics* 32 (2000) reviews the EF concept and methods.

More comprehensive approaches of measuring sustainable development include, besides the SDI (described in the text), the direct *greening of the Human Development Index* (HDI), or at least supplementing the index with environmental indicators (Neumayer, 2001; Vemuri & Costanza, 2006). The idea is to assess the environmental sustainability of the HDI, which so far features economic and social concerns only. A technical note of UNDP's (annual) human development reports describes the HDI calculation method (http://hdr.undp.org/reports/global/2004/pdf/hdr04_backmatter_2.pdf).

Sustainability Now (2006) presents a synopsis of the IUCN/IDRC adaptation and results of Prescott-Allen's (2001) *Well-being Index.*

The *Environmental Performance Index* assesses the current environmental performance of countries. It is a weighted average of 25 indicators expressed in distance-to-(policy)target values (http://epi.yale.edu/2008EPIOverview).

Review and Exploration

- Do the indicators and faces of Table 5.1 suggest improvement or deterioration of the state of the European environment?

- Compare the strengths and weaknesses of different aggregation methods.
- Do more indicators in an index improve its validity?
- What do the theme equivalents (Table 5.2) tell us about the state of the environment in the Netherlands?
- Is averaging (and equal weighting) of indicators a practical way of index calculation? Compare the results of SDI, WI, ESI and HDI (Table 5.4).
- Do you agree with van Kooten and Bulle (Section 5.3) that the EF 'is less a scientific measure than one designed to raise public awareness and influence policies'?

Chapter 6
Energy and Material Flow Accounting

Physics provides the theory for physical accounting. Thermodynamic laws, applied not only to energy but also to matter, combine with economic bookkeeping to create the framework for energy and material flow accounts, and physical input-output tables. Explaining life on earth as a matter of energy supply suggests using energy values and accounts for assessing the state of the environment and the sustainability of human activity.

The relative ease of compiling material flows and interpreting them as pressures on natural systems makes material flow accounts (MFA) the favourite of both ecological economics and official statistics. The purpose of the MFA is to assess the possible de- or rematerialization of the economy, catering implicitly to ecological sustainability. In practice, physical flows alert to actual and potential environmental pressures. They fail in measuring the environmental and economic significance of different materials and substances. Part III will explore monetary valuation and accounting to this end.

6.1 Rationale: Social Metabolism and Environmental Sustainability

Ecological economists summon natural laws and limits for restricting economic performance and growth. To operationalize and promulgate these limitations they describe the interaction between nature and human activity as a case of – disturbed – industrial, social or societal 'metabolism'.[1] Fischer-Kowalski (2002) even credits social metabolism with cutting across 'the "great divide" ... between natural and social sciences'. Insight into the material and energetic metabolism of society provides thus the rationale for physical, non-monetary accounting [FR 6.2].

[1] No such distinction is made here, as the focus is on the interaction between nature and all economic activity. Also, industrial metabolism appears to be more microeconomic in nature aiming at eco-efficiency in production (see Section 13.3.1).

P. Bartelmus, *Quantitative Eco-nomics*,

The social view of metabolism extends the original concept of biological metabolism, i.e. the biochemical processes of energy generation in a living organism, to the economy or even society. At the national level this means observing energy and material flows through a country's economy, rather than measuring chemical reactions in an organism. The question of whether to perform this aggregation in terms of energy flows and energy units of measurement, or material flows and their weight, divides physical accounting into schools of *energy* and *material flow accounting*.

Both approaches set out from the first thermodynamic law of conservation of energy and mass, according to which energy and mass cannot be created or destroyed. This law justifies and facilitates double-entry accounting of the quantity of inputs and outputs of energy and material flows into and out of an ecosystem, region or economy. Environmental and sustainability problems show up when scrutinizing the *quality* of the flows after their use, i.e. on the output side. The second law of thermodynamics refers to the increase of entropy (disorder) of energy after its use/transformation, and hence to the reduced usefulness of dissipated energy. The first thermodynamic law provides the means of input-output accounting whereas the second law refers to the grounds for accounting, i.e. evaluation of the use and usefulness of energy.

Georgescu-Roegen (1979) extended the entropy law beyond energy to matter. Since matter can also be dispersed as non-reusable waste, he famously claimed 'matter matters, too'. Georgescu-Roegen consequently criticized the 'energetic dogma', which proclaims that energy alone poses the 'ultimate limitation' (Slesser, 1975). This is also the idea behind the 'spaceship earth' metaphor (Boulding, 1966), which assumes that all material waste can be recycled with sufficient energy supply.

In reality the bulk of matter and energy cannot be mutually converted. Georgescu-Roegen (1979) suggested, therefore, to 'keep two separate books – one for matter and one for energy'. Bookkeepers seem to opt, however, either for energy accounts or for material flow accounts. The energy accountants believe that their approach permits to assess environmental sustainability in terms of the availability of useful energy. On the other hand, material flow accountants argue that their accounts include energy 'carriers'. The aggregate flow of materials reflects, in their view, total pressure from both energy and material throughput on nature's carrying capacities of human populations – an indicator of ecological (non)sustainability (cf. Sections 2.2.3, 2.4.1). The following sections present the concepts and methods of the two basic physical accounting approaches and discuss their results with regard to sustainability concerns.

6.2 Energy Accounting

'Practically everything on the earth can be considered to be the direct or indirect product of past and present solar energy' (Costanza, 1980). The quote reflects the above-mentioned energetic dogma, assessing society's sustainability in terms of its energetic metabolism. It might also be behind the view of energy-related emission as a surrogate for overall environmental deterioration (Section 4.3).

Looking back far enough into the past and standing back far enough for a global view of our planet, this argument is certainly valid. Solar energy generated our fossil fuels over millions of years; it is also the only input into an otherwise closed planetary system in which the use of energy and matter is ruled by entropy increase. Plate 6.1 shows the overall energy balance for the earth according to the first thermodynamic law: whatever influx of energy is not reflected back into space is either absorbed by oceans and land or trapped in the atmosphere to keep the planet at a comfortable temperature. In recent times human activity has increased the presence of heat trapping greenhouse gases in the atmosphere. The result is an upset of the earth's energy balance that can be held responsible for global warming of the atmosphere (Section 4.3). This human-induced greenhouse effect is an important, if not the most important, finding of global energy accounting.

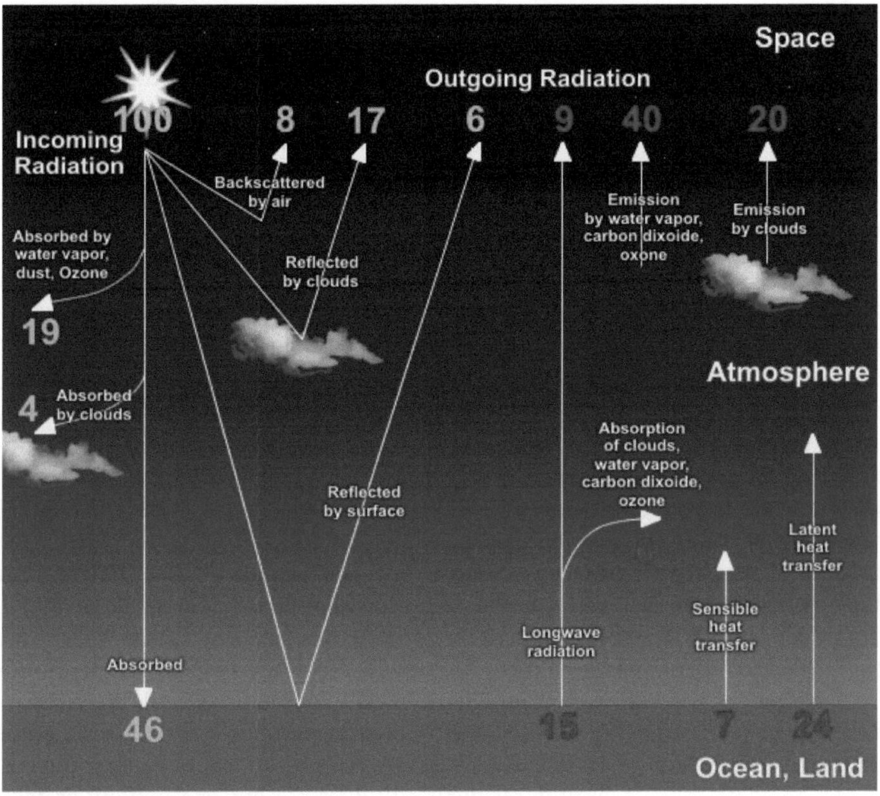

Plate 6.1 Global energy balance
Source: US National Weather Service, JetStream – Online School for Weather (http://www.srh.weather.gov/srh/jetstream/atmos/energy_balance.htm) (*See Colour Plates*).

The energy dogma considers energy not only as the ultimate (non-substitutable) source of human life but sees the availability of useful energy as a limitation for meeting ever-increasing human needs and wants. The human-needs oriented concept of useful energy is called *exergy*. The reason for its limited availability is the second thermodynamic law, which reduces exergy through dissipation in human production and consumption processes. The exclusive focus on energy or exergy flows implies that in principle all matter can be recycled or converted into energy. As a consequence, the energy theory of value claims that all product, resource and waste flows can be accounted for in terms of their 'natural' and 'scientific' energy value. [FR 6.1]

Two different energy accounting methods apply their particular concepts of available energy:

- ***Ex**ergy accounts* express all energy, resource, product and residual flows from the environment through production processes to final consumers in terms of *potential* energy available for doing work; their focus is on the (energy) *efficiency* of economic activity.
- ***Em**ergy accounts* measure the same physical flows, as well as their 'stores' of natural capital, in terms of their *embodied* energy, i.e. directly and indirectly used energy that would have to be expended for creating any particular product; their focus is on the *sustainability* of economic activity.

6.2.1 Exergy Accounting

Plate 6.2 presents the exergy system of Sweden. It shows the loss of exergy from the stage of extracting primary natural resources, through various processes of production and high-quality energy (notably electricity) generation, to their final use by industries and households. The white rectangles at different stages of the exergy flows symbolize the numerous and complex conversion processes. The overall efficiency of the energy sector comes to less than 15% as indicated by the diminishing widths of the exergy flows towards final use. Note that in comparison to energy carriers other material flows (of timber, ores and iron scrap) are quite insignificant in terms of their (internal) exergy content. This is an indication that the energetic dogma might not do full justice to the role of natural resources in the economy.

Ayres et al. (1998) point out that comprehensive exergy accounts should not only measure exergy inputs but also 'unexpended' exergy outputs. The reason is the potential of residual exergy in emissions and wastes for generating environmental damage. Estimating residual exergy content requires knowledge about the chemical composition of wastes and emissions in order to calculate their internal exergy contents. Also, exergy in residuals still does not indicate toxicity or environmental damage.

Together, input and output flows of exergy could give an indication of the 'eco-efficiency' of economic activity in terms of resource use and emission per

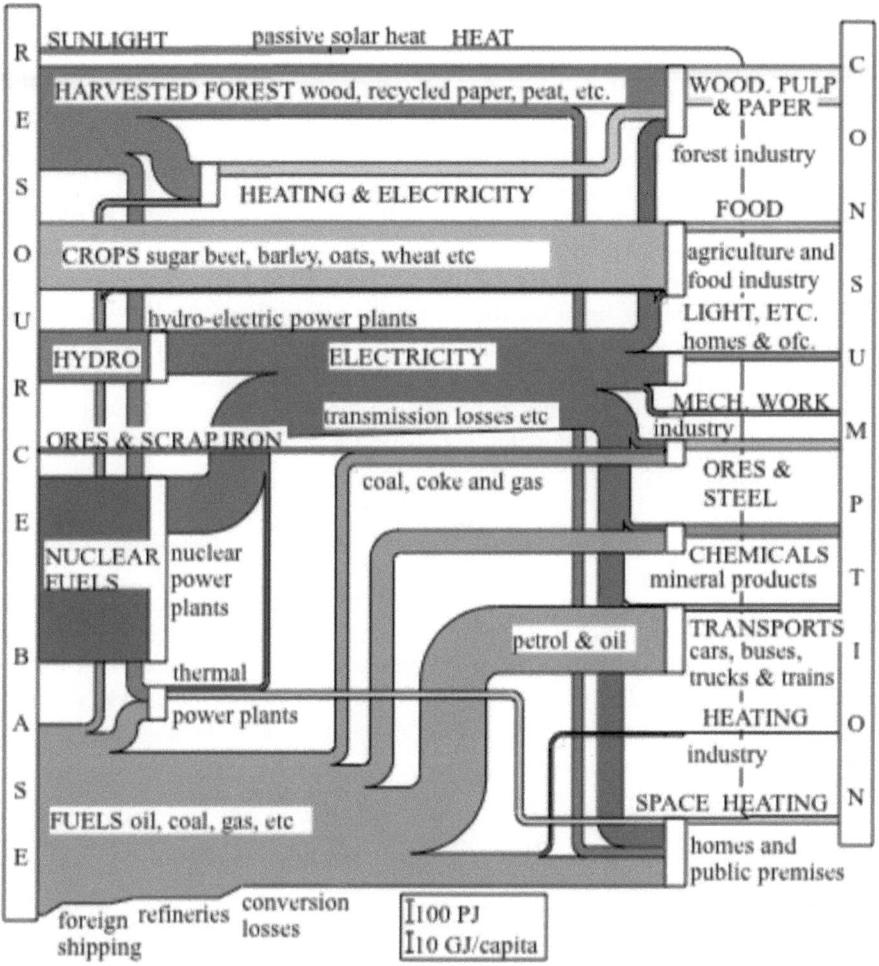

Plate 6.2 Exergy flow system, Sweden 1994
Source: Wall (2001b), The use of natural resources in society, plate 30; Copyright Eolss, with permission from Eolss (*See Colour Plates*).

economic output (Section 13.3.1). However, for assessing the long-term sustainability of production and consumption, exergy flows would have to be related to the funds (resource stocks) available for future use. On the output side, sustainability would be ensured by zero-exergy content of the residuals, or at least an exergy amount that could be safely handled by nature's absorptive sinks. So far, measurement problems of exergy stocks and flows prevented the regular use of exergy accounts for assessing sustainability at the national level.

6.2.2 Emergy Accounting

Emergy accounts present the total amount of energy required in all production stages and sectors that finally create a particular product. It is the embodied energy, or 'energy memory', of a product. Emergy measures actual or potential use of energy in solar energy joules (sej) for making the product (Odum, 2002). Emergy 'stores' of natural resource stocks are an indicator of natural wealth in physical rather than conventional money units.

Provided that we have a complete input-output table for i = 1…n sectors or products and the energy input coefficients ε_i for all products are known, we can use input-output analysis (Section 10.2.2) to calculate the direct and indirect (embodied) emergy for the sector j, $\varepsilon_j x_j$, from its energy balance (Costanza, 1980). Equation 6.1 presents this balance as the sum of the direct (external) energy inputs E_j into the sector plus the total embodied energy in all inputs required for the production of x_j, i.e. $\Sigma\varepsilon_i x_{ij}$.

$$E_j + \Sigma\varepsilon_i x_{ij} = \varepsilon_j x_j \qquad (6.1)$$

As mentioned, emergy accounts can assess the embodied energy not only of flows of materials and products but also of stocks of natural resources. Such capital measurement assesses the sustainability of economic activity when there is a risk of running down the capital stock, in this case of available energy. Of course, comprehensive sustainability analysis would require measuring also the energy embodied in produced capital and its consumption.

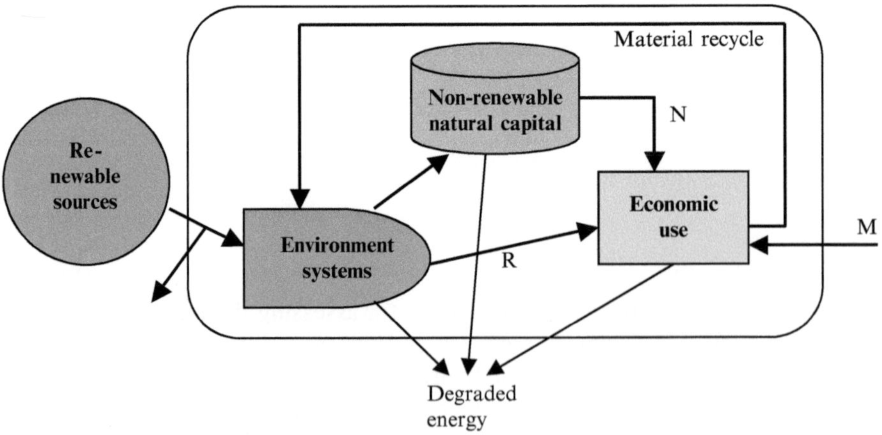

Fig. 6.1 Simplified emergy accounting system
Source: Brown and Ulgiati (1999), *Ambio* vol. 28, no.6, fig. 5, modified.

Figure 6.1 shows a simplified system of emergy flows for a country or region including their economy as 'economic use'. The figure shows renewable energies (R) such as sun or wind power coming in from outside the country. Various kinds of energy are also imported from other countries (M), while degraded energy is 'exported'. Within the system, renewable (R) and non-renewable (N, from long-term build-up of national deposits) emergy flows for economic use are distinguished since their relative share affects the long-term sustainability of the system.

The emergy accountants suggested a number of indices that are essentially ratios of renewable and non-renewable emergies. Neglecting the emergy content of imports one can derive from the figure an emergy yield ratio (EYR) and environmental loading ratio (ELR) as:

$$EYR = (R+N)/N \tag{6.2}$$

where R and N are the emergies of renewable and non-renewable energy sources provided for economic use

$$ELR = N/R \tag{6.3}$$

The EYR indicates the emergy benefit for the economy (or economic activity) per use of non-renewable energy. It can thus be interpreted as a sustainability-efficiency indicator for different economic processes or regions. ELR on the other hand is a stress indicator showing how much non-renewables are used by the economy in relation to the use of renewables.[2]

Global EYR declined from 3.7 in 1950 to 1.6 in 1995, with ELR showing a corresponding increase (Brown & Ulgiati, 1999). The same source also indicates that the main culprits are industrialized countries with an ELR of up to nine times that of some developing countries. As to the long-term sustainability of the biosphere, a comparison of emergy flows with emergy stores, i.e. natural capital, might be more meaningful. The total emergy value of natural capital is estimated at 739.8 E25 sej with freshwater making up 40% (ignoring oceans!). The authors thus estimate that global non-renewable natural emergy stores are about 360 times the present annual emergy flows to the economy. If past trends continue we might expect running out of non-renewable energy supply in about 300 years. Of course, such predictions would have to be refined by appropriate assessments of substitutability with renewable energy sources and by modelling technological and demographic trends, and production and consumption patterns (see Part IV).

---

[2] A further 'energy sustainability index' (Brown & Ulgiati, 1999), defined as the ratio of EYR over ELR is murky. Presumably it is a benefit-stress indicator; however the resulting $(R^2+NR)/N^2$ quotient is difficult to interpret.

6.2.3 Energy Theory of Value

Exergy and emergy accounts convert different material and energy flows into the common *numéraire* of energy units of joule or watt. The purpose is to compare the relative importance of these flows for different final uses in particular production processes, natural systems, society or the biosphere. There is however a major difference between exergy and emergy accounting with regard to the reach of energy valuation.

Exergy analysis can link up with economic activity by including a further primary production factor, energy, in addition to labour and capital in the production function of the economy (Ayres & Warr, 2002). Exergy accounting thus serves productivity, efficiency and growth analyses of particular production processes or the whole economy. It does not attempt to replace economic valuation and analysis with a theory of energetic value. Consequently, exergy analysis appears to have found an accepted niche in economic-environmental analysis.[3]

In contrast, emergy accounts adhere to the above-described energetic dogma. From this point of view, expressing the value of goods and services in energy terms permits to avoid bias in economic preferences and valuation. As a consequence, 'human-centred' (utility) values are rejected in favour of a 'valuation system free of human bias': 'emergy is a biosphere value, it is the energy the biosphere invests in its goods and services (including the goods and services of society)' (Brown & Ulgiati, 1999). Science should thus overrule short-sighted human preference when its findings indicate that the ultimate limits of energy supply are about to be reached.

This deeply environmentalist reasoning assumes that the biosphere's 'priority' is its own preservation. Accounting for the – physical – energetic inputs needed for maintaining natural systems would determine the necessary minimum level of preservation and sustainability. There is thus less interest in the outputs of natural systems for meeting human needs and wants. This becomes quite obvious when describing emergy flows in terms of economic accounting principles and techniques such as input-output tabulation. Costanza (1980) showed that treating final economic consumption as just another energy input into the 'human ecosystem' justifies expressing all product flows in embodied energy values. As a result the only final output remaining in this expanded system is net capital accumulation.[4]

It is no surprise that economists and policymakers did not accept this paradigm of human activity, which sees capital accumulation, rather than welfare-creating goods and services, as the only output (purpose?) of economic activity. Still, such accounting presents an interesting, albeit extreme ecological perception of humans and their environment. The validity of this view depends of course on the closeness of mankind to environmental disaster: imminent catastrophe would indeed override any economic valuation in favour of nature's goal of self-preservation.

---

[3] However, official statistics still ignores energy accounting in both conventional and green accounting (Chs. 7, 8).

[4] Modified physical input-output tables take a similar approach, in terms of the mass (weight) of material flows (see Section 6.3.4).

Still, energy accountants seek to construe some links to economic analysis by couching energy accounts in money values. The idea is to relate circulating money or GDP to emergy flows. Assuming constancy of the emergy/money ratio across economic sectors, sectoral emergy use could be expressed in so-called em-dollars (em$) by dividing a sector j's emergy E_j by the E/GDP ratio (Wall, 2001a, Odum, 2002). One study does indeed indicate a high correlation between sectoral output in dollars and embodied energy, except for primary energy use (Costanza, 1980). However, the interpretation of the national em$ flow as the 'money circulation, whose buying power is supplied by use of a quantity of emergy' (Odum, 2002), or (globally) as 'the amount of GWP [gross world product] that results from the emergy flow' (Brown & Ulgiati, 1999) is rather obscure.[5] In the final analysis, the purpose of em$ valuation seems just to raise the popularity of emergy accounting 'since people do not think in emergy units' (Odum, 2002).

Apart from possible ignorance about energy accounting and analysis by social scientists and national accountants, there are three major obstacles to making such accounting a standard tool of assessing sustainability:

- Inconvertibility (in practical measurement) of matter and different energy sources into a common energy unit
- Lack of knowledge about, and differing measurement methods for, a multitude of energy transformation processes, and most importantly,
- Incompatibility of energy value theory with economic choice and preferences.

The following section introduces material flow accounts as an alternative to energy accounting. Material flows are easier to observe and measure while covering comprehensively both energy and non-energy materials. On the other hand, their use of weight units raises questions similar to those of energy valuation.

6.3 Material Flow Accounting

The relative ease of definition and measurement of material flows appeals to national statistical offices under pressure to launch environmental accounting. Physical accounting can be added to their statistical work without affecting the centrepiece of official statistics, the national accounts (cf. Section 7.3). Ecological economists agree with this focus on physical accounts. As discussed in Section 2.4.2, they advocate material throughput analysis for assessing the ecological non-sustainability of economic activity.

European countries were the first to implement material flow accounts (MFA). In response, the statistical office of the European Commission, Eurostat (2001), prepared a methodological guide as a first step towards harmonizing concepts and

[5] Is it the one, or the other? How does GWP 'result' from the emergy flow? What is the connection of purchasing power with money circulation?

methods. More ambitious physical input-output tables connect material flows to economic sectors and provide thus a direct link to economic accounting and analysis [FR 6.2].

6.3.1 Concepts, Methods and Indicators

Georgescu-Roegen's admonition that matter – just as energy – is ruled not only by the first but also by the second thermodynamic law provides the incentive to account for both energy and material flows. Ayres and Kneese (1969) first proposed material and energy balances (MEB). Later, Ayres (1976) presented these balances as a framework for environment statistics to the international community. The MEB seek to measure material and energy inputs from the environment to the economy, their transformation in economic production and consumption processes, and their return to the environment as wastes and residuals.

The United Nations Statistical Commission rejected this presentation as 'a good paper for the long term' (United Nations, 1977a). The rejection delayed the further development of MEB for almost 20 years that is until the United Nations Statistics Division advanced the System for integrated Environmental and Economic Accounting (SEEA) (United Nations, 1993). The SEEA incorporated material and residual flows in somewhat aggregated form in its physical accounts. The revised SEEA-2003 (United Nations et al., in prep.) takes up the 'economy-wide MFA' but takes a dim view of their policy use (Section 8.4.1).

The main difference between MFA and MEB is that the MFA treat the economy as a black box. The MFA thus focus on the big picture of an economy's sustainability, in terms of overall natural resource supply to and disposal of residual output from the national economy (or a particular region). This allows ignoring the myriad of intra-economy processes, summing up the primary inputs and imports, and residual outputs and exports, with accumulation of materials in the economy as the balance. To this end the MFA apply a common measuring rod, the weight of materials. The MEB, on the other hand, use different units of measurement for different material inputs and outputs at different transformation stages.

Plate 6.3 depicts the material throughputs through the (blue-coloured) black box of the economy as inputs and outputs from and to the environment. The plate also shows 'translocations' of unused primary natural resources that had to be moved in generating the national product. These 'ecological rucksacks' do not become a part of a product but their movements may create considerable environmental disturbance. Plate 6.4 shows that an environmental rucksack can exceed by far the weight of the product itself. According to the Wuppertal Institute for Climate, Environment and Energy, we need a 2,000 kg rucksack of moving earth and sand to produce a 5 g gold wedding band.

Plate 6.3 refers to Total Material Requirement (TMR) as the measure of total input during an accounting period, including the movement of unused materials. As indicated in the plate, this key indicator of MFA can be related to economic output

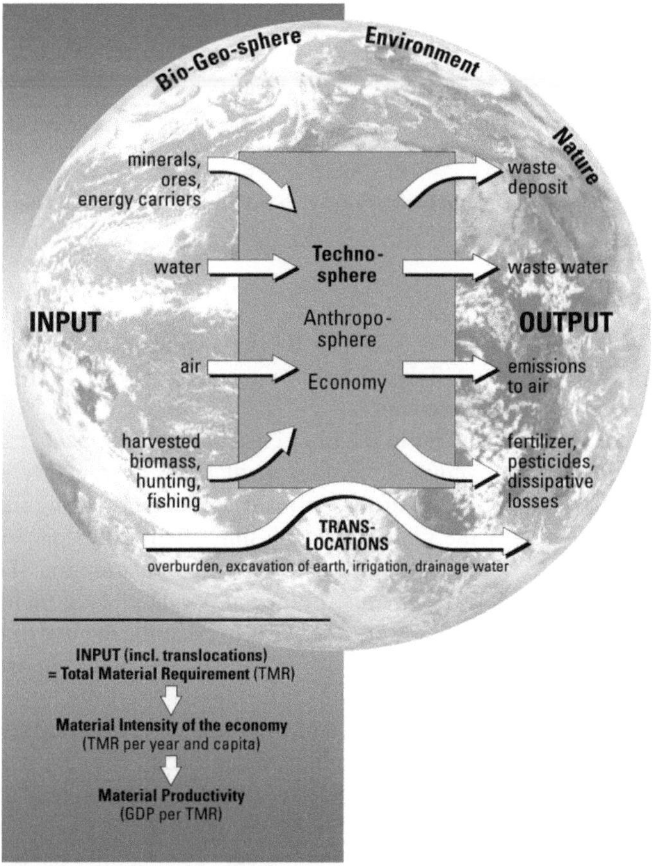

Plate 6.3 Material flows through the economy (*See Colour Plates*)
Source: S. Bringezu (2000). *Ressourcennutzung in Wirtschaftsräumen*. Berlin: Springer, cover page (translated by the author); with permission by the author, VisLab/Wuppertal Institute for Climate, Environment and Energy, and Springer Science and Business Media.

Plate 6.4 Ecological rucksack of a wedding band: 'too heavy to marry?' (*See Colour Plates*)
Source: Seppo Leinonen, with permission by the artist.

Table 6.1 Material flow balance and derived indicators

Inputs	Outputs
Domestic extraction + import	Emissions and waste
DMI (Direct Material Input) + Ecological rucksack (hidden flows) = TMR (Total Material Requirement) = DPO + X + NAS (Net Additions to Stock)	DPO (Domestic Processed Output) + Disposal of ecological rucksack + Export (X) = TMO (Total Material Output)

as material or resource productivity (GDP/TMR), or material intensity of a population's resource use (TMR p.c., p.a.). Table 6.1 shows the accounting definitions of TMR and other input and output indicators.

All these indicators endeavour to measure environmental impacts from material throughput. However, the weight of ecological rucksacks frequently overwhelms the weight of material inputs and of the ultimate products. Ignoring the rucksacks contained in TMR allows the compilation of simplified indicators, in particular Direct Material Input (DMI) and Domestic Material Consumption (DMC). DMI consists of domestic extraction and import of primary materials; it includes materials that are exported. DMC deducts these exports to describe the use of materials in the national economy. Both indicators are in fact more consistent with national accounts conventions.

TMR measures the overall pressure from the use of raw materials on natural systems by adding up the inputs and ecological rucksacks of primary materials in tons. Such a pressure is deemed to be indicative of actual and potential environmental impacts of natural resource use. Reducing the pressure by decreasing material inflows into the economy represents the ecological sustainability concept of dematerialization (Section 2.4). Dematerialization reflects thus a precautionary approach, which anticipates potentially disastrous and largely unknown environmental effects (Hinterberger et al., 2000).

On the output side, Total Material Output (TMO) measures the generation of waste and residuals. The measurement of particularly noxious substances as selected outputs could assess the success or failure of the 'detoxification' of production and consumption by pollution control. The EU's strategy of sustainable natural resource use (Commission of the European Communities, 2005) views detoxification as a supplement to dematerialization in a combined strategy of 'double decoupling' (see Section 13.3.1). However such a view looks like overkill since dematerialization eventually decreases wastes and residuals on its own. TMO and other output indicators seem thus to be less relevant for anticipatory and comprehensive sustainability analysis. Still, they may help check MFA balances and ensure the consistency and comprehensiveness of environmental (emission and waste) statistics.

Net Additions to Stock (NAS) are the balancing item in the accounts. The meaning of the NAS is controversial. They represent the materials stored in inventories or durable goods such as buildings, machines and infrastructure. As an environmental pressure index NAS reflect to some extent increased land use through built-up areas. However, area statistics are probably better (direct) measures of land use. Another interpretation views the accumulation of materials in the economy as physical growth of the economy (Bringezu & Moriguchi, 2002). Daly (1996), on the other hand, considers aggregate throughput as the physical growth indicator when calling for zero growth in a 'steady-state economy'.

6.3.2 Results

Figure 6.2 presents, for the region of the European Union (EU), the material flows depicted in Plate 6.3. Total material input of 18.5 billion tons exceeds total output by 3.7 billion tons, i.e. by the accumulation of materials in the region. The highly aggregated flows do reveal some structural characteristics of material flows in and out of the EU (Bringezu, 2002):

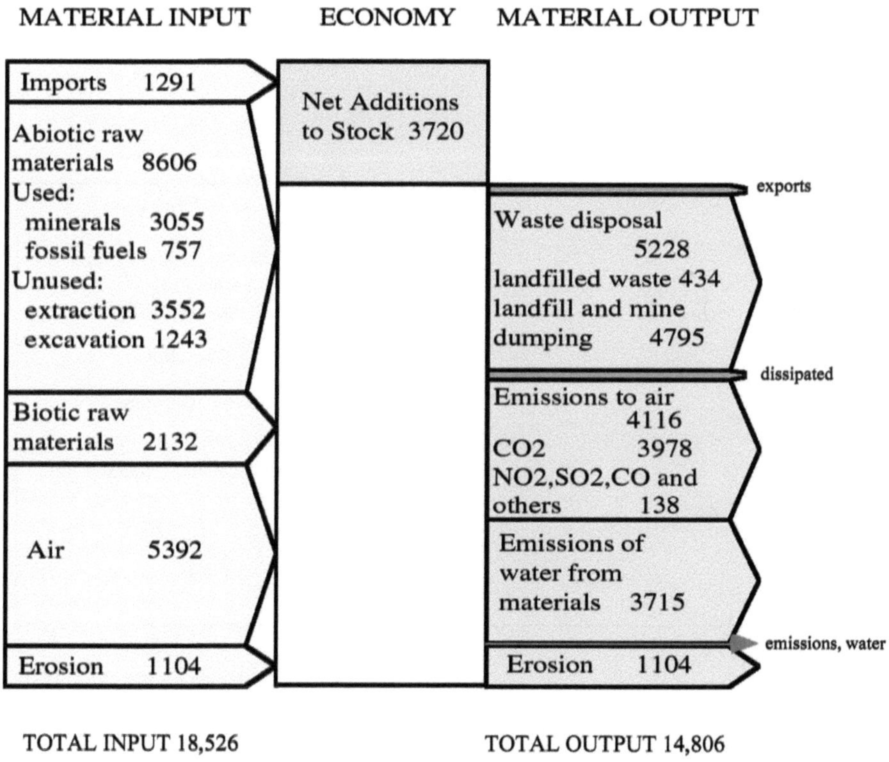

Fig. 6.2 Material flow balance of the European Union 1996 (million tons)
Source: Bringezu (2002), fig. 2.1.

- The movement of abiotic (non-renewable) raw materials is four times the flow of biotic (renewable) materials.
- The ecological rucksack of intra-EU extraction of abiotic resources exceeds the used-in-production part of these resources by 11%.
- Biotic agricultural resources are associated with an ecological rucksack of 0.5 tons erosion per ton of biomass.
- NAS are about 20% of total material input, indicating physical economic growth of the region.
- Waste dumping is 11 times the amount of controlled waste disposal.

Note that water inputs and (waste water) outputs are excluded because the inclusion of their huge amounts would indeed 'drown' the TMR by several orders of magnitude.

The purpose of dematerialization is to delink economic growth from the consumption of primary materials and its potential environmental impacts. Figure 6.3 plots the changes of TMR per capita against growth of GDP per capita for selected countries. TMR per capita seems to be levelling off for the industrialized countries at about 80 tons p.a., except for Japan, at 40 tons, due to its low per capita energy consumption. Low TMR per capita in Poland and China reflects these countries' relatively low levels of per capita economic output (in the 1990s). Upward-pointing arrows indicate that these (and probably other developing and transition countries) might well catch up with the high-material-intensity economies of industrialized nations.

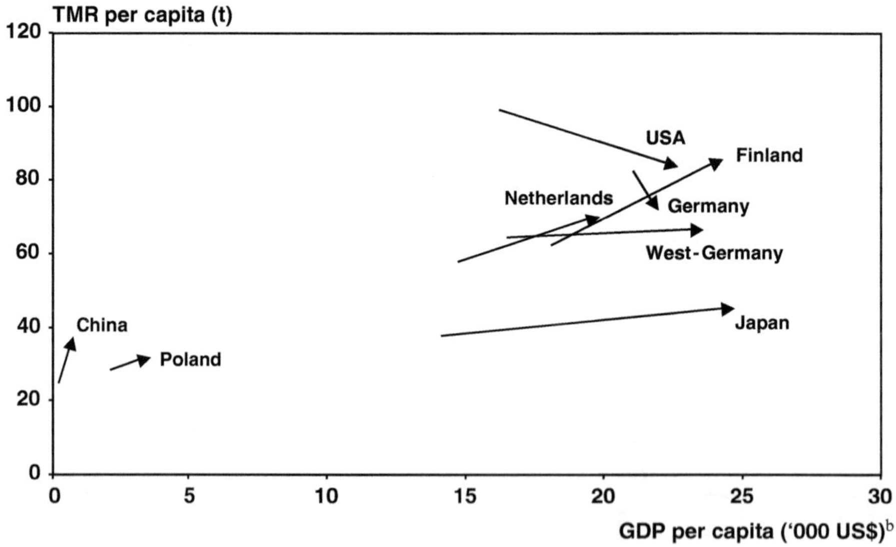

Fig. 6.3 Material use and economic growth in selected countries[a]
Notes: [a]China 1989–1996, Germany 1991–1996, Japan 1975–1994, Poland 1992–1997, USA 1975–1994, West Germany 1975–1990. [b]GDP in 1990 prices and exchange rates.
Source: Bringezu (2002, fig. 2.3).

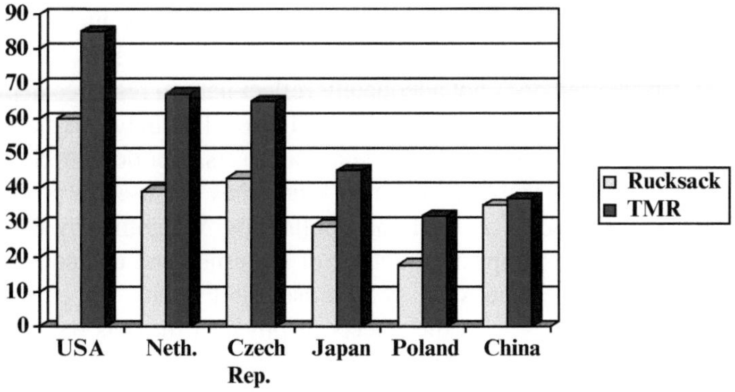

Fig. 6.4 TMR per capita and ecological rucksack (tons)
Source: Bringezu et al. (2004), table 2, p. 102; with permission by the copyright holder, Elsevier.

In general, there is some *relative* delinkage from growing GDP. However, relative dematerialization implies growing material use and pressure on natural systems with faster growing GDP. Most of the economies presented in Fig. 6.3 have thus not dematerialized in *absolute* terms and are still a far cry from sustainability standards such as Factors 4 or 10. Japan seems to comply with the Factor 4 standard of about half the other industrialized countries' TMR per capita and growing GDP. However, an upward trend of material use signals increasing 'materialization' of the economy. Germany is also an exception for more political reasons. Diminished overburden from the closure of unprofitable lignite mines in the new States (*neue Länder*) after the country's reunification explains the absolute decrease in TMR per capita. Converging arrows of Germany and West Germany indicate that West Germany's production and consumption patterns and concomitant constant high level of material use might soon prevail.

Figure 6.4 illustrates the huge inflation of material use if one accounts for ecological rucksacks (about two thirds of TMR in the average). Import from other countries generates much of these rucksacks. Except for the USA and China, whose TMR is largely domestic (Bringezu et al., 2004), at least part of the economic growth has been facilitated by importing sustainability, possibly from developing countries. Such 'burden shifting' of environmental pressure to other countries is particularly relevant in the EU, whose member states appear to rely increasingly on foreign resources (op. cit.). Globalization, together with domestic resource depletion, are significant factors in this outsourcing of natural resource supply (Section 14.1).

6.3.3 Critique: Ton Ideology, Early Warning or Policy Guide?

In their aggregate form, MFA provide better warning about environmental trends than indices based on averages of selected indicators (cf. Section 5.3). The grounding on physical laws and comprehensive accounting makes the MFA internally consistent and applicable to a wide range of environmental concerns. However, the total weight of primary materials used does not adequately reflect natural resource depletion and environmental degradation. There is no clear and direct link between material inputs and the depletion of stocks of (renewable) natural assets. Nor do material flows create uniform and equally hazardous damages when they are used in production and consumption and disposed of as waste and pollutants. Weighting by the weight of materials ignores different impact potentials of materials and excludes other environmental functions and effects such as land use, biodiversity, the ethical and aesthetic appreciation of nature, and the effects of physical impacts on human health and well-being. There are at least two attempts at countering this criticism.

One is to modify the MFA indicators by weighting them with environmental impact factors. A report commissioned by the EU (van der Voet et al., 2005) introduces an Environmentally weighted Material Consumption (EMC) indicator. The EMC calculates the weights for different materials consumed by means of life cycle analysis, which assesses the environmental impacts of each material from its cradle (extraction, import) to its grave (disposal) (see Section 9.1.2). Admittedly, the EMC faces a number of 'obstacles', which include

- Distinguishing raw materials from finished materials and a corresponding risk of double-counting materials and impacts
- Tracing and combining (by equal weighting!) the different impacts for each material into one impact factor
- Omission of depletion, i.e. impacts of permanent losses of renewable resource stocks
- Uncertainties about toxicity.

The reference to toxicity indicates that physical impact weights still do not capture health and welfare effects – neither in terms of physical damage to humans and non-humans, nor by some kind of valuation or evaluation by those suffering the damage. Finally, correlation of the weighted EMC with unweighted direct material consumption (DMC) is low ($R^2 = 0.56$); the easier-to-compile DMC of the standard MFA does not accurately present, therefore, actual or potential environmental impacts.

Another attempt at ensuring the validity of MFA softens the analytical use of its indicators. MFA-derived radical prescriptions such as the tenfold reduction of material inputs (Factor 10 Club, 1994) drew fears of 'eco-dictatorship' and 'ton ideology' (Gawel, 1998).[6] Later interpretations of Factors 4 or 10 view these targets as 'guard rails' rather than strict policy objectives (Section 2.4.2). Such guard rails refer to an environmental corridor, within which economic activities can be played

[6] There has been some heated argument on the risks and merits of material flow analysis vs. neoclassical environmental economics in Germany: see in particular Gawel (1998) criticizing 'material flow economics' for its inefficiency and interventionist ideology, with Hinterberger et al. (1999) presenting the counter-critique.

out without harming the environment. The Factor X authors, or at least their disciples, seem thus to have grown doubts about the use of MFA as a decision-making tool; they appear to favour using MFA indicators for warning about violations of a largely unspecified environmental space (cf. Section 2.4.1).

A first step towards throwing some light on the environmental space available for national economic performance is to find out what is happening within the black box of material flow accounting. Linking the material flows to different production and consumption processes requires disaggregation, i.e. greater detail in primary resource input and residual output. This would permit tracing environmental impacts back to their causes, and possibly forward to the damage on humans and ecosystems. Physical input-output tables achieve this linkage of environmental impacts with economic activities.

6.3.4 Physical Input-Output Tables

A physical input-output table (PIOT) fills the black box of the economy with details on material flows, in consistency with the national accounts.[7] The PIOT's usually large number of economic sectors shows sectoral supply (output) and use (input) of materials and products.

Compared to a conventional PIOT, a greened PIOT introduces MFA categories of raw materials as primary inputs into the economy and residuals of wastes and pollutants as final outputs of the economy. For instance, the German PIOT [FR 6.2] presents 58 branches of economic activity, 9 raw materials, 49 categories of products and 11 residuals. On the other hand, the physical tables, usually measured in tons, do not account for non-material flows of labour or other non-material services. The focus of an environmentally expanded PIOT is thus – just like the MFA – the flow of material throughput but with added detail on their transformation in production and consumption. In this they resemble the originally advanced MEB (Section 6.3.1), but with reduced and hence more manageable detail of production and consumption processes.

As in the MFA, the law of conservation of matter also holds for the individual sectors of the PIOT, with inputs equalizing outputs. Of course, the corresponding sectoral balances need calibration since residual outputs might not necessarily occur in the accounting period, which recorded the original material inputs. One could either assume in this case that all discharges and uses of materials take place in the same period, or one would have to introduce a balancing NAS item.

Table 6.2 is the aggregated PIOT for Germany with the economic sectors defined in consistence with the monetary input-output system and the national accounts.

---

[7] The standard system of national accounts, the SNA, treats (monetary) input-output tables as an integral part of its supply and use accounts. The PIOT represent the physical counterpart of the monetary tables.

Table 6.2 Physical input-output table, Germany 1990 (million tons)

Output \ Input	Transformation P	(HH)	Δ Cap C	Produced assets	Non- produced natural assets	X	Total material use
P	7,577	←	3,075	713	48,295	208	59,868
HH	2,645			11	700		3,356
Δ CAP	49,252	←	281[b]	20	56	0	49,609
Total material supply[a]	59,474	←	3,356	744	49,051	208	112,833

Notes: [a]Total material supply = direct material input (DMI). [b]Consumption of non-produced natural resources by households.
Source: Stahmer et al. (1998, table 12, modified and aggregated).

The supply rows include imports of materials and products as input to the different sectors. Exports (X), on the other hand, are a separate final use category, shown together with final consumption (C) and capital formation (Δ CAP). The production sector P consists of the 58 branches, which supply physical products to each other and final use. Both households (HH) and industries produce residual outputs. The residuals are either recycled or captured in environmental protection, a production activity, or dumped into the natural environment. Discharges to the environment generate an increase in the natural (non-produced) 'assets' part of Δ CAP.

The German PIOT ejects the primary production factor labour as a non-material service, but retains the final demand categories of the national accounts. A more radical eco-centric approach – as encountered above in emergy accounts (Section 6.2.3) – treats household consumption as an input into the 'national ecosystem'. This 'endogenizes' (Strassert, 2000) all household activity into the production sector. Such endogenization treats both enterprises and households as producers. The expanded production or material 'transformation' process absorbs final consumption of households as indicated by arrows in Table 6.2. Exports, capital formation and the accumulation of wastes and residuals remain as the only final outputs of the domestic transformation processes.

As a result, Germany's physical GDP of 3,603 million tons (C + Δ CAP, produced + X-M) declines drastically down to 528 million tons (Δ CAP, produced + X-M)[8], unless the new final output of mostly waste and residuals from conventional (enterprise) production (48,295 million tons) and residuals and natural resources from household production (700 million tons) is added. The overwhelming load of residuals creates a far-from-desirable final output of the economy. The remaining physical GDP amounts to only about 6% of total material output, and about 7% of the (weight of) residuals.

The meaning of these overall tonnages and their respective sizes is opaque. It is obviously not an indication of the value or significance of economic production, con-

[8] With M = –393 million tons (not shown in Table 6.2).

sumption and (produced or natural non-produced) capital formation. This might explain why some ecological economists, while shunning the 'human bias' of market valuation, revived Sraffian system-inherent pricing as the dual solution of a linear programming system (Strassert, 2001; Friend, 2004). Such pricing avoids market valuation by reflecting the inherent technology and preset ecological and economic restrictions of the linear programming model (see Section 12.2). It is unable and decidedly unwilling to take human preferences for goods and services into account.

Apart from these radical changes in production and consumption concepts, PIOT of the more traditional kind serves a number of statistical and analytical uses (Eurostat, 2001), including

- Statistical checks for balances and consistency with accounting identities
- Assessment of material use efficiencies
- Intermediate tabulation of resource use and emissions for monetary environmental accounting (Section 7.3)
- Assessment of direct and indirect environmental impacts of products and services, covering the whole production chain
- Decomposition analyses, which assess the influence of overall economic growth and technological and structural change on environmental impacts
- Modelling of trends and scenarios of environmental impacts for different production and consumption patterns and economic growth rates.

The fourth part of the book will explore the analytical uses of input-output modelling.

Further Reading

FR 6.1 Energy Concepts and Accounting

The analysis of available energy or *exergy* for the efficiency of economic processes has tradition (e.g. Soddy, 1933; Slesser, 1975; Gilliland, 1978; Martinez-Alier, 1987; Szargut et al., 1988). More recently, Ayres et al. (1998) and various contributors to the Encyclopedia of Life Support Systems (Tolba, 2001; Brodianski, 2001; Frangopoulos, 2001; Szargut, 2001; Wall, 2001b) advanced exergy for valuation and sustainability analysis in energy accounts. Assigning an economic value to exergy flows is the task of 'thermoeconomic accounting' (Wall, 2001a). Odum (1996, 2002) and his disciples, Brown and Ulgiati (1999), suggest similar monetary valuation for *emergy* flows.

The United Nations Statistics Division (http://unstats.un.org/unsd/energy/balance/default.htm) and the International Energy Agency (http://www.iea.org/Textbase/stats/index.asp) prepare more narrowly defined *energy balances*. These balances use a common energy unit, i.e. a fuel combustion heat equivalent, which refers directly to the underlying energy carriers. They can also be linked to the physical material flow accounts for measuring the 'energetic metabolism of societies' (Haberl, 2001). Bartelmus (2004) describes the relationship of such accounting with greened national accounts and physical input-output tables.

FR 6.2 Material Flow Analysis and Accounting

Fischer-Kowalski (1998), Fischer-Kowalski and Hüttler (1999) and Ayres (1989) made societal or industrial metabolism the underlying principle of material flow analysis. Ayres and Kneese (1969) pioneered material and energy balances (MEB) that showed not only inputs into and outputs from the economy but also the transformation of material and energy inputs within the economy. Steurer (1992), Bringezu (1993) and Schmidt-Bleek (1994) simplified the approach by covering only those flows in the MFA that cross the boundaries of the economy. International guidelines (Eurostat, 2001) and applications (Adriaanse et al., 1997; Matthews et al., 2000) indicate growing interest in material flow analysis, sometimes under the umbrella of 'industrial ecology' (Ayres & Ayres, 2002). Numerous country projects of MFA have been carried out under the auspices of the European Union (van der Voet et al., 2005).

The German Federal Statistical Office pioneered physical input-output tables as a sectoral extension of the MFA (Stahmer et al., 1998). See [FR 10.1] on input-output tables and their connection with the national accounts.

Review and Exploration

- Would you consider social metabolism as just a metaphor for environment-economy relationships or as the basic principle of material and energy flow accounting? How do thermodynamic laws apply in such accounting?
- What are the reasons for replacing market valuation by energy values in regional or countrywide accounts?
- Does measurement in weight and energy units reflect the environmental significance of material and energy flows? How do these flows assess the (non)sustainability of economic performance and growth?
- How does Fig. 6.3 indicate delinkage of economic growth from environmental impacts? Do we need a sustainability standard to assess the ecological sustainability of economic growth?
- How do the MFA measure the import/export of sustainability (burden shifting) from/to other countries?
- Is NAS a good indicator of physical economic growth? How does it compare to Daly's concept of physical (throughput) growth?
- Compare the approaches of PIOT and MFA and their sustainability assessments. What is the meaning of a physical GDP in PIOT?
- Can the monetization of energy and material flows in emergy-dollars or by input-output inherent pricing assess the economic and environmental significance of physical flows?

Part III
Greening the Economic Accounts

This part follows up on the ecological/physical vs. economic/monetary dichotomy in environmental accounting and analysis. The economic approach extends economic valuation to physical environmental impacts within the worldwide-adopted System of National Accounts, the SNA. Figure 7.1 illustrates the dichotomy in environmental-economic accounting. It distinguishes between the physical accounts described in Part II, and hybrid (mixed physical and monetary) and monetary accounts of this part. The figure also displays the links of environmental accounting to data sources (Ch. 4) and uses (explored in Parts IV and V).

Chapter 7 first examines ad hoc monetary index calculations. The indices attempt to measure economic welfare by adding positive effects and deducting negative ones from conventional accounting indicators. The flaws of these corrections make the case for embedding the costing of environmental impacts in

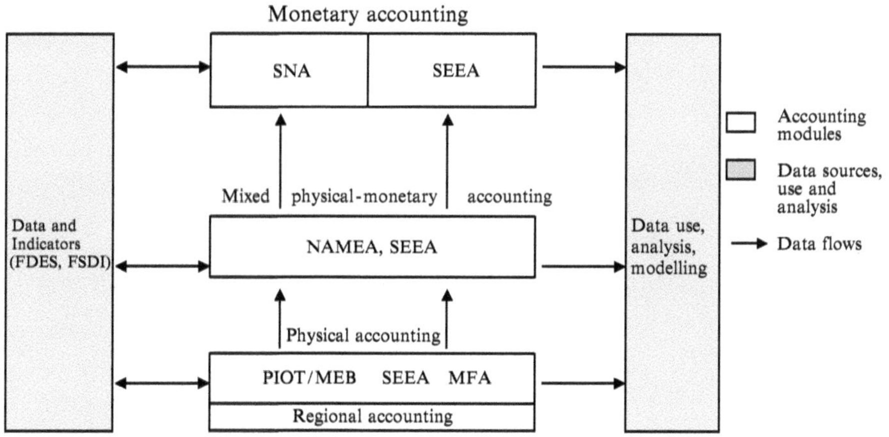

Fig. 7.1 Framework for environmental and economic accounting
Acronyms: see text.
Source: Bartelmus (2001), Accounting for sustainability: Greening the national accounts, fig. 2; Copyright Eolss, with permission from Eolss.

the more consistent national accounts system. The chapter also describes hybrid accounts, which one can see as an intermediate step towards monetary environmental-economic accounting, the topic of Ch. 8. The hybrid accounts thus bridge the physical-monetary dichotomy by connecting key variables of ecological and environmental economics. Official statisticians take a different view in the revised version of the System for Environmental and Economic Accounting (SEEA). They tend to consider the hybrid accounts as the ultimate goal, with a view to avoiding controversial monetary valuation of environmental impacts (Section 8.4).

Corporations, concerned with their responsibility for environmental impacts, have more and more taken up environmental accounting and reporting. Chapter 9 discusses the use of corporate environmental accounts for improving environmental management and demonstrating social responsibility. The chapter also examines the usefulness of a 'micro-macro link' in green accounting and accounting analysis.

Chapter 7
Linking the Physical and Monetary Accounts

Popular indices of sustainable income, welfare and wealth are precursors to greening the national accounts. They take a short cut by correcting accounting indicators for environmental and social effects without grinding the corrections through the accounting rules and definitions. Their flaws make the case for a more systematic introduction of environmental concerns into the national accounts framework.

The physical input-output tables of the national accounts overcome some of the conceptual and procedural differences between material flow and economic accounts. They remain, however, in the physical realm and do not connect the physical data with their monetary counterparts. *Hybrid accounts* make this connection by embedding the physical accounts in the monetary national accounts.

Introducing asset accounts into the hybrid accounts and expanding the asset boundary of produced capital to non-produced natural assets generates the physical accounts of the United Nations System for integrated Environmental and Economic Accounting (SEEA). These accounts provide rigorous definitions of environmental impacts in terms of accounting concepts of capital consumption and accumulation. The actual measurement of natural capital and its consumption in a fully integrated green accounting system requires, however, monetary valuation of environmental impacts, described in Ch. 8.

7.1 Measures of Economic Welfare and Wealth

Chapters 5 and 6 dealt with the problem of aggregating environmental indicators by converting them into a common scale of relative deviation from maximum or average targets, or into common physical units of area, energy or mass. The resulting measures cannot answer, however, the basic question whether we are really better off, and if so, whether this had been achieved by living off (unsustainably?) our produced and natural capital endowment. To this end, environmental economists suggested amending the economic (flow) indicators of the national accounts by incorporating social environmental costs and benefits, and by expanding the coverage of wealth beyond produced capital.

7.1.1 *Welfare Indicators*

Section 2.3 introduced non-declining welfare as a rather abstract concept of economic sustainability. Reacting to the critique of using national income, product or consumption as economic welfare measures, mainstream and environmental economists set out to adjust these aggregates for missing welfare effects [FR 7.1]. Nordhaus and Tobin (1973) were among the first to compile a Measure of Economic Welfare (MEW), which adds 'desirables' of leisure and non-market outputs and deducts 'regrettables' to and from GNP. Desirables include leisure and non-market products; regrettables consist of instrumental (defensive) expenditures, requirements for production and capital from population growth, and environmental externalities.

Ecological economists criticized the MEW for reflecting preconceived faith in economic growth. Large positive imputations for leisure and non-market subsistence, and relatively low deductions for environmental and social deterioration understated in their opinion the social costs of economic growth. They advanced, therefore, their own improved measures, presenting a broader list of defensive expenditures and environmental damages. Daly and Cobb (1989) incorporated these and other detractions from the human quality of life in a final-consumption-based Index of Sustainable Economic Welfare (ISEW).

As could be expected, the ISEW, later called Genuine Progress Indicator (GPI) after some modification for leisure and work, differed dramatically from conventional GDP: America appeared to be 'down' despite an upward trend of GDP (Cobb, Halstead & Rowe, 1995). This scissor movement of GDP and ISEW (see USA in Fig. 7.2) seems to confirm Max-Neef's (1995) 'threshold hypothesis'. The hypothesis claims that after attaining a certain level of economic output, its further growth brings about a decline in the quality of life. Considering that environmental quality is an important aspect of the human quality of life, the threshold hypothesis inverts the EKC hypothesis of environmental improvement through economic growth (Sections 2.2.2, 3.2.1).

Unrealistic assumptions behind the valuation of impacts and their welfare effects (Neumayer, 2000) might explain why such a threshold was not found for several other countries. For instance, the German ISEW rose by 30% between 1979 and 1992, while GDP increased by 50% (Diefenbacher, 1995). The differing trends of Fig. 7.2 might therefore be the result of differences in coverage, concepts and methods of valuation, changes in the quality of life, or both.

ISEW and GPI claim to assess the ultimate goal of economic activity, the generation of human welfare. They attempt to solve the equal weighting problem of indicator averages by applying monetary price or cost weights to social and environmental welfare effects. However, the indices do not clearly connect physical impacts to effects on health and well-being. Making this connection would require an assessment of the exposure of humans and non-humans to environmental and other social impacts, and of the final dose-response reactions of those affected by the impacts.

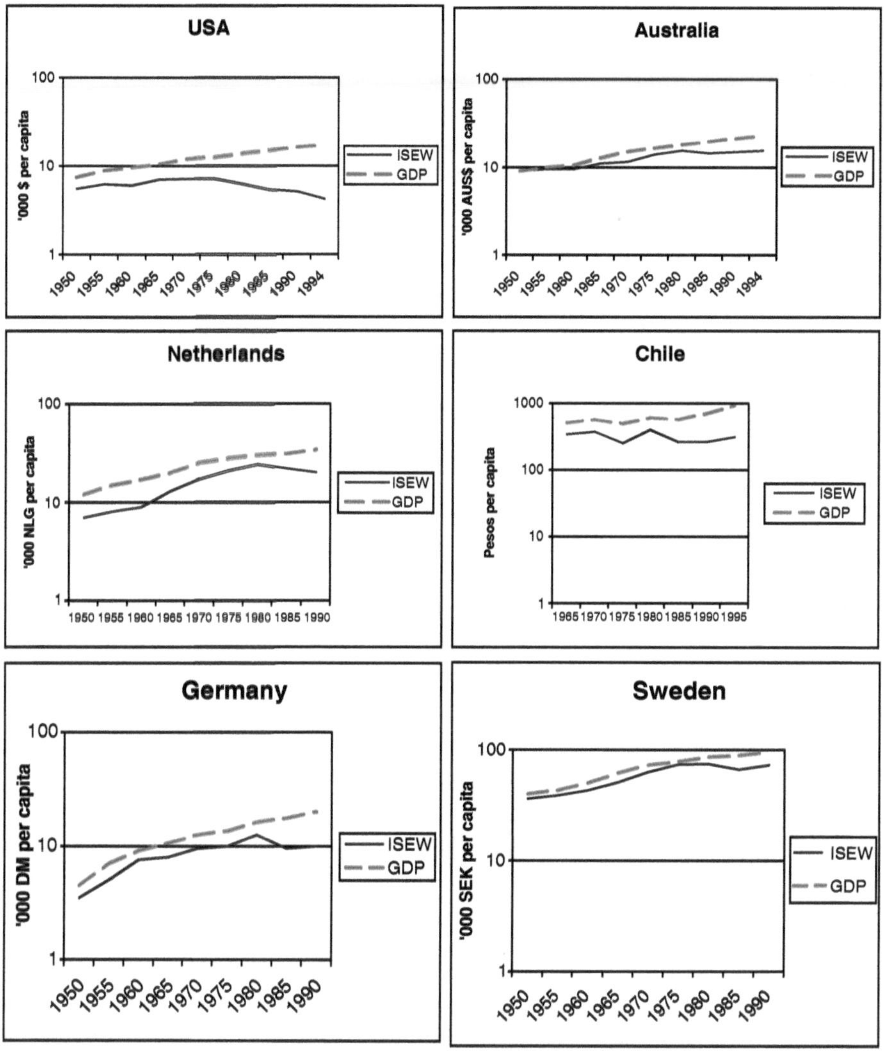

Fig. 7.2 ISEW and GDP in selected countries
Source: http://www.foe.co.uk/campaigns/sustainable_development/progress/international.html (modified).

Unsurprisingly, the index is mired in problems of welfare definition, measurement and valuation, including

- Confusion about combining *sustainability of* economic *production*, conceived as the maintenance of forms of (human, social, natural, financial and produced) capital, with *welfare* effects *from* the *consumption* of welfare generating goods and services

- Arbitrary definition (and exclusion) of private and public consumption categories, deemed to be defensive with reference to some ideal period (usually in the 1950s) when all consumption supposedly generated only positive welfare
- Consideration of all capital formation as defensive, ignoring current and future welfare effects from deferred consumption, i.e. saving
- Muddled use of national accounts concepts of final consumption, capital formation and consumption (including consumer durables)
- Mixing economic costs and outputs in market prices with widely differing welfare (damage) valuations (cf. Section 8.1)
- Biased focus on welfare losses (e.g. from reduced leisure for gaining higher income, or forced leisure from unemployment) at the exclusion of welfare gains (e.g. from increased leisure for recreation).

All these conceptual inconsistencies and questionable estimates in actual index compilations tend to make the ISEW/GPI more of an instrument of advocacy than scientific measurement. Admittedly, the index 'would blast away the obfuscatory polemics of growth – and the devious politics that goes along with it' (Cobb et al., 1995).

7.1.2 Comprehensive Wealth Measures

Kenneth Boulding (1966) famously asked: 'Is it ... eating that is a good thing, or is it being well fed?' In his view, the 'spaceman economy' with limited stocks of natural assets makes minimizing production and consumption *flows*, while maximizing the *state* of well-being, a much better option for generating human welfare than maximizing consumption. Three decades later, the World Bank seems to have responded to Boulding's complaint that 'the economics profession has neglected with astonishing singlemindedness' the state and stock aspects of human welfare. Interpreting sustainable development as 'preserving and enhancing the *opportunities* open to people', the World Bank (1997) calls for the 'portfolio management' of the stocks of environmental resources, produced assets and human resources.

The reason for shifting the focus from flow indicators such as income, production or consumption to wealth measures differs, however, from Boulding's arguments. An updated version of the 1997 report (World Bank, 2006) describes comprehensive wealth as an easier-to-measure *proxy* for utility and welfare measurement, compared to any direct measure of the state of well-being. The justification is derived from 'economic theory', which postulates that 'current wealth should equal the present value of future consumption'. The admonition 'should' points to the assumption that wealth is used optimally, so as to realize maximum welfare-creating consumption over an extended future period. A further assumption is that consumption can be modified sufficiently to make up for the depletion of different (substitutable) types of capital. Such argument, based on economic theory, shows the power of economics to give its indicators (in this case wealth) meaning and context. On the other hand, it reveals crucial deviations of modelled variables from real-life statistics and indicators.[1]

The World Bank defines total wealth as the economic value of three real (non-financial) capital categories, viz.

- Produced capital, measured by the perpetual inventory method of adding capital formation and deducting depreciation to/from an initial capital stock
- Natural capital, mostly valued as the net present value of natural resource rents (cf. Section 8.1.1)
- Intangible capital, defined as the residual after deducting produced and natural capital from total wealth; total wealth in turn is calculated as the present (discounted) value of current and future final consumption.

The intangible wealth residual contains all kinds of non-natural and non-produced aspects of human, institutional and social capital that are expected to contribute to current and future consumption. Further analysis suggests that the residual is dominated by human capital (op. cit., ch. 7).

Ranking countries according to these wealth estimates shows high correlation of total wealth with GDP (both per capita). This is not surprising since total wealth represents the potential for future economic growth.

More interesting is the composition of wealth. Figure 7.3 shows the relatively high endowment of developing countries with natural capital, including agricultural land. The World Bank interprets the low share of natural (economic) assets in rich countries as an indication of their relatively low significance in these countries. On the other hand, the absolute value of natural capital has been increasing with economic growth. The World Bank (2006) asserts that this 'contradicts the common assumption, that development necessarily entails the depletion of the environment and natural resources'. Both conclusions are questionable, considering that (1) the 'burden shifting' or import of sustainability (see Section 6.3.2) does not absolve rich countries from demand-side responsibility for imported natural capital services, and (2) the increase in value of natural capital could be the result of increasing demand, which does not reflect actual physical depletion and degradation of natural assets.

Given the assistance of sustainability-exporting developing countries, rich nations seem to be able to rely on their (higher share of) knowledge, skills and institutions, i.e. intangible capital, to generate their ever-growing output. But there is another international aspect of sustaining economic development, which the comprehensive wealth estimates ignore: the increase in net wealth through transfer of financial capital as part of foreign aid. Such a potential for investment clearly contributes to the sustainability of economic performance. A special case of linking

---

[1] Cf. Section 8.4.1 for a similar controversy of using Hicksian (ex ante) income to describe the sustainability of national income generated during a past accounting period. See also Section 12.3.2 for the necessary assumptions and explanatory power of optimal growth analysis.

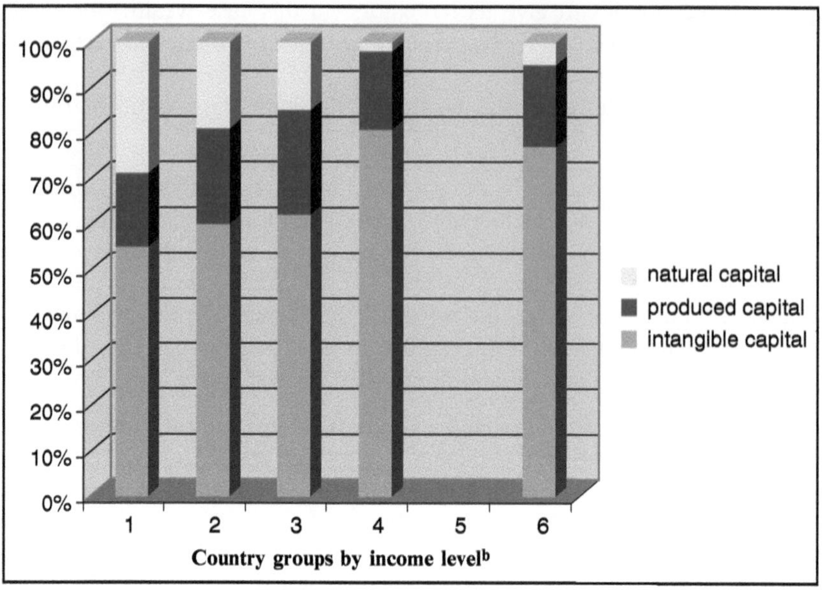

Fig. 7.3 Total wealth and its composition, 2000
(low-, mid- and high-income countries[a])
Notes: [a]Excluding oil exporting countries. [b]1: lower-income, 2: lower-mid-income, 3: upper-mid-income, 4: high-income, 6: World.
Source: World Bank data (2006, Table 2.3).

such transfer to environmental concerns is foreign debt relief for environmental preservation. The purpose of such 'debt-for-nature swaps' is to substitute negative wealth (the debt liability) by the conservation of a natural asset.

Ad hoc estimates of the economic value of wealth draw attention to the limited availability of exhaustible non-produced capital for sustaining economic growth and development. However, compiling macroeconomic aggregates outside the national accounts bears the risk of inconsistencies with standard economic indicators. Most wealth estimates thus ignore the definitions and links of national (disposable) income, savings and 'net worth' (representing the wealth of an economic agent in terms of his assets and liabilities) in capital accounts and balance sheets.[2]

The results are wealth estimates, whose wide differences remain unexplained. What are we to make of a decrease of total world wealth by over 30% between 1994 and 2000 (in the World Bank 1997 and 2006 reports)? Or how can we relate nature's *annual* services of $33 trillion (Costanza et al., 1997b) to nature's capital

[2] Note in particular that the national (disposable) income definition, from which savings and consumption are derived can only be connected with the net worth or net wealth concept of the national accounts if several 'erratic' changes in net worth are excluded (see Box 8.3).

stock of $42 trillion (as calculated from World Bank 1997 data)? Extending consistently the rigorously defined national accounts *system* appears to be the only way to assess the sustainability of economic activity with compatible stock and flow indicators.

7.2 Extending the National Accounts: Incorporating Nature's Assets

The risk of undermining economic growth by over-mining natural capital (Section 2.3.1) is the reason for greening the national accounts and their income, output and capital indicators. The United Nations (1993) advanced to this end the System for integrated Environmental and Economic Accounting (SEEA).[3] Its basic approach is to extend the asset boundary of the System of National Accounts (SNA) (United Nations et al., 1993) by

- Including nature's resource stocks and sink capacities in the asset accounts and balance sheets of the SNA
- Costing their depletion and degradation as natural capital consumption in the production, income and capital accounts.

The incorporation of environmental assets and asset changes is at the heart of environmental-economic accounting. It determines the additional environmental information that is to be incorporated in the national accounts. Box 7.1 defines and explains environmental assets as non-economic natural assets, using the SNA definition of economic assets. At the same time, the SEEA maintains that these assets do provide economic benefits since their products and services have become scarce due to increased human use.

Box 7.2 presents the basic categories of non-financial assets of the SNA used in the SEEA. Both treat cultivated (managed) natural assets of agriculture, fishery and forestry as produced assets. The non-produced assets include natural resources of wild animals and plants (including wood from uncultivated forests), subsoil reserves, land and water. Air is an additional non-produced asset in the SEEA, allowing for its overtaxed environmental sink function.

At first sight, the scope and classification of produced and natural assets appear to coincide, with the exception of air, in the SNA and SEEA. However, as discussed in Box 7.1, non-produced natural assets may hold economic functions (in the SNA sense: as raw materials for production), as well as environmental functions of sink and other ecological services. Accounting for environmental functions distinguishes the SEEA

---

[3] The SEEA has now been revised, relaxing some of its compatibility with the SNA; Section 8.4 reviews critically the revised version of the SEEA. Unless otherwise indicated, 'SEEA' refers to the original 1993 version, rather than the revised SEEA-2003 (United Nations et al., in prep.).

Box 7.1 Asset definition in the SNA and SEEA

The *economic asset* definition of the SNA includes all natural assets, 'over which ownership rights are enforced by institutional units, individually or collectively, and from which economic benefits may be derived' (United Nations et al., 1993). These natural economic assets can be produced such as agricultural products or non-produced such as land, mineral deposits or fish in the ocean (for which fishing rights are established). The SNA accounts for changes in the availability of economic non-produced assets, resulting from depletion or degradation, as 'other changes in volume' in asset accounts. These changes are recorded only in the asset accounts; they do not affect the key economic flow indicators of income and product. In contrast, the SEEA shifts the value of depletion and degradation of these assets as a cost of natural capital consumption into the production accounts, modifying the main economic indicators (Section 8.2.2).

Implicitly, *environmental assets* are all those non-produced natural assets, for which no property rights exist and which do not function as providers of natural resource inputs into production or as stores of wealth (the 'economic benefits'). They supply environmental services of waste absorption, ecological functions such as habitat or flood and climate control, and other amenities of recreation, public health or aesthetic values. Their loss is valued (costed) therefore in the SEEA (Section 8.1.2), further adjusting the conventional economic indicators. Particular natural assets such as a forest can of course exhibit characteristics of both, economic assets (e.g. timber) and environmental assets (e.g. habitat or CO_2 absorption).

Box 7.2 Classification of produced and non-produced assets in the SNA/SEEA

1. Produced assets
   - 1.1 Fixed assets
     - 1.1.1 Tangible fixed assets, including cultivated assets
     - 1.1.2 Intangible fixed assets, including mineral exploration
   - 1.2 Inventories of products, including cultivated ones
   - 1.3 Valuables
2. Non-produced assets
   - 2.1 Tangible non-produced assets
     - 2.1.1 Land
     - 2.1.2 Subsoil assets
     - 2.1.3 Non-cultivated biological resources
     - 2.1.4 Water resources
     - *2.1.5 Air*
   - 2.2 Intangible non-produced assets, including permits of resource use and pollution

Source: United Nations et al. (1993); United Nations (1993).

from the SNA which excludes them explicitly. The SNA considers these services as externalities, i.e. 'unsolicited services, or disservices', which are neither a market transaction nor economic production; rather, they are seen as 'a purely natural process'.

Related ecological assets of terrestrial and aquatic ecosystems could be viewed as a separate category of non-produced natural assets. However, the assessment of the quality of ecosystems and their complex interactions of species and abiotic components is probably best left to special ecological statistics and analysis, rather than forcing them into an accounting framework.[4] As shown in Ch. 8, assessing the sustainability of economic activity should focus on the immediate interaction between environment and economy; it does not require and cannot deal with detailed physiological and ecological processes and health effects, especially at the national level.

Figure 7.4 describes how the SEEA extends the conventional stock (asset) and flow (supply and use) accounts of the SNA. An additional (shaded) column illustrates the inclusion of environmental assets and asset changes. Economic natural resources are actually part of the SNA asset accounts; their use reaches therefore into the economic asset column. As mentioned in Box 7.1, the SNA accounts for their depletion as 'other asset changes' outside the supply and (capital) use accounts. In contrast, the SEEA, at least in its original (1993) form, considers the depletion of natural resources as natural capital consumption. Figure 7.4 shows, therefore, the depletion of economic non-produced natural assets as a (shaded) change in both the flow and asset accounts.

The asset accounts measure the value of opening and closing stocks. They overlap with the flow accounts as consumption and formation of fixed and non-produced natural capital. Certain other asset changes result from disasters, unmanaged natural growth or discovery. Both SNA and SEEA exclude these changes from capital formation and consumption since they are the result of non-economic (natural or political) events. These asset changes are therefore recorded as other asset changes in the asset accounts only; they do not affect the conventional and greened income and output accounts and their indicators (see Section 8.2.2). The purpose is to remain as consistent as possible with the established national accounts definitions and balances.

7.3 Hybrid Accounts: Expanding the Production Boundary

Some national accountants contest the valuation of non-marketed environmental impacts as modelling – rather than statistical measurement. They opt, notably in the revised SEEA-2003, for opening the monetary national (flow) accounts to physical material inputs and residual outputs, shunning monetary valuation of environmental impacts. In this way they modify the accounting frame by expanding the production boundary to include nature's 'production' of source and sink services.

[4] The SEEA-2003 distinguishes, however, natural resources from land and surface water and ecosystems (United Nations et al., in prep.). Besides a confusing overlap, these categories are measured in different units of area, volume, weight, etc. and are thus impossible to add up in aggregate accounts.

		OPENING STOCKS	Economic assets	Environmental assets	
			+		
	DOMESTIC PRODUCTION	*FINAL CONSUMPTION*	*CAPITAL FORMATION*	*CAPITAL ACCUMULATION*	*REST OF THE WORLD*
SUPPLY of products	Output				Imports
USE of products	Intermediate consumption	Final consumption	Gross capital formation		Exports
CAPITAL use	Capital consumption		Capital consumption		
NATURAL ASSET use	Environmental cost	Environmental cost	Natural capital consumption		
			+		
			Other asset changes	Other asset changes	
			=		
		CLOSING STOCKS	Economic assets	Environmental assets	

Fig. 7.4 Incorporating environmental assets in the national accounts
Source: Bartelmus (2001), Accounting for sustainability: Greening the national accounts, fig. 3; copyright Eolss, with permission from Eolss.

The Dutch National Accounting Matrix including Environmental Accounts (NAMEA) is the prototype of a hybrid approach, which presents physical data of natural resource inputs and residual outputs of wastes and emissions next to the monetary aggregates of the national accounts [FR 7.2]. The SEEA-2003 considers the implicit avoidance of any effects on the – monetary – SNA as a 'strength' of such hybrids. Table 7.1 illustrates how physical resource inputs and residual outputs (in italics) are simply placed next to the core of the conventional monetary national accounts. The result is a hybrid physical-monetary input-output table.

Difficulties of tracing materials and residuals from their origin to their destination of receiving sectors reduces the NAMEA in most cases to tabulations of wastes and emissions and resource use in the standard economic activity classification. Table 7.2 confirms this reduction, even by the highly developed Dutch statistical services. The table classifies residuals mostly by origin (economic sector), except for waste generation and treatment (and phosphorus and nitrogen, not shown in the table), and natural resource discovery and use. CO_2 emissions create the heaviest burden (in weight units). Environmental clean-up services capture only a small portion of waste (10%); they are part of the industries, excluding 'for the time being' recycling.

Table 7.2 also shows, as does Table 7.1, the contribution of materials and residuals to environmental 'policy themes'. The categorization of policy themes as a 'destination' seems to be out of place since 'themes' are not a receiving or absorbing sector. As discussed in Section 5.1, the application of theme equivalents to environmental indicators is a method of aggregating physical flows into more compound indices. Skirting monetary valuation, NAMEA thus simply imported – from outside the accounting framework – both, the environmental indicators and their weighting scheme (Adriaanse, 1993). The purpose of calculating theme contributions is thus to regain some of the aggregative and comparative power lost with the restriction to physical accounting. As pointed out in Section 5.1, this approach still fails in making inter-theme and overall economic-environmental comparisons.

The NAMEA does not include any asset accounts, whether extended or not. It cannot, therefore, assess the depletion of natural capital and hence the sustainability of economic activity. NAMEA does facilitate, though, the linkage of physical environmental impacts with their economic causes. This is a necessary step towards costing these impacts and thus truly greening the national accounts. NAMEA also stands on its own as the database for analytical applications such as decomposition analysis and scenario modelling (see Part IV).

Further Reading

FR 7.1 Monetary indices of sustainability

Eisner (1988) reviews early attempts at modifying monetary accounting indicators for welfare measurement. Nordhaus and Tobin's (1973) *Measure of Economic Welfare* (MEW), later renamed Net Economic Welfare (NEW) (Samuelson

Table 7.1 Simplified structure of a NAMEA

	Monetary accounts (use by)				Physical Accounts	
Supply (of)	Industries 1, 2, …	Households (public and private)	Capital formation and accumulation	Rest of the world (ROW)	*Material flows (environmental source and sink functions)*	*Policy themes*[a]
Outputs (excl. capital goods)	Intermediate consumption	Final consumption		Exports	*Emissions from industries*	
Capital goods	Capital consumption		Gross capital formation	Capital transfer to ROW	*Other volume changes (discoveries etc.)*	
'Outputs' of households					*Emissions from households*	
Income generation	Value added, NDP					
Rest of the world (ROW)	Imports		Capital transfer to ROW	Balances	*Imports of natural resources and residuals*	
Material flows (natural resources, residuals)	*Natural resource inputs, residuals received*	*Natural resource use, 'consumption' of residuals*	*Net accumulation of materials and substances*	*Exports of natural resources and residuals*	*Balances*	*Contributions to policy themes*

Note: [a] Greenhouse effect, ozone layer depletion, acidification, eutrophication, waste, loss of natural resources.
Source: based on Bartelmus (2004), table II, p. 49; with permission by the copyright holder, Elsevier.

Table 7.2 NAMEA 1997 (Netherlands) – origin and destination of material flows

	CO_2 (million kg)	SO_2 (million kg)	...	Waste (million kg)	Natural gas (pj)	Crude oil (pj)
Origin						
Households	36,790	2.05		5,120		
Industries	164,230	234.10		10,050		
...						
Discoveries, etc.[a]					3,364	250
Imports of residuals		82.40				
Total	201,020	318.55		15,170	3,364	250
Destination						
Households						
Industries[b]				4,460	2,541	88
...						
Accumulation of materials						
Exports of residuals		222.85				
Environmental themes	201,020	95.70		10,710	823	162
Total	201,020	318.55		15,170	3,364	250

Notes: [a] Other volume changes in natural resources.
[b] Environmental cleansing and sanitary services.
Source: de Haan and Kee (2004, tables 1 and 2, reduced with some modifications).

& Nordhaus, 1992), has now been sidelined by the propagation of the *Genuine Progress Indicator* (GPI) (Venetoulis & Cobb, 2004). Less ambitious is the calculation of a *Sustainable National Income* indicator, adjusted for the cost of restoring the environment to a sustainable level (as to be found in some earlier pristine period; Hueting 1993). Ekins and Simon (2001) propose a similar *sustainability-gap index*, which would measure the total cost of reducing environmental impacts to desirable standards. Contrary to Hueting, they do not deduct these costs from national accounts aggregates, considering such costing as 'not commensurable' with accounting indicators. The World Bank's (1997, 2003) *Genuine Saving* corrects the sum of saving and human capital formation for environmental depletion and degradation (see Section 8.2.2).

FR 7.2 Hybrid Accounting

Statistics Netherlands advanced the NAMEA in the 1980s; it is now an annual publication (Keuning & de Haan, 1998; de Haan & Kee, 2004). An attempt to extend the NAMEA for the inclusion of social accounting matrices in a System of

Economic and Social Accounting Matrices and Extension (SESAME) (Keuning & de Haan, 1998) appears to be overambitious in gathering all types of statistics under the umbrella of the national accounts. The revised SEEA-2003 made NAMEA a (if not *the*) key element of environmental-economic accounting (see Section 8.4).

Review and Exploration

- Is America really down, while GDP is up?
- Which is more plausible: Max-Neef's threshold hypothesis or the Environmental Kuznets Curve (EKC) hypothesis? See also Section 11.1 on EKC testing.
- What is the basic approach to incorporating environmental concerns into the national accounts?
- Can hybrid accounts measure the economic sustainability of economic performance and growth?
- Explain the use of NAMEA in linking environmental impacts with the values of production and consumption; are weight units and theme equivalents useful tools of aggregating environmental concerns in national accounts?
- Why are national accountants so keen on avoiding the full integration of environmental concerns into the (monetary) national accounts? See also Section 8.4.

Chapter 8
SEEA – The System for Integrated Environmental and Economic Accounting

US national income accounting set the tone for first attempts at incorporating environmental and social concerns into the national accounts [FR 8.1]. The purpose was to modify national income and product, considered to be imperfect measures of national welfare. Similar to the above-described welfare indices (Section 7.1.1), extended income and product accounts and derived measures of economic welfare deducted 'defensive' (welfare maintaining) expenditures, and added or subtracted environmental and social externalities to/from the conventional accounts indicators. National statistical offices dismissed welfare measurement as 'more suitable for research than for statistical compilation' (United Nations, 1977b). The System of National Accounts (SNA) later confirmed that 'GDP is a measure of production', and 'changes in the value of consumption are not the same as changes in welfare' (United Nations et al., 1993).

The statisticians did not succeed in muting criticism of the conventional accounts. 'Political unease with GNP as a bellwether' (Ward, 2004) increased with growing awareness of the role of social and environmental issues in national and international policies. The environmental critique included the

- Neglect of scarcities in the availability of natural resources, threatening the sustained productivity of the economy
- Exclusion of environmental degradation as an externality of economic activity
- Accounting for expenditures of environmental protection as increase in national income and product although such outlays could be considered as a maintenance cost of society.

In response, the United Nations Statistics Division issued a handbook on a System for integrated Environmental and Economic Accounting, the SEEA (United Nations, 1993). At first sight, such broad integrated accounting looks like a good replacement for narrow economic accounting. Conventional national accounts have, however, a large variety of micro- and macroeconomic uses. Modified standard indicators of market transactions would obscure the assessment of short-term economic disequilibria and medium- and long-term business cycles. Conventional economic and fiscal policy uses suggest, therefore, developing integrated environmental

P. Bartelmus, *Quantitative Eco-nomics*,

and economic accounting as a *satellite* (parallel) system of the SNA, rather than a substitute for the conventional accounts.[1]

After a decade of further testing and discussion, the United Nations Statistical Commission charged the so-called London Group of natural resource and environmental accountants with the revision of the SEEA. The revision process brought out the natural reluctance of official statisticians to embark on new and controversial issues such as the pricing of non-marketed environmental services. The result is some bias against aggregation and the modification of economic accounting indicators, even in satellite accounts.

8.1 Pricing the Priceless

Acceptance or rejection of monetary valuation in green accounting is at the heart of the ecological-economic dichotomy. Physical accounts acquire some systemic features by using physical thermodynamic laws in formatting the accounts for flows of materials, energy and residuals. Aggregating these flows in mass or energy units meets the requirements of analysing biophysical ecosystems, but not those of the economic system. To this end, the national accounts cater to economic preferences for goods and services, expressed in market prices.

Imputing a proxy market value on natural assets and asset services, which are not traded in markets, is a prerequisite for maintaining the system character of the national accounts. It is also the only possibility of fully integrating environmental concerns into the economic accounts. The SEEA ensures in this manner the comparability of environmentally adjusted with conventional economic indicators. Monetary valuation faces its own problems, though. Environmentalists, but also more conservative national accountants, criticize the imputation of monetary values on non-market transactions and processes. This section reviews therefore the main valuation techniques before discussing SEEA concepts and methods.

8.1.1 Market Valuation of Depletion and Degradation

Markets may trade, in some cases, natural resource stocks such as land, timber tracts or water wells. In these cases, natural assets fetch observable market prices. Frequently, natural assets are however either owned by governments as a public good or are situated outside national jurisdiction. In the absence of markets for such common property or common access resources (cf. Annex I.1), income from the sale of harvested or

[1] The Rio Earth Summit confirmed this approach in its Agenda 21: '... systems of integrated environmental and economic accounting ... to be established in all member States at the earliest date should be seen as a complement to, rather than a substitute for, traditional national accounting' (United Nations, 1994).

otherwise extracted resources can be used for estimating the economic value of the resource stocks. The generic approach is to calculate the – discounted – sum of economic benefits or 'net returns' that can be ascribed to the exploitation of a natural resource over its lifetime. One can argue that it would be at this *net present value* that a natural asset would be traded if a market existed for the asset.

The concept of economic rent is central to assessing the value of a scarce, non-priced category of natural 'economic' wealth *in situ* (Section 7.2); it is also crucial for defining and understanding the value of the non-sustainable use of natural wealth, i.e. its *depletion* (see Annex II). Based on the national accounts concepts of capital earnings and capital consumption, depletion D is the difference between the rent (or gross earnings) R_{nc}, obtained from natural capital use in production, and the net return to natural capital NR_{nc}.[2] Expressed in net present values, the depletion of a natural resource reflects the change in value of natural capital from the direct use in production during the accounting period; it is thus also the difference between the discounted values of the opening stock $OpSt_{nc}$ and the closing stocks $ClSt_{nc}$:

$$D = R_{nc} - NR_{nc} = OpSt_{nc} - ClSt_{nc} \tag{8.1}$$

Besides the direct use of a resource in production, other changes in its availability and value stem from natural regeneration, discovery, revision of previous estimates of resource availability and quality, natural and political events, and changes in the prices of the resource products. The national accounts include these effects in their asset accounts as 'other asset changes'. In line with the (production) cost concept of the national accounts, the SEEA does not count these 'other asset changes', which are not brought about by production, as depletion cost; on the other hand, it calculates natural resource depletion (and its cost) net of the natural regeneration of renewable resources.

In the absence of market prices for natural resource stocks, their valuation has to discount their future net rents for calculating their net present values. Environmental accounts typically use a 'normal rate of return' to capital of similar industries, which do not use natural capital, for estimating the portion that should be allocated to natural capital. The difficulties of predicting and discounting future net returns from natural resource use (Section 2.3.2) are the reasons for suggesting several simplified valuation techniques (United Nations et al., in prep., ch. 7).

The *net price method* makes use of the Hotelling rent assumption of compensating net price and discount rate increases, so as to dispense with discounting future net returns. Hotelling (1931) showed that in perfectly competitive markets the net

[2] The accounting category of rent should not be confounded with the rent concept of economic theory, i.e. payment to a production factor in fixed supply (such as land). The earnings from such a production factor can be conveniently taxed away without losing allocative efficiency. On the other hand, most natural capital can be depleted and augmented (if renewable), and its taxation would therefore affect its use and availability.

price of an exhaustible resource, defined as the difference between the market price p_{nc} and the unit marginal cost of extraction c_{nc}, will rise at the rate of interest of alternative investments, offsetting the discount rate. The value-depressing effect of the discount rate would thus neutralize the future price increase, brought about by increased scarcity of the exhaustible resource. In this case, the net-price based depletion value D_{np} of the resource approximates its economic rent generated during the accounting period, especially if the average unit value $\bar{p}_{nc}$ and the average cost $\bar{c}_{nc}$ are used as a further simplification (Bartelmus, 1998):

$$D_{np} = (\bar{p}_{nc} - \bar{c}_{nc})Q \approx R_{nc} \tag{8.2}$$

where Q is the physical amount of depletion or extraction (in case of a non-renewable resource) during the accounting period.

Comparing the depletion value D_{np} to the basic definition (8.1) reveals that D_{np} tends to consider all the rent as depletion. In other words, the extraction or harvest of the resource does not create itself any profit or net return – beyond a normal return to produced capital. Natural capital may be necessary for income generation, but all the credit goes to produced capital. As a consequence, the results of the net price method tend to exaggerate depletion from income-generation and -allocation points of view. Moreover, the validity of the various assumptions underlying the Hotelling rule and net price method can be questioned. Nevertheless, because of its relatively easy calculation, requiring only knowledge of the resource price and average production cost, most case studies of green accounting (in particular those described in Section 8.3) apply this convenient method.

In response to the criticisms of the net price method, El Serafy (1989) advanced another simplified method of calculating depletion, the *user cost allowance*. He questioned, in particular, the obliteration of income generated by natural resource exploitation: 'Countries with marketable natural resources are evidently better off than those without such resources'. His method suggests reinvesting a constant part of the rent generated, the user cost, so as to create a 'perpetual stream of true income' NR^{∞}_{nc}. The user-cost method is a special case of the generic definition of depletion as the change in net present values of a natural resource; it assumes that the discount rate and rents remain constant over the lifetime of n years of the resource (Hartwick & Hageman, 1993). In this case, the perpetual income element is the difference between the current and last (discounted) rent, and the depletion (user cost) allowance D_{uc} is equal to the discounted last (at the end of its lifetime) rent generated by resource exploitation:

$$D_{uc} = R_{nc} - NR^{\infty}_{nc} = R_n/(1+r)^n \tag{8.3}$$

Comparing (8.2) and (8.3) shows that the user cost allowance is indeed only a fraction – of $1/(1+r)^n$ – of the depletion allowance calculated by the net price method. It is therefore useful to calculate both allowances for an estimate of the range of natural capital loss from natural resource exploitation.

8.1.2 Maintenance Costing of Environmental Degradation

Dealing only with natural resources, which conveniently supply marketable and hence priced products, reduces drastically economic analysis. Economics is indeed concerned with scarce goods and services, whether traded in markets or not. There is no a priori reason why scarcities in environmental services should be excluded from the analysis. Notably in industrialized countries, environmental externalities of pollution are usually of far greater importance than natural resource depletion. The SEEA proposed, therefore, maintenance costing as a way of valuing the losses of environmental functions that are usually not traded in markets. Natural resources, whose products are traded, may provide some of these functions (e.g. erosion control or habitat by forest plantations). Note, however, that the market prices of their products (e.g. for timber) do not include the value of these services.

The SEEA defines maintenance costs as those that 'would have been incurred if the environment had been used in such a way as not to have affected its future use'. The maintenance costs refer to the – missed – opportunity of mitigating or avoiding the environmental impacts caused during the accounting period. Of course, these costs are generally hypothetical since environmental impacts did occur and no money was spent on actually tackling the impacts. This is in fact the most persistent criticism raised against such costing. It is also the argument for excluding environmental degradation from the monetary accounts and their indicators.

There is, however, nothing hypothetical about the environmental impacts themselves. In order to incorporate these impacts in the national accounts system, maintenance costing monetizes the impacts by weighting them according to society's obligation and capacity for dealing with environmental concerns. This is probably the only realistic way of an *ex post* assessment of environmental social costs, in consistency with the (similarly hypothetical) 'replacement cost' of conventional capital consumption. In practice, one would have to cost compliance with society's environmental (emission) standards where high marginal costs make the total avoidance cost unrealistic.[3]

Markets can in fact provide a maintenance value for nature's source and sink functions. Capping emission and natural resource use, and trading permits for the capped environmental services create a socio-economic scarcity value for the maintenance of these services. This value reflects the standard (cap) setter's and the market's evaluation of the social cost of environmental degradation. The annualized

[3] Baumol and Oates (1971) recommended a similar approach for measuring the optimal level of externalities in practice. Cf. Annex I.1 (Fig. I.1, part D) for such costing of standard compliance. Contrary to the marginal cost assumption in justifying the net price method of market valuation (Section 8.1.1), maintenance-cost weighting of actual physical impacts works with average cost observed in the (past) accounting period. Warnings against modelling marginal cost (Radermacher, 1999; United Nations et al., in prep.) in environmental accounts miss the point.

cost of purchasing the permits could and should be treated as a capital depreciation cost for deduction in environmentally adjusted indicators.[4]

Environmental costs are 'externalities' that by definition were not internalized, i.e. actually budgeted, by households and enterprises. However, these costs *should* have been accounted for, from society's social caring and costing points of view. It is certainly a valid question to ask what would happen to the economy if these costs were indeed internalized. Such analytic modelling is not the task of descriptive accounting. Chapter 12 (Section 12.1) will take up this question of modelling a 'greened-economy GDP'.

On the other hand, individual corporations might actually account for the depletion and degradation of their natural assets (including the annualized cost of environmental permits), make provisions for potential liabilities for environmental damage, or show (and pay for) 'social responsibility' (Section 9.1.1) for impacts on their neighbourhood. The conventional national accounts would in these cases overstate the net value added generated by these corporations. Green accounting would then simply correct the distorted information about the (net) production level and structure of the economy by incorporating environmental costs in the production accounts of the different industries.[5]

Maintenance costing could also apply to the social costs exported to or imported from other countries by transboundary pollution. The SEEA largely ignores accounting for environmental impacts across national borders. In analogy to capital transfer accounting in the SNA, one could treat 'international' externalities as free services or disservices, modifying national income rather than domestic product. Export of pollutants would thus increase national income and import would decrease it because of the transfer of a social cost to/from abroad.

Maintenance costing refers to environmental impacts of the current accounting period. Extending the costing to the restoration of the environment to desirable levels, experienced in previous pristine periods, muddies such costing by referring to a difficult-to-determine situation in the past. The accumulation of environmental impacts and the cost of recreating this ideal situation (apart from any 'irreversibilities') can be viewed as an *environmental debt* of the current and past generations to future generations (Section 8.3).

---

[4] The SNA treats the creation of tradable permits of pollution and natural resource use as the 'appearance of an intangible non-produced asset'. As these permits approach their expiry date they lose value and should therefore be 'amortized' over their lifetime. Both, SNA and SEEA record the creation and amortization of the permits as 'other asset changes', which do not affect production and capital formation (see Section 8.1.1).

[5] In the case of so-called diseconomies, i.e. producer-producer external effects, the receiving enterprise will in all probability account for the external cost borne (for instance, in the classical case of industrial smoke impacts on a laundry). However, this would still represent a distorted cost allocation according to the polluter-pays principle (Section 13.3.2).

8.1.3 Contingent and Related Damage Valuation

In theory, from an economic optimality point of view, welfare losses or gains, rather than imputed environmental protection costs, should be internalized. Box 8.1 contrasts production-oriented, supply-side environmental costing with demand-side valuation of environmental welfare effects. Cost-benefit analyses (CBA) of programmes and projects (Section 2.3.2) use demand-side valuations of environmental benefits to assess welfare effects, including environmental damage as loss of benefits. Demand-side valuations include directly stated preferences for environmental benefits in contingent valuations and indirectly revealed preferences by defensive expenditures and surrogate markets.

Contingent valuation uses interviews of stakeholders to determine their willingness to pay or be compensated for environmental services or service losses, contingent upon hypothetical situations of welfare gains or losses. Other methods such as hedonic pricing of land and property with different environmental characteristics, or the travel cost to benefit from protected environmental sites, are indirect indicators of preferences for environmental amenities. The closeness of welfare valuations to economic (utility/welfare maximizing) theory makes these valuations the favourite of textbooks on environmental economics [FR 2.2].

The theoretical desirability of demand-side valuation for internalizing externalities from a welfare-maximizing point of view should not hide the fact that these valuations are proxies for utility measurement. Utility estimates are difficult to apply at the project level and hardly possible to compile at sectoral and national levels. Furthermore, stress-response reactions frequently delay the generation of environmental damage. Current accounts face therefore the problems of discounting future effects and tracing them back to causing agents. Contingent valuations of the willingness to pay or to be

Box 8.1 Supply- and demand-side valuations in accounting and cost-benefit analysis

A. Supply-side valuation (costing):

- *Natural capital consumption*
  - Market valuation of natural resource depletion
  - Maintenance costing of environmental impacts
- *Environmental protection expenditures*
  - Capital expenditures
  - Operating cost

B. Demand-side valuation (benefits foregone):

- *Expressed preferences (contingent valuation)*
- *Revealed preferences*
  - Hedonic pricing
  - Travel cost method
  - Defensive expenditures

compensated are also inconsistent with market prices, the basic valuation principle of the national accounts. The inconsistency stems from the inclusion of consumer surplus in willingness to pay declared by individuals. Interview-based valuations also face problems of free-rider attitudes and consumer ignorance. These are the reasons why the national accounts do not consider welfare measurement as their main objective, and focus instead on the market values of goods and services.

A few environmental accounting studies applied contingent and related damage valuations with questionable results (Section 8.3). The original SEEA focuses, therefore, on supply-side valuations, considering the use of welfare valuations as exploratory and experimental (United Nations, 1993). The SEEA-2003, on the other hand, deals extensively with CBA valuations, since 'damage-adjusted income clearly says something about the country's revenue-creation capacity under prevailing conditions' (United Nations et al., in prep.). There is no explanation, however, how these conditions (including environmental ones) relate to production and income-generation capacities.

The practical solution for including environmental impacts in environmental accounting – beyond economic resource accounting – is, therefore, maintenance costing. 'Costing the maintenance of environmental "capital" is the anchor, which prevents environmental accounts from drifting away into the realm of welfare measurement and analysis' (Bartelmus, 1998).

8.2 SEEA Objectives, Structure and Indicators

In response to the above-mentioned criticisms of the national accounts, the original SEEA set the following objectives for greening the accounts (Bartelmus, 2001):

- Segregation and elaboration of all environment-related flows and stocks of the conventional national accounts, including environmental protection expenditures as part of a broader concept of 'defensive expenditures'
- Linkage of physical with monetary environmental accounts and balances, with a view to overcoming the ecological-economic dichotomy
- Accounting for the maintenance of tangible wealth by covering not only human-made but also non-produced natural capital and its consumption
- Assessment of hitherto ignored costs of (1) depletion of natural resources and (2) impacts on environmental quality, in particular from pollution
- Definition and measurement of indicators of environmentally adjusted product, income and capital formation, accounting for the costs of environmental depletion and degradation as capital consumption.

All these objectives cater to the overall goal of assessing the environmental sustainability of economic performance and growth. Figure 8.1 shows the accounting indicators as they emerge from their respective accounts. The figure elaborates on Fig. 7.4, which illustrated the basic approach of incorporating environmental assets and asset changes in the conventional national accounts.

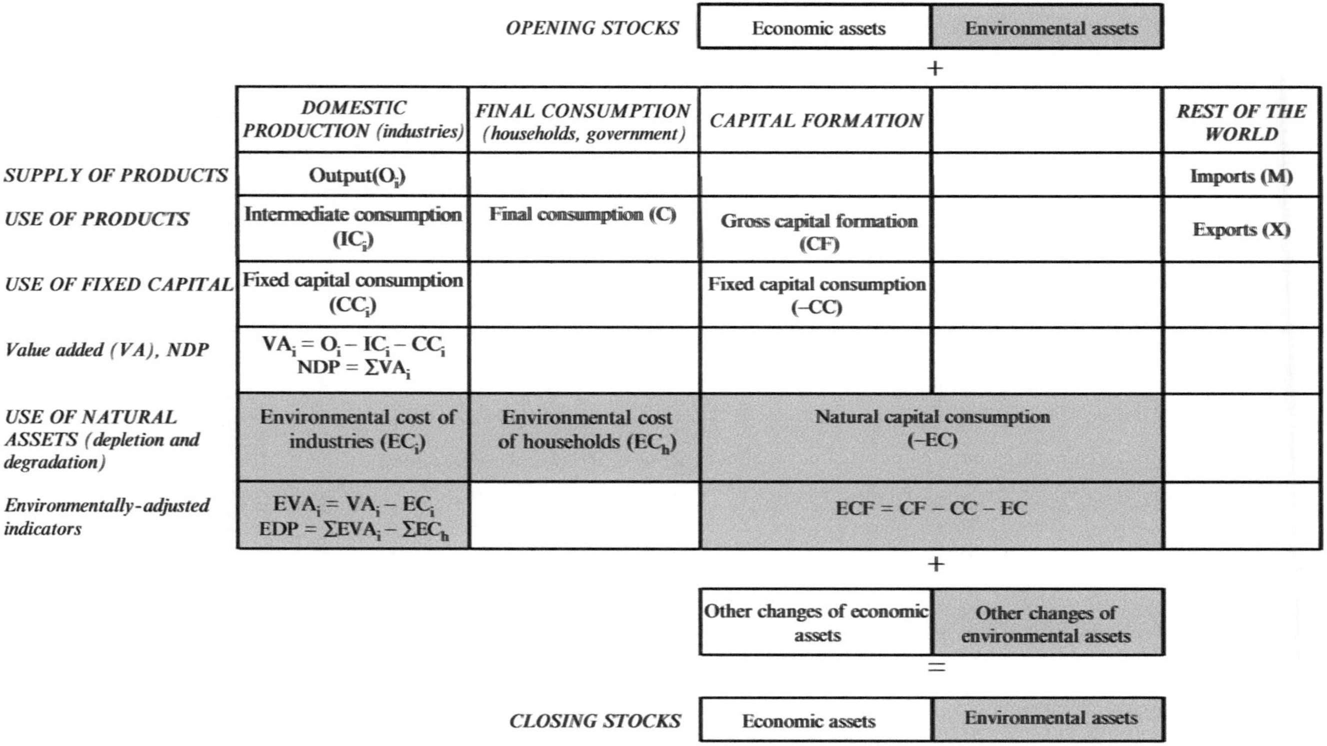

Fig. 8.1 SEEA structure and indicators
Source: Bartelmus (2004), fig.3, p.50; with permission by the copyright holder, Elsevier.

8.2.1 Accounting for Sustainability

Chapter 7 discussed the linkage of physical and monetary accounts by extending the asset definition of the conventional accounts. Broader concepts of capital and national wealth are the results. Changes in these capital categories in terms of capital consumption and formation may indicate compliance or non-compliance with minimum conditions for sustainable economic growth, i.e. capital maintenance (Section 2.2.3).

Accounting for natural capital consumption and maintenance expands the sustainability notion that is built into the conventional *net* indicators of value added, income and capital formation. In analogy to the wear and tear, i.e. the ultimate destruction, of capital goods in the production process, one can define natural capital depletion and irreversible degradation as the permanent loss of parts or all of natural resource stocks and waste absorption capacities. Accounting conventions thus clarify the contents of physical depletion and degradation as a process of natural capital consumption by economic activities – beyond regeneration and replenishment and excluding other non-economic impacts on natural capital. The regeneration of nature can be seen as a cost-free natural repair process, recorded outside the production and income accounts as other changes of assets (Section 8.1.1). In contrast, capital consumption creates a private cost of produced capital loss for the owners and a social cost of environmental depletion and degradation for society.

One could also see the non-sustainable use of a natural resource in production as the reduction of nature's 'inventory' of (primary) materials. The SNA would treat the resulting negative change in the value of an inventory of goods as negative capital formation. The corresponding increase in intermediate consumption and its deduction in net value added would then obtain the same environmentally adjusted net indicators as the natural-capital-consumption costing of depletion. Since the loss of absorptive capacities is difficult to conceptualize as a decrease in the 'inventory' of environmental services, the inventory-loss concept is not further explored here.

As discussed in Section 8.1.1 and Annex II, the depletion value represents a loss in the income/value added generation capacity of a natural asset. Depletion cost allowances reflect therefore a weak sustainability concept, calling for the reinvestment of these allowances in any income-generating activity. At first sight, maintenance costing of environmental services, discussed above, looks like aiming at the preservation of environmental functions. However, the strength of sustainability created by such valuation and accounting depends, of course, on the actual use of the cost allowance. Investing in the restoration of depleted and degraded natural capital would indeed reflect strong sustainability. If such use is not possible because of 'complementarities' in capital use (Section 2.3.1) or is ignored, investing in any other income-generating source would cater to weak sustainability.

Ultimately the strength of sustainability depends on (1) actual cost internalization or absorption (e.g. by governmental eco-taxation) and (2) on the actual use of the cost allowance made or tax revenue obtained (cf. Section 13.3.3). Given that such cost internalization or absorption did not actually take place, it is probably safe to interpret the adjusted accounting aggregates as indicators reflecting *potentially weak sustainability*.

Attempts at accounting for other non-produced capital categories, in particular human and social capital, have not reached the same levels of conceptualization and measurement as natural capital. Treating education expenditure as capital formation (as in the genuine savings indicator of the World Bank: see Section 8.2.2) is problematic. Education has benefits of private consumption, and health expenditure would also have to be considered as contributing to human capital formation and maintenance. Furthermore, the notion of human capital 'consumption' as a cost of a production process is not very enticing. Even more difficult is the measurement of social capital, i.e. social coherence and networking within a more or less 'civil' society.

At least for now, definition, measurement and valuation problems consign human and social capital accounting to research rather than recurrent accounting. One should not forget, though, that determining natural resource rent by deducting the earnings of produced capital from gross operating surplus generates a residual, which includes, besides natural capital, other intangible influences on corporate earnings and profits from production. Note also that assessing the role of financial wealth in contributing to the sustainability of economic growth needs still further clarification in analysis and accounting (see Box 8.3, below).

8.2.2 Environmentally Adjusted Macroeconomic Indicators

Figure 8.1 illustrates how the inclusion of natural capital consumption as environmental cost affects the main accounting identities. Most of the environmentally adjusted economic indicators can be calculated as sum totals and elements of the following equations:

- Value-added identity for industry i:

$$EVA_i = O_i - IC_i - CC_i - EC_i = VA_i - EC_i \tag{8.4}$$

describing Environmentally adjusted Value Added EVA_i generated by an industry i as the difference of its output O_i and cost, including intermediate consumption IC_i, fixed capital consumption CC_i, and environmental depletion and degradation EC_i

- Net domestic-product identity for the whole economy:

$$EDP = \Sigma EVA_i - \Sigma EC_h = NDP - EC = C + CF - CC - EC + X - M \tag{8.5}$$

defining Environmentally adjusted net Domestic Product (EDP) as the sum of environmentally adjusted value added of industries, with a further deduction of environmental costs generated by households EC_h.[6] Alternatively, and as in the conventional accounts, EDP can also be calculated as the sum of final uses, including final

[6] Deducting the (maintenance) cost of household pollution from NDP treats these emissions as negative production or natural capital consumption of a sector whose activity is otherwise limited by definition to (final) consumption.

consumption C, Environmentally adjusted net Capital Formation ECF and the balance of exports X and imports M; ECF is defined as gross capital formation CF minus produced and natural capital consumption:

$$ECF = CF - CC - EC \quad (8.6)$$

- Supply-use identity:

$$O + M + EC = IC + C + EC + CF + X \quad (8.7)$$

indicating that the supply of goods and services produced ($O = \Sigma O_i$), imported (M) and provided by nature (EC, valued at replacement cost) equals their use in intermediate consumption (IC, ΣEC_i) and final consumption (C, ΣEC_h), capital formation CF and export X, with $\Sigma\ EC_{i,h} = EC$

- Asset balance:

$$OpSt + CF - CC - EC \pm OC = ClSt \quad (8.8)$$

explaining the changes in the value of stocks – from the beginning of the accounting period (opening stocks OpSt) to its end (closing stocks ClSt) – as gross capital formation CF, produced and natural capital consumption (CC, EC), and other changes in assets OC.

Other asset changes play an important role in greening the conventional accounts. The SEEA shifts part of the 'economic disappearance of non-produced assets' as the depletion cost of natural resources from SNA's asset accounts to the production accounts. This rejects the notion of somehow vanishing natural assets, as the responsible users of environmental source and sink services are charged with the cost of depleting and degrading these assets. All other asset changes remain outside the production accounts, since natural disasters, the creation of subsoil resources or unmanaged natural growth are not the result of an economic production process (Section 8.1.1). Such changes should not affect, therefore, the value of product, income and capital formation.

There is some controversy about accounting for natural resource discoveries ('economic appearance of a non-produced asset' in SNA terminology). US national accountants (Landefeld & Howell, 1998) argue that the discovery of subsoil resources turns them into 'developed natural assets'. Consequently they account for discoveries as capital formation in the supply and use accounts, thus largely offsetting their depletion.[7] This argument ignores, on the one hand, that the SNA actually

[7] Despite this 'self-effacing' treatment of natural resource depletion, the coal-mining lobby succeeded in convincing the US Congress to suspend further work on green accounting for an external review by the National Academy of Sciences (NAS). As a result of this suspension, work on green accounting by the Bureau of Economic Analysis was effectively halted, notwithstanding the positive recommendations by the NAS panel (Nordhaus & Kokkelenberg, 1999).

accounts for resource development as fixed capital formation (in the case of mineral exploration). On the other hand, the creation of *in situ* mineral deposits is obviously more in the nature of a cost-free gift by the 'creator' (nature) than the result of economic production.

One green accounting indicator attempts to assess sustainable development in terms of 'enhancing human well-being through time' (World Bank, 2003). Genuine Savings S_g, which now runs under the name of 'adjusted net savings', sets out from national income NI and final consumption C to calculate 'education enhanced' ($+C_e$) and environmentally adjusted (–EC) net savings as

$$S_g = NNI - C + C_e - EC \tag{8.9}$$

However, the presumed relations of welfare-generating consumption with savings (a source of finance for capital formation) and, partially, capital formation (for including education as human capital) obscures the indicator's meaning for sustainability – of welfare, wealth or income?[8] In the end, using the SEEA's ECF indicator (Equation 8.6) would be clearer with regard to capital maintenance. It would also be more consistent with national accounts conventions of capital formation and consumption.

8.2.3 *Accounting for Policy Performance*

At first sight, environmental expenditure by governmental and non-governmental actors seems to indicate society's willingness to take environmental action. These outlays are part of the conventional accounting indicators of output, input, consumption, capital formation, and exports and imports. In Fig. 8.1 environmental protection expenditures could therefore be shown as 'thereof' subcategories of the conventional flow accounts. Consequently, these outlays do not require any basic changes of the system structure. National accountants readily embraced environmental protection and related expenditures as a major part in greening the national accounts. The segregation of environmental activities is a matter of relatively uncontroversial expansion of classifications and data collection; it is extensively discussed in the SEEA-2003.

Environmental expenditures are, however, not a good indicator of environmental performance. They depend crucially on a country's particular environmental conditions and the efficiency of its regulative and legislative institutions. Still, environmental

[8] Besides the general problem of reflecting utility by public and private consumption, genuine savings does not clearly define environmental cost for depletion (with regard to the treatment of other volume changes such as discoveries or natural disasters), and takes \$20 per ton of carbon emission as the basis for calculating a placeholder value for total environmental damage. The savings indicator seems also to ignore capital transfers from other countries as a source of potential investment and a factor in the generation of 'net worth' in the national balance sheet (United Nations et al., 1993, ch. XIII).

expenditures can assess the significance of an emerging environmental industry in terms of conventional indicators such as sales, value added and investment (OECD and Eurostat, 1999). More questionable are proposals to deduct such expenditures and other regrettables from gross or net national product as a defence against the deterioration of environmental and social conditions (Leipert, 1986, Daly, 1996). Box 8.2 shows the wide range of defensive expenditures including, besides the cost of environmental protection, those of maintaining health, security and other social standards.

The SEEA presents only environmental protection expenditure accounts (and their classification) and refrains from deducting such expenditure from national accounts aggregates. The reason is that such deduction would destroy the coherence of the accounting system (United Nations et al., in prep.). From a more substantive point of view, it seems hardly possible to distinguish defensive from 'real' welfare creating outlays. When, for instance, does defence increase security rather than maintaining it, or when is food improving, maintaining or damaging human health and well-being? Moreover, any deduction of a particular expenditure would have to trace – and exclude – all antecedent industries' contributions to this expenditure. Such assessment of direct and indirect outlays is, however, more a matter of modelling than accounting or index calculation. As discussed in Section 7.1.1, the deduction of defensive expenditures may be part of ad hoc index calculations of human welfare but should not be included in systemic accounting of economic activity.

Specific environmental policy measures, in particular those using 'market instruments', are probably of greater relevance for environmental policy. Accounting for the costs and revenues generated by these instruments is one of the highlights of the revised SEEA. Somewhat hidden under 'accounting for other environmentally related transactions', Ch. 6 of the SEEA-2003 (United Nations et al., in prep.) explains

Box 8.2 Categories of defensive expenditures

Expenditures for

- Environmental protection and damage compensation
- External costs of production and consumption
- External costs of spatial concentrations and urbanization (noise protection, rent increases, security and commuting costs)
- Risks in the industrial system (provisions for hazardous industries, crime, defence etc.)
- Costs of car transport (accidents and emission control)
- Health costs from unhealthy consumption patterns, and living and working conditions.

'Minimum' estimates of defensive expenditures for Germany (excluding, in particular, health costs) amounted to about 10% of GNP.
Source: Leipert (1986, 1989).

- The nature of fiscal (dis)incentives as production taxes and subsidies that affect value added and domestic product (in the income-generation accounts)
- Ecological tax reform as 'hypothecated' (earmarked for reducing labour cost) eco-taxes
- Environmental fees or charges for governmental environmental services such as waste disposal as intermediate or final consumption
- Natural resource rent absorption through royalties and other resource taxes as a significant source of governmental property income (shown in the primary income distribution accounts)
- The acquisition of tradable emission and resource use permits as an increase in intangible non-produced wealth (cf. Section 8.1.2 as to the accounting of amortized outlays for tradables).

Chapter 13 describes the objectives of different policy instruments and evaluates their ecological and economic efficiency, in particular as part of an ecological tax reform. Predicting the success or failure of these instruments is, of course, a matter of modelling, addressed in Ch. 12.

8.3 Case Studies

Integrated environmental and economic accounts translate the concept of economic sustainability into environmentally adjusted indicators of non-declining net output or positive net capital formation. Net domestic product (NDP) and capital formation play key roles in conventional economic accounting and analysis. A similar significance can be expected for their environmentally adjusted counterparts, EDP and ECF, in long-term sustainable growth analysis and policy.

Policymakers usually refer to a 'green GDP', rather than green NDP.[9] The reason is that GDP calculation avoids the difficulties of estimating capital consumption. Interpreting the environment as an 'inventory' of nature's goods and services that enter production as intermediate consumption (Section 8.2.1), might justify ignoring capital consumption in an environmentally adjusted GDP. Since sustainability requires the maintenance of natural and produced capital, green GDP is misleading, however: omitting fixed capital depreciation ignores the need to replace worn-out capital goods. Crumbling infrastructure has been a significant cause of non-sustainability of economic development, not only in poor countries but also in industrialized ones; the spectacular collapse of a highway bridge in Minneapolis is a case in point.

Table 8.1 presents EDP as the overall result of pilot studies of natural resource and environmental accounting [FR 8.2]. The studies show the significance of natural capital in production and income generation by comparing EDP with NDP. An effort was made to adjust those indicators, which were compiled outside the

[9] For instance, China's leadership endorsed (but later refuted) the idea of compiling a green GDP as the scientific approach to assessing economic development [FR 8.2].

Table 8.1 NDP and EDP in case studies of green accounting (lowest and highest percentages)

Country	EDP 1[a]/NDP(%)	EDP 2[b]/NDP(%)
China (1992)		94
Costa Rica (1970–1989)[c]	89–96	
Germany		96–97
Ghana (1991–1993)[c, d]	85–89	
Indonesia (1971–1984)[c]	69–87	
Japan (1985/1990)	98/99.6	97/98
Korea, Republic of (1985–1992)[d]	100	96–98
Mexico (1985)	94	87
Papua New Guinea (1986–1990)	92–99	90–97
Philippines (1988–1992)[d, e]	96–99.5	75–83
United Kingdom (1980–1990)[f]	95–100	
USA (1987)[g]	98.5–99.6	

Source: Bartelmus (1997b, table 1) and updates.
Original sources: China: Akita and Nakamura (2000); Costa Rica: Solórzano et al. (1991); Germany: Bartelmus (2002); Mexico: van Tongeren et al. (1991); Indonesia: Repetto et al. (1989); Japan: Oda et al. (1996); Korea: Kim (1998); Papua New Guinea: Bartelmus et al. (1992); Philippines: Domingo (1998); Ghana: Powell (1996); United Kingdom: Pearce (1994); USA: Landefeld and Howell (1998).
Notes: [a] EDP 1 is NDP, adjusted for natural resource depletion only.
[b] EDP 2 is NDP, adjusted for natural resource depletion *and* environmental quality degradation.
[c] Concept adjusted to United Nations (SEEA) methodologies.
[d] Preliminary estimates.
[e] Soil erosion not yet covered.
[f] Oil and gas depletion only.
[g] Depletion of subsoil assets, range of estimates (valuations).

national accounts framework (Indonesia, Costa Rica, United Kingdom), to SEEA concepts. However, as indicated in the table notes, comparability still suffers from remaining differences in concepts, methods, valuations and coverage of environmental concerns. Several studies stopped short of estimating environmental degradation cost, compiling only EDP 1, which accounts for natural resource depletion only. EDP 2 calculations include additional maintenance costs of pollution.

All SEEA applications took a cautious approach, leading to undercoverage and underestimation. This could explain the rather modest shares of depletion and degradation cost (the difference between NDP and EDP), especially in the industrialized countries of USA, Germany, Japan and the Republic of Korea. Japan, Korea and Germany hardly extract or harvest domestic natural resources. The USA, on the other hand, limited its study to the depletion of subsoil resources and assigned only a placeholder value of actual environmental expenditure to environmental degradation (Landefeld & Howell, 1998). Other (developing) countries show more significant effects on their natural capital. At a time, Costa Rica and Indonesia exploited their natural resources at rates of 10% and 30% of their NDP, respectively.

Of course, most industrialized countries depleted their natural resources in the past and accumulated thus an environmental debt to future generations. The SEEA does not account for such debt because current production and cost measures do not recognize costs incurred in previous accounting periods. Still, environmental debt estimates (Hueting & Bosch, 1994; Azar & Holmberg, 1995) point to the need for assessing the environmental sins of the past, and also those against other countries through 'burden shifting' (Section 6.3.2).

One way of looking at the sustainability of economic performance and growth is to assess a nation's capability of generating new capital after taking produced and natural capital consumption into account. Figure 8.2 presents ECF in per cent of NDP. Only Indonesia, Ghana and Mexico appear to have performed non-sustainably, showing a disinvestment of negative ECF. Non-negative ECF reflects the fact that natural capital consumption did not offset the net increase of fixed capital. The countries maintained or increased in this case the *total value* of capital during the accounting period, achieving weak sustainability of economic performance.

World Bank estimates of adjusted net savings, which is similar to ECF, seem to indicate widespread non-sustainability for Africa (Table 8.2). However, as pointed out in Section 8.2.2, the indicator is not strictly comparable with national accounts categories of income, savings, NDP or changes in net worth.

For structural and sectoral policy and management, overall environmental cost and the affected indicators need to be disaggregated by economic sectors. The case studies of Mexico and Thailand show that the depletion costs incurred by forestry and mining reduce the conventional value added of these industries by over 70%.

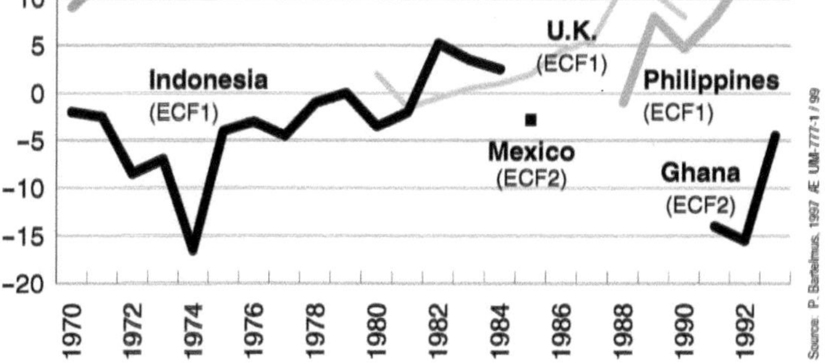

Fig. 8.2 ECF in selected countries (% of NDP)
Note: ECF 1 is net capital formation minus the cost of natural resource depletion; ECF 2 covers both depletion and environmental degradation cost.
Source: Bartelmus (1997b, fig. 2).

Table 8.2 Adjusted net savings, world regions 1999 (% of GDP)

Countries	Gross domestic savings	Adjusted net savings (including education expenditure)	Adjusted net savings (excluding education expenditure)
Low income	20.3	7.8	4.9
Middle income	26.1	14.3	10.8
High income	22.7	13.5	8.7
East Asia and Pacific	36.1	25.2	23.5
Europe and Central Asia	24.6	11.9	7.8
Latin America and the Caribbean	19.2	9.6	5.5
Middle East and North Africa	24.2	**–1.3**	**–6.0**
South Asia	18.3	8.3	5.2
Sub-Saharan Africa	15.3	3.9	**–0.8**

Explanation: The World Bank definition of adjusted net savings differs from national accounts definitions of saving and capital formation. To make the indicator more comparable with ECF education expenditures are excluded in the last column.
Source: World Bank (2003, table 2.1).

Table 8.3 Green accounting indicators, Germany 1990, 1991 and 1995 (provisional estimates)

		1990			1991	1995
	Total (1990)	Agriculture, fishery and forestry	Iron and steel	Energy supply	Total	Total
NDP(VA) (billion DM)	1,943	24.8	14.4	39.6	2,527	3,002
EDP(EVA) (billion DM)	1,884	16.6	11.7	26.3	2,444	2,926
EC/NDP(VA) (%)	3.0	33.4	18.9	33.6	3.3	2.5
NCF/NDP (%)	11.1				12.0	9.2
ECF/NDP (%)	8.1				8.7	6.7

Source: Bartelmus (2002, table II.2); with permission by the copyright holder, Springer.

In Germany, pollution costs amounted to about one third of value added in both, the agriculture/forestry/fishery and the energy supply sectors (Table 8.3).

One of the arguments against compiling an environmental satellite account is cost. Using the software of the SEEA operational manual (United Nations, 2000a) the author carried out a test application for Germany with two assistants within three months. Annex III shows the result of this test for the year 1990. The annex also presents a synoptic view of the greened accounts, which demonstrates their consistency with the standard national accounts. The admittedly rough study indicates sustainable performance in this year in terms of positive ECF (Table 8.3: ECF/NDP > 0).

Table 8.3 also presents EDP and EVA estimates for West Germany (1990) and for the unified country (1991, 1995). EDP estimates for 3 years (and moreover in current prices) can obviously not assess any trends in the environmental sustainability of economic growth. Still, the table shows a distinct increase of environmental cost (NDP – EDP in current prices) from 59 billion DM (3% of NDP) in 1990 to 83 billion DM (3.7% of NDP) in 1991 because of the coverage of East Germany's wasteful and polluting industries in 1991. Thereafter (by 1995), adaptation or elimination of these industries in competition with West Germany's modern production methods appears to have largely offset the relative increase in environmental cost after unification. The table also identifies agriculture and energy supply as the most environmental-cost-intensive industries (per unit of value added).

The good news of the case study is that the avoidance or immediate mitigation costs of attaining weak sustainability are relatively low at 3% of net product. The bad news is that actual and potential damage from these impacts could be considerably higher. As discussed in Section 8.1.3, such damage costs are near-impossible to estimate at the national level. In fact, those brave enough to do so came up with values ranging from about the same as our natural capital consumption value to twenty times this value. Damage estimates for the late 1980s and early 1990s in Germany thus vary between 100 billion and 1,000 billion DM (Wicke, 1993). On the other hand, a DM 50 billion result from a European Union project of green accounting (Markandya & Pavan, 1999) is even lower than the maintenance cost estimated by the author for the same year (1990). In an accounting project in the Philippines (Delos Angeles & Peskin, 1998),[10] the costs of 'complete' water pollution and household wastes control exceeded the estimated damages (foregone earnings and medication expenses for disease and premature death). The authors suggest, therefore, that in these cases 'complete control is unwise'. The question is, how much control would be wise and, of course, how much is the value of human life?

8.4 SEEA Revision

In 1993 the United Nations Statistics Division published the SEEA as an 'interim version'. For 10 years, the London Group of national accountants worked on improving the SEEA for broader international acceptance. The Group intended to change the interim status into more permanent guidelines, which would last at least another decade [FR 8.1]. Since we will have to live for some time with this document and an operational version of the SEEA-1993 might continue to be used in practice, this section discusses critically the revision in comparison to the original SEEA. The purpose is to facilitate an informed choice of concepts and methods from both handbooks in future country applications.

The revision process also revealed, but failed to resolve, core questions of environmental-economic analysis and accounting. The questions include the various above-discussed physical-monetary, income-welfare and accounting-modelling dichotomies in the measurement and evaluation of sustainability. There was also

[10] This project was conducted by the Department of Environment and Natural Resources; it was not coordinated with, and is indeed quite different from, the SEEA application carried out by the National Statistical Coordination Board (presented in Table 8.1 and Fig. 8.2).

little participation by data users and developing countries, which could be a source of further dissent, this time between national accountants and policymakers.

The result of the revision is a bulky report, well-researched in part but often ambivalent in its recommendations. In the end, the report admits that the SEEA is still 'very much ... work in progress'.[11] The ambivalence of the SEEA stems from opaque analytic concepts and contradictions, in particular with regard to the

- Sustainability of economic growth vs. development as the main objective of the SEEA
- Aggregation of environmental impacts in physical units vs. monetary values
- Incorporation of basic environmental statistics in the accounting system.

As a result, the SEEA looks sometimes more like a framework for environmental and economic data than an integrated economic-environmental accounting system.

8.4.1 Accounting for Economic Sustainability?

'The purpose of this handbook is ... investigation and analysis of the interaction between the economy and the environment. Only by integrating the two areas can the implications for sustainability of different patterns of production and consumption be examined ....' This statement of the SEEA seems to focus on the sustainability of economic activity and growth, rather than development. However, reference to Hicksian income presumably brings in 'sustainable development', deemed to be 'closely related to the long-standing economic concept of income'.

Hicks (1946) himself argued that the *ex ante* notion of income cannot be meaningfully aggregated. The SNA also makes it quite clear that Hicks' income definition of 'the maximum value ... [a man] can consume during a week, and still expect to be as well off at the end of the week as he was at the beginning' is not compatible with the national accounts conventions of income and wealth (Box 8.3). Nor can income be considered as a measure of welfare as suggested by the SEEA, due to the incompatibility of damage/welfare valuations with market prices (Section 8.1.3). The revised SEEA thus fails to clearly specify its objective of assessing the sustainability of economic performance and growth vs. socio-economic development. Right at the outset, there is confusion about the scope and coverage of the SEEA.

The SEEA addresses in principle two main operational categories of environmental sustainability, classifying them as weak and monetary, and strong and physical. This comes close to advocating the assessment of economic and ecological sustainability

[11] Unless otherwise stated 'SEEA' and citations in this section refer to the (revised) SEEA-2003 (United Nations et al., in prep.). Direct references for most of these citations can be found in Bartelmus (2007).

Box 8.3 National income vs. Hicksian (sustainable) income and wealth

The main difficulties of adopting the widely accepted Hicksian income definition in the SEEA stem from the following accounting conventions:

- Hicksian income is an *ex ante* notion (based on expectations), which, for measurement purposes, and as pointed out by Hicks (1946), would have to be translated into 'realized', i.e. actually received, income.
- The 'well-off' or wealth notion would have to be defined in accounting terms as maintaining 'real net worth' (the net value of financial and non-financial assets and liabilities: United Nations et al., 1993). It is far from clear to what extent the SEEA's 'capital base', which might include human and social capital, refers to real net worth – at individual and national levels.
- Assessing net worth maintenance for defining income would change the national (disposable) income concept by accounting also for changes in net worth due to (1) capital transfers (from/to other countries), (2) other changes in volume of assets (from natural disasters, war, discoveries and depletion of non-produced natural assets, i.e. 'windfall profits or losses' according to Hicks (op. cit.), and (3) real holding gains.

The connections between income generated (net domestic product), national income and the treatment of other volume changes therein are crucial for assessing the sustainability of natural capital use (Section 8.2.2). The SEEA mixes up or misinterprets these concepts when it distinguishes between a damage-based stock concept of sustainability of Hicksian income, and a cost- and income-based concept of the very same income notion. The objective seems to use Hicksian income for defining sustainability of both, welfare and income generation, interpreting 'being well off' as well-being and (non-declining) income.

as defined in Ch. 2. However, a persistent wavering between, on the one hand, the necessity of monetary valuation for assessing overall capital maintenance and, on the other hand, rejecting such valuation as research or hypothetical modelling prevents specifying the sustainability categories in terms of accounting indicators.

Obviously such ambivalence in valuation carries over into ambivalence towards calculating environmentally adjusted indicators. 'Adjustment accounts' first seem to present a large variety of greened aggregates. Soon enough we are warned, however, against carrying out these calculations: 'there are theoretical, practical and institutional reasons why a statistical office may not implement this part of the SEEA or at least not yet'. In fact, the adjustment accounts seem to conceal the key aggregates of EDP and ECF by focusing on the modification of a little-known and

-used sub-item, 'operating surplus'. EDP ('eaNDP') is listed under a bewildering list of indicator options.[12] ECF (excluding environmental degradation) is mentioned in passing as 'depletion-adjusted measure of capital formation'. The SEEA's unusual focus on the adjustment of saving in its 'captial' account seems to be a concession to the World Banks's promotion of 'adjusted net saving' (Section 8.2.2).

Without fully modifying the key monetary national accounts indicators, the SEEA cannot directly compare the 'goods' of production and consumption with their 'bads' of pollution and depletion. Consequently, proclamations on accounting for sustainable growth or development remain largely rhetoric. In fact, after an introductory discussion of natural capital and sustainability, the concept of natural capital and the role of its consumption in sustainability measurement are studiously avoided. Only the last chapter refers briefly to total national wealth as an indicator of sustainability, cautioning against its use because of the difficulties of assessing substitution among all capital categories.

8.4.2 Accounting for Ecological Sustainability?

The SEEA looks much better in physical accounting. Physical and hybrid accounts show material flows and stocks underlying the monetary transactions and the value of natural assets. Unfortunately, the SEEA is again ambivalent with regard to measuring comprehensively the inputs and outputs (throughput) of materials and substances in material flow accounts (MFA). A few paragraphs address the problem of using a common physical unit for various types of natural resources and emissions. In the end, the ambiguous conclusion is either to 'aggregate all materials on the basis of weight and … use caution in the interpretation of the results', or to 'build accounts on a material-by-material basis and avoid altogether the creation of potentially misleading measures…'.

The ambiguity towards aggregation carries over into discussing the ecological sustainability concept of dematerialization. According to the SEEA, the purpose of the MFA is to show the 'decoupling of economic growth from materials use' as 'an important sustainability goal for environmentalists'. However, setting standards for such dematerialization, notably of Factor 4 (cf. Section 2.4.2), is disparaged as 'rather vague for use as guides to policy…'. The physcial accounts fail therefore to provide comprehensive indicators for sustainability *policies*. They do include particular natural resource accounts and environmental statistics for the *management* of resources and residuals.

The SEEA does present the maintenance of critical capital as an alternative notion of strong ecological sustainability. This ecological sustainability concept

[12] What are we to make of the different versions for dpOS, dpS, dpNDP, daNI, daS, eaGDP, eaNDP and geGDP, where dp stands for depletion adjusted, da for damage adjusted, ea for environmentally adjusted (including depletion cost), and ge for greened-economy (modelled) indicators?

would indeed provide a justification for measuring irreplaceable environmental assets, selectively and in different units of measurement, i.e. without forcing them into the straightjacket of tonnage. The opportunity to show how the physical accounts could capture the strong sustainability concept of complementarity by proper definition and classification of critical capital categories is not seized, though. There are, however, promising attempts at defining and monitoring critical capital in terms of importance and vulnerability (de Groot et al., 2003), and by means of safe minimum (sustainability) standards (Ekins et al., 2003). These criteria should be further examined as to their compatibility with SEEA objectives and conventions.

The SEEA also includes ecosystems and their inputs into production and consumption, at least 'conceptually', while admitting to 'limited knowledge and experience', and measurement problems. The half-hearted inclusion of ecosystem accounts cannot provide a thorough discussion of the need for assessing ecosystem services and resilience as a measure of ecological sustainability (cf. Sections 2.4.1, 3.3.1). This draws the ire of ecological economists. Box 8.4 summarizes their critique and argues that, after all, welfare valuation of ecosystem services and modelling of system resilience do not fit in a national environmental-economic accounting system. The Millennium Ecosystem Assessment (2005) appears to confirm this view: it does

Box 8.4 Accounting for ecosystem services?

A special issue of *Ecological Economics* (2007, 61/4) confronted ecological economists with the revised SEEA-2003. With regard to ecosystem accounting, their critique focused on the SEEA's deficiencies in covering:

- The *spatial dimension*: land and ecosystem accounts of the SEEA need further development (Weber, 2007).
- Measurement and welfare valuation of *ecosystem services*: these services are 'Nature's public goods' and must be included in comprehensive welfare measures, notably a green GDP (Boyd, 2007).
- *Resilience: the* SEEA needs to address 'key ecological issues, such as system dynamics and … vulnerability' (Walker & Pearson, 2007).

However, these criticisms look more like arguments for removing ecosystem accounts from the SEEA. When it comes to accounting for ecosystem health, diversity and resilience, Weber's (2007) 'accounts' turn into 'counts', i.e. indicators and classifications. Boyd's (2007) suggestion of extending the production boundary of the national accounts upsets accounting identities and balances and introduces welfare valuations that are incompatible with the market valuations of the national accounts (Sections 7.3, 8.1.3). Finally, the modelling of potential welfare effects of changes in resilience (Walker & Pearson, 2007) blurs both *ex post* accounting and predictive modelling (Sections 8.1.2, 3).

not attempt a systematic accounting or presentation, but answers a range of 'key questions' on how ecosystems change, affect well-being and how they can be managed sustainably. Other frameworks, notably for environmental or ecological statistics and statistical ecology are indeed better suited for assessing and modelling the benefits and damages of particular ecosystems (cf. Ch. 4 and FR 4.1).

8.4.3 Revising the Revision

In summary, the revision process sought to minimize changes to the conventional national accounts. The idea is to elaborate on physical accounts and their underlying statistics, and using monetary values for those transactions that need only limited adjustment (as part of the SNA). These transactions can either be shifted around (from other asset changes to the production accounts) or presented in greater detail (environmental expenditures, taxes, permits and licenses).

The revised SEEA makes, therefore, most progress in physical accounting. The price is a loss of much of its systemic character by dealing with difficult-to-aggregate physical data. Meaningful aggregation is however a prerequisite for assessing and comparing the significance of environmental impacts and economic benefits. An opportunity for operationalizing the opaque notion of environmental sustainability with the help of environmentally adjusted accounting indicators is missed. The revision also misses a chance of overcoming, or at least assessing, the persisting environmental-economic dichotomy discussed in Ch. 2.

The necessary next revision will have to tackle, among others, the following issues:

- Defining clearly the goal of assessing the environmental sustainability of economic performance and growth, in produced and natural capital terms and corresponding physical and monetary indicators
- Streamlining a voluminous and difficult-to-read handbook by concentrating on aggregative physical and monetary accounts; separate handbooks could present the databases for natural resources and residuals with reference to other frameworks of environmental data and indicators (Sections 4.1,2)
- Reassessing the need for costing environmental externalities, required for cost internalization and full-cost pricing (Sections 2.3.2, 13.3)
- Describing the use of tradable pollution permits for market-price valuation of environmental degradation (Section 8.1.1)
- Reviewing critically the need for introducing welfare (damage) valuation into a system geared toward measuring economic performance (Section 8.1.3)
- Exploring neglected aspects of sustainability accounting and analysis, including
  - The maintenance of human, social and financial capital categories and their substitution (Section 8.2.1)
  - Environmental debt owed to future generations and other countries, whence sustainability is 'imported' (Section 8.1.2)

- Definition and measurement of critical capital in physical accounts (Section 2.3.1)
- Accounting for goods, services and pollutants in energy (exergy) units (Section 6.2.3)
- The feasibility of subnational accounting for regional environmental pressures and ecological capacities
- The treatment of transboundary pollution as transfers in environmentally adjusted national income (Section 8.1.2)

- Establishing guidelines, software and training material for the implementation of green accounting projects, building upon the operational manual of the SEEA (United Nations, 2000a) and experience gained in case studies.

The implementation of these proposals requires a greater involvement of the research and user communities – beyond the narrow views of official statistics. Decision-makers need to learn about the analytic capabilities of green accounting for both, the environmental management of particular natural resources and pollutants, and the formulation and evaluation of national sustainability policies. Statisticians should familiarize themselves with data uses in sustainability analysis and policy.

Satellite accounts can assess progress towards long-term sustainability of economic performance, without changing the basic principles of the conventional accounts. Why not use the satellites for what they are intended, namely to present and test alternative assessment tools for new concepts and paradigms? At present, there is a distinct risk that green accounting will be ignored as yet another – complex and costly at that – indicator framework. In 1992, the Rio Earth Summit proposed 'a programme to develop national systems of integrated environmental and economic accounting in all countries' (United Nations, 1994, ch. 8). Ten years later, the 2002 Johannesburg Summit did not mention environmental accounting but 'encourage[d] further work on indicators for sustainable development' (United Nations, 2003). It remains to be seen if a new United Nations Committee of Experts on Environmental-Economic Accounting [FR 8.1] will be able to achieve its declared objective of raising the SEEA from a technical report to a 'statistical standard'.

Further Reading

FR 8.1 SEEA History and Revision

Ward (2004) devotes a chapter of his book to the 'environmental dimension' of statistical work by the United Nations. Bartelmus and Seifert (2003, Introduction) present a concise history of green accounting. Their reader also selects key works of the methods and use of green accounts at national and corporate levels. The new Earth Portal to the *Encyclopedia of Earth* provides an overview article on green accounting: http://www.eoearth.org/article/Green_accounting.

Bartelmus et al. (1991) developed the basic system of integrated environmental and economic accounting. The same authors also prepared a draft handbook, which was published by the United Nations (1993) after submission to the Rio Earth Summit. Using experience gained in pilot case studies, the United Nations (2000a) later issued an 'operational' manual. The SNA presents the SEEA as part of its satellite accounts (United Nations et al., 1993, ch. XXI).

The London Group of national accountants, named after the place of its first meeting, revised the original SEEA. The draft revised version, the SEEA-2003, is available from the web site of the United Nations Statistics Division: http://unstats.un.org/unsd/envaccounting/seea.asp. The successor of the London Group, the United Nations Committee of Experts on Environmental-Economic Accounting seeks to coordinate the further development of concepts and methods and to promote the use of the SEEA (http://unstats.un.org/unsd/envAccounting/ceea/default.asp). Inexplicably, the SEEA-2003 is still not published.

Hecht (2005) is a largely textual presentation of the SEEA, presenting useful summaries of its modules and national accounting in general. A special edition of *Ecological Economics* (2007, 61/4) presents a first outside review of the revised SEEA.

FR 8.2 Case Studies of Green Accounting

The sources of Table 8.1 refer to pilot studies of green accounting. Some of these studies can be found in Uno and Bartelmus (1998). The operational SEEA manual (United Nations, 2000a, annex) describes software available for a step-by-step implementation of case studies. The Institute of Advanced Studies of the United Nations University conducted case studies on green GDP in China, Japan and Indonesia (Akita & Nakamura, 2000). Following a call by China's President, the State Statistical Bureau and the State Environmental Protection Administration carried out a case study of green accounting (http://english.gov.cn/2006-09/11/content_384596.htm); as in the USA (cf. note 7), the recent halting of China's green accounting project (see Box 4.4) reflects fears of revealing the economic significance of environmentally hazardous activities. Markandya and Pavan (1999) and delos Angeles and Peskin (1998) attempted to apply welfare valuation to environmental damages in green accounting for selected European countries and the Philippines, respectively. The United Nations Statistics Division is building a database on environmental accounting mostly by governmental agencies (http://unstats.un.org/unsd/envaccounting/ceea/archive/Introduction.asp).

Review and Exploration

- Why should we impute a money value on the use of natural resources (depletion) and environmental sinks (degradation)? Do we need to adjust the national accounts for costing environmental impacts?

- Explain the pros and cons of different valuation techniques for greening the national accounts.
- How do green accounting indicators assess the sustainability of economic performance and growth? Do they account for the sustainability of development? Compare the monetary indicators with the physical aggregates of material flow accounts.
- Find the key green accounting aggregates in Germany's SEEA matrix (Annex III). What do they say about the sustainability of Germany's economy?
- Does the deduction of defensive expenditures turn net national product into a welfare measure?
- Should we replace the conventional national accounts with greened ones?
- Does the SEEA revision address the different accounting dichotomies? Do we need a revision of the revised SEEA?

Chapter 9
Corporate Accounting: Accounting for Accountability

Corporate environmental accounting mirrors national environmental-economic accounting at the enterprise level. Corporations picked up the messages of the Earth Summits, presenting environmental management as a sign of corporate social responsibility. They are more reticent, though, to account publicly for their environmental impacts, in particular if it comes to providing a cost value for these impacts. If at all, corporate accountants favour physical eco-balances and life cycle analyses over full-cost accounting. Dissenting voices introduced the physical-monetary dichotomy into a debate of the accountancy profession.

International guidelines promote cost-saving environmental management, or at least the cost-efficient implementation of environmental rules and regulations. Mostly, they ignore the need for standardizing the monitoring of environmental impacts and their costs. The SEEA could provide a framework and standards for extending the management guidelines to green accounting. The result would be harmonized micro- and macro-level environmental accounting and analysis, establishing the so-called micro-macro link.

9.1 From Accountability to Accounting[1]

9.1.1 Corporate Social Responsibility

The social indicator discussion of the 1970s in Europe triggered the inclusion of non-economic social concerns in corporate accounts. However, the widely propagated Swiss-German 'social balances' (*Sozialbilanzen*) (Hoffmann-Nowotny, 1981) were short-lived. The reasons were measurement and aggregation problems of quality of life components, on the one hand, and conflicting interests between corporate-economic and social objectives, on the other hand. Nonetheless, social accounting can be seen as the 'Trojan horse', which opened the walls of conventional corporate

[1] This section draws on Bartelmus and Seifert (2003), Introduction, section 4.

accounting to the accountability of corporations for their social and environmental impacts (Gray, 1992). Scholars at the university of St. Gall (Switzerland) used the breach made into economic accounting by extending the social balances into 'ecological bookkeeping' (Ullman, 1976; Müller-Wenck, 1978).

The failure of assessing the quality of life by social indicators [FR 4.3] stalled the further development and implementation of green corporate accounting. It took a long time and inspiration from the international environmental and sustainable development movements for the accountancy profession to acknowledge the relevance of environmental concerns. By now, the idea of corporate social responsibility (CSR) seems to have 'won the battle of ideas' (Crook, 2005). Business, government, civil society and international organizations all advocate the need for catering not only to the economic gain of the company's shareholders but also to the welfare of its stakeholders, i.e. the neighbourhood community and society at large. Globalization (cf. Ch. 14) contributed to this general acceptance of corporate accountability and good corporate citizenship: multinational corporations, which got mired in human rights violations, corruption, social conflicts and environmental disasters in some countries, are ready to take up and flaunt social responsibility.

The United Nations and other international governmental and non-governmental organizations promote CSR in all dimensions of sustainable development [FR 9.1]. At the same time, there are voices questioning the wisdom of letting the boardroom decide about social and environmental concerns. There is no general electorate to legitimize the formulation and implementation of social and environmental policies by companies; nor should companies compromise their obligation to shareholders for maximizing profitability. Moreover, banking on public-private partnership for fostering sustainability may be a sign of governments shirking their responsibility for improving environmental and social conditions.[2]

The general drive for CSR puts pressure on enterprises to move beyond rhetoric by monitoring the implementation of proclaimed social objectives. Actual efforts at changing the established accounting procedures reveal how far enterprises and their accountants are willing to go in subjecting the CSR ideals to scrutiny. For instance, the United Nations (2002a) programme on the promotion of environmental management accounting sees the main benefits of such accounting in cost-saving waste management, reduction of environmental liability and improvement of corporate image.

This is a far cry from showing responsibility for improving the quality of life of community and society. On the other hand, it is a sign of good corporate management that could enhance profitability and environmental goodwill. It remains to be seen if, possibly in reaction to accounts manipulations such as exaggerating oil

[2] The 2002 Johannesburg Summit advanced so-called type-2 partnerships. UNEP's former Executive Director, Klaus Töpfer, maintained that these partnerships 'threaten to mask the failure of governments to agree on meaningful action' (the type-1 partnership) and 'could … result in "greenwash" by polluting companies wanting to divert criticism' (as cited by the Friends of the Earth, an NGO: http://www.foe.org/WSSD/partnerships.html).

reserves, we will now enter a new 'age of transparency' for share- and stakeholders (Tapscott & Ticoll, 2003).

9.1.2 Getting Physical or Monetary?

Changing accounting rules and regulations seems to be easier at the national level as national accountants have some advantages in this regard over their corporate counterparts: they are less confined by accountancy laws and rules, they are not directly affected by their own calculations, and their macroeconomic vantage gives them a broader view and earlier recognition of changing socio-economic priorities. This may explain why corporate 'financial' environmental accounting has lagged behind national accounting in addressing environmental and human quality-of-life concerns. On the other hand, environmental 'management' accounts (EMA) have been widely propagated, even at the international level [FR 9.2].[3]

However, EMA face the same physical-monetary dichotomy as their national counterparts. Gray (1990, 1992) has been among the first to call for introducing notions like carrying capacity and capital maintenance into corporate accounts. He recognizes the value of both physical impact accounting and 'sustainable cost' accounting in monetary 'shadow accounts'. He stops short, though, of advancing an accounting system to this end, considering the difficulties of doing so 'monumental'.

Schaltegger and Burritt (2000) tackle the monumental task. In their seminal book they distinguish between financial (monetary) and ecological (physical) accounting; they also suggest to 'take the two together' in a modular presentation of an environmental accounting framework. This is indeed similar to the conservative modular approach of the revised SEEA (United Nations et al., in prep). Schaltegger and Burrit also adopt a cautious valuation approach,

- Including only 'internal costs' of outlays for environmental protection in the monetary accounts
- Assessing environmental impacts through physical input-output accounts
- 'Integrating' economic and environmental data by means of eco-efficiency indicators as the ratio of (monetary) value added and (physical) environmental impact.

9.1.2.1 Physical Accounting

Physical accounting of natural resource use and residuals is the most popular way of meeting stakeholders' demand for environmental information. Depending on the scope of the analysis, *eco-balances* assess the physical environmental impacts of

[3] Financial accounts are typically subject to strict legislative regulation to ensure consistent disclosure of the firm's performance to regulators, investors and stakeholders. Management accounts serve the internal cost analysis of a firm's activities according to its particular needs and priorities.

corporations or local plants, while *life cycle analyses* focus on product-specific impacts at different production and consumption stages.

Table 9.1 shows the internationally acclaimed eco-balance of a German corporation[4] as an example of physical input-output accounts. Contrary to conventional input-output systems, the eco-balances also present assets and asset changes of equipment, buildings and land – the latter with environmental categories. The flow accounts show material and energy inputs and residual outputs (in addition to product outputs) – similar to the national material flow accounts (MFA) (Section 6.3).

Applying impact analysis to a particular product or production process over the lifetime of the product (from 'cradle to grave') is the approach of life cycle analysis (LCA) [FR 9.3]. Plate 9.1 illustrates the production process of jeans from

Table 9.1 Eco-balance, Kunert AG

	Stocks (12/31/93)	Input (1994)	Output (1994)	Stocks (12/31/94)
Stocks				
1. Land (sq. m)[a]	649,143	12,931	9,602	646,960
1.1 Sealed	68,606	636	2,692	65,750
1.2 Green	448,659	938	340	448,386
1.3 Built-over	131,878	11,357	6,570	132,824
2. Buildings (sq. m)[a]	178,473	14,447	17,923	185,369
3. Equipment (piece)	16,542	1,436	1,263	16,715
Flows				
4. Materials/products (kg)		1,055,912	8,492,704	
4.1 Raw materials	697,183	3,558,124		
4.2 Goods		2,082,292		
4.3 Auxiliary materials		3,936,325		
4.4 Ancillary materials		1,479,171		
5. Waste (kg)			2,357,988	36,398
5.1 Hazardous			62,883	3,910
5.2 Other			660,225	32,488
6. Energy/waste heat (kWh)	n/a	118,986,313	118,986,313	n/a
7. Water/waste water (cu. m)	n/a	428,770	339,277	n/a
8. Air emission (kg)	n/a			n/a
8.1 NO_x			100,548	
8.2 SO_2			170.132	
8.3 CO_2			36,109,594	
8.4 Steam			96,895,400	

Note: [a]Imbalances in stocks 1993/1994 are due to improved data collection in Tunisian and Moroccan factories.
Source: Kunert AG (1994/1995, pp. 14/15).

[4]In 1995 the Kunert AG's environmental report was chosen as the 'world-best' by SustainAbility Ltd., a London-based research institute, on behalf of the United Nations Environment Programme.

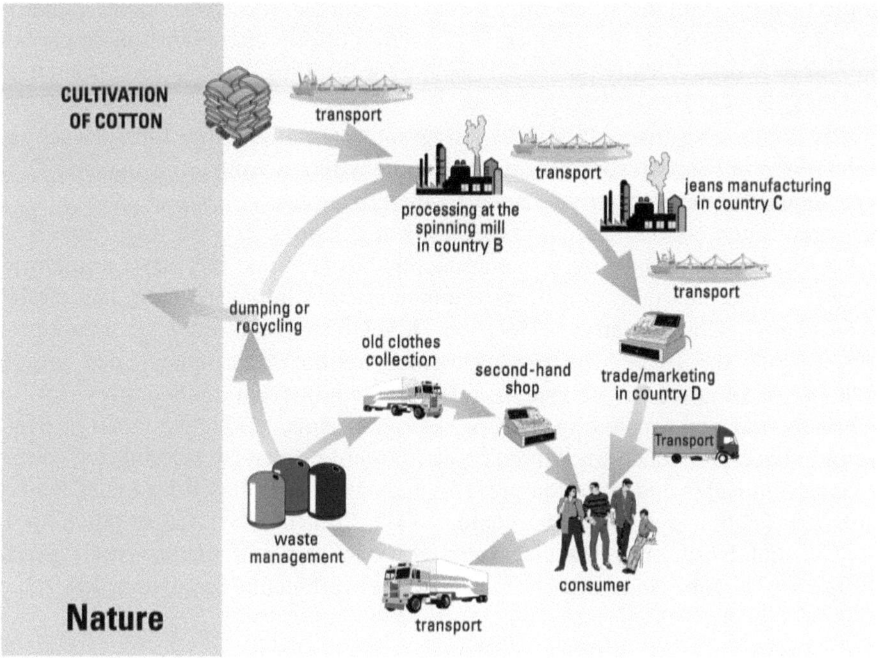

Plate 9.1 Life cycle of jeans
Copyright VisLab/Wuppertal Institute for Climate, Environment and Energy; with permission by the copyright holder (*See Colour Plates*).

the production of cotton to the use and disposal by consumers, with recycling loops back to the consumers as second-hand goods or to reprocessing in cloth manufacture. Detailed analyses could and should assess the environmental impacts at all stages of production and transport, especially with regard to emissions and fuel use.

Physical accounting faces of course the problem of comparing the significance of impacts assessed in different measurement units. As in the MFA, the closest physical corporate accounts can come to combining environmental impacts with economic output are resource productivity or eco-efficiency ratios. At the same time, the detail and knowledge available at the micro-level of the enterprise permit a more valid intuitive evaluation of environmental impacts than at the national level. However, full integration is possible only by costing environmental impacts in monetary accounts.

9.1.2.2 Monetary Accounting

On the monetary side of corporate environmental accounting, the less problematic assessment of internal environmental protection expenditures has made greater strides than the valuation of environmental externalities. It is clearly more attractive for a firm to present its environmental protection efforts than to cost its impact on the outside world. Vividly put: 'who could expect turkeys to vote for Christmas?' (Bebbington et al., 2001).

It is no surprise that calls for assessing and internalizing the full (private and social) costs of the corporation's activities come typically from policymakers. Their expectation is that voluntary initiatives by the private sector might obviate unpopular market interventions such as eco-taxes or regulations (see Ch. 13). Agenda 21 of the Rio Earth Summit urges 'governments, business and industry … [to] work towards … the internalization of environmental costs into accounting and pricing mechanisms' (United Nations, 1994, ch. 30). Under the heading of 'getting the prices right', the EU's Fifth Environmental Action Programme called for the 'redefinition of accounting concepts, rules, conventions and methodology' for full environmental cost accounting (European Commission, 1993). Not much progress seems to have been made since then, except, possibly, when considering accounting for emission rights and emission prevention as assets and liabilities under the EU Emission Trading Scheme (Casamento, 2004). Still, professional associations in the UK and North America elaborated the concepts and methods of *full-cost accounting*, possibly in anticipation of further governmental regulation [FR 9.2].

9.1.3 Micro-Macro Link

The national accounts are based on double-entry bookkeeping of enterprises. Micro-level corporate accounting that is fully consistent with aggregate national accounting would facilitate statistical data compilation. It would also support economic analysis, in particular of the distribution of income and wealth. One of the SNA handbooks thus explores the relationships between micro- and macro-accounts (United Nations, 2000b). The handbook also reveals numerous differences in accounting concepts, procedures and indicators such as depreciation by firms (for tax purposes) and capital consumption in the national accounts (for assessing the wear and tear of fixed capital).[5]

Despite these differences, corporate environmental accounting takes approaches that are similar to the greening of the national accounts. They include, in particular,

- Corporate 'parallel' or 'shadow' accounting for externalities (Bebbington et al., 2001), comparable to the SNA's satellite accounts for the SEEA

[5] As the national accounts record transactions between different economic agents, they frequently expand double-entry accounting (for internal production and financial flows) of enterprises into quadruple-entry accounting, adding the same transaction for buyers and sellers (United Nations et al., 1993). The SNA also describes the micro-macro links between business and national accounting and underlying economic theory.

- The dichotomy of physical vs. monetary accounting in physical eco-balances and full-cost accounts
- The segregation of environmental protection expenditures from corporate overhead costs, and from the SNA's economic activity classifications in the SEEA.

Corporate and national accountants could indeed learn from each other about their respective methods and the use and usefulness of harmonized green accounting at micro-, meso- and macro-levels. The benefits of this micro-macro link would be

- Enhanced compatibility of physical material flow and monetary environmental cost accounts at enterprise, household, regional and national levels
- Consistent micro- and macroeconomic strategies and policies, addressing the sustainability of production and consumption patterns of economic sectors, corporations and households, and of the overall economic development of regions and countries
- Identification and measurement of critical capital maintenance, the key ingredient of strong sustainability (cf. Section 8.4.1), notably through LCA and with a view to exploring aggregation at sectoral and national levels
- Improved quality of aggregated environmental stock (ledgers, assets) and flow (input, output) data from harmonized data sources.

The integrated – physical and monetary – accounting system of the SEEA appears to provide the best available framework for further developing the micro-macro link in the fields of environmental-economic accounting and analysis.

9.2 From Accounting to Management

Corporate environmental accounts provide direct data input for corporate environmental management. However, the main international management guidelines of the ISO (International Organization for Standardization) 14000 and the European Union's EMAS (Environmental Management and Audit Scheme) [FR 9.3] do not clearly link environmental accounting and management. Some connections can be envisaged, though, between accounting data and *performance indicators* proposed by ISO and EMAS for environmental management. Both management guidelines categorize these indicators as

- Operational performance indicators of material inputs and outputs
- Management performance indicators of programme costs, and internal safety and health
- Environmental condition indicators of environmental quality and effects on human health and other socio-cultural amenities.

Based on these indicators ISO and EMAS suggest internal and external *audits* for the evaluation of environmental performance. Such audits serve the information

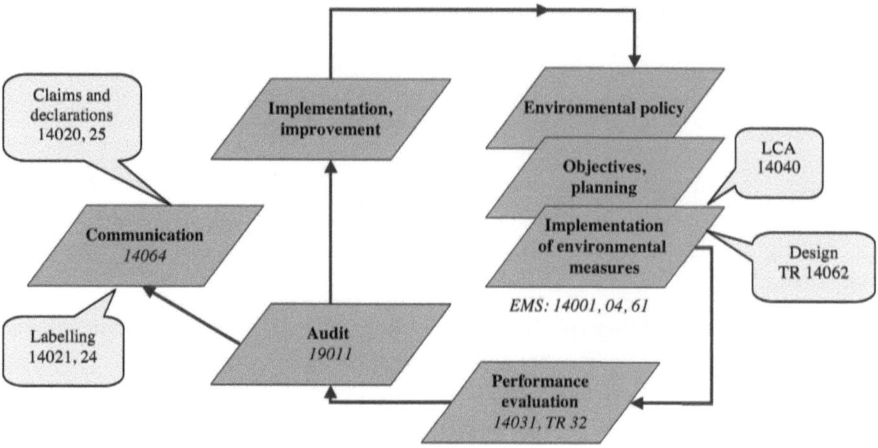

Fig. 9.1 ISO 14000 standards for environmental management
Source: Based on Wohlfahrt (1999).

needs of stakeholders and the improvement of environmental management in the organization. Figure 9.1 depicts a cycle of continuous performance evaluation and improvement for the ISO 14000 series with an indication of more specific recommendations (by numbers of ISO standards). The company's environmental policy, planning and measures form its 'environmental management system' (EMS). The evaluation of the EMS by performance indicators and audits may warrant further improvement in environmental management, possibly changing environmental policy. The basic goal of this cycle is to encourage organizations to move from reactive treatment of environmental damage to proactive damage prevention.

ISO 14000 and EMAS are quite similar in their scope and coverage, owing to the incorporation of ISO 14001 ('Environmental Management Systems – Specifications with Guidance for Use') into the revised EMAS II. There remain, however, important differences, in particular

- The regional validity: EU member states for EMAS, and worldwide coverage for ISO 14000
- An environmental 'declaration' under the authority of the EU vs. a less specific environmental 'policy statement' proposed by the non-governmental ISO
- The EMAS logo, which can be used on-site and on stationary, but not for product advertising (Plate 9.2).

The global scope and the less stringent supervision of ISO explain its greater popularity: as of January 2007 there were 129,031 ISO certifications as compared to 5,389 for EMAS.[6]

[6] http://www.ecology.or.jp/isoworld/english/analy14k.htm.

Plate 9.2 EMAS logo
Source: http://europa.eu.int/comm/environment/emas/index_en.htm; with permission by the copyright holder, Stora Enso Kabel Mill, Germany (*See Colour Plates*).

Both guidelines are voluntary. Despite the clamorous advocacy of corporate social responsibility, they found only limited application. It remains to be seen whether actual or perceived economic benefits of environmental management will foster greater use, notably by small and medium-sized enterprises. Like the United Nations programme of environmental management accounting (Section 9.1.1), the ISO and EMAS management guidelines advertise their benefits as

- More efficient environmental management
- Natural resource (cost) savings
- New business opportunities and innovations
- Reduction of liabilities for environmental hazards
- Improved staff-management relations
- Better credit conditions and credibility
- Improved image of the corporation.

Catering to a broad notion of CSR, a coalition of business, accountants, investors and stakeholders advanced further guidelines on *sustainability reporting*. The Global Reporting Initiative (GRI) aims to extend environmental performance evaluation and reporting, covering contributions to all three dimensions of sustainable development. To this end, the GRI also presents economic, social and environmental performance indicators [FR 9.3].

Further Reading

FR 9.1 Corporate Social Responsibility

The World Business Council for Sustainable Development seeks to bring about sustainable development through eco-efficiency, innovation and corporate social responsibility (CSR) (http://www.wbcsd.ch/templates/TemplateWBCSD5/layout.asp?type=p&MenuId=NjA&doOpen=1&ClickMenu=LeftMenu). Agenda 21 of the Rio Summit promotes 'cleaner production' by full-cost pricing, life cycle analysis and 'responsible entrepreneurship' (United Nations, 1994, ch. 30). The Secretary General of the United Nations, after addressing the World Economic Forum in 1999, launched a Global Compact of United Nations agencies, business, labour and civil society to take stakeholder concerns into account through 'responsible corporate citizenship' (http://www.un.org/Depts/ptd/global.htm). The Johannesburg Summit (United Nations, 2003) stresses in its Political Declaration 'the duty' of companies 'to contribute to the evolution of equitable and sustainable communities and societies' and the 'need … to enforce corporate accountability'. Its Plan of Implementation promotes 'corporate responsibility and accountability', among others through 'public-private partnerships'. The European Union developed a European Strategy on CSR, whose 'centerpiece' is the European Multistakeholder Forum. The Forum is to promote 'transparency and convergence' on CSR (http://europa.eu.int/comm/enterprise/csr/index_en.htm).

The Journal of Corporate Citizenship presents special theme issues on the theory and practice of CSR. The CSR Newswire is a source for 'press releases, reports and news' on corporate responsibility and sustainability (http://www.csrwire.com/).

FR 9.2 Environmental Management Accounting and Full-Cost Accounting

The United Nations organized a series of workshops to assess governments' role in promoting Environmental Management Accounting (EMA) (http://www.un.org/esa/sustdev/sdissues/technology/estema1.htm). The United Nations also surveyed national and international EMA efforts, recommended exploring the relationships of environmental management systems and national green accounting (United Nations, 2002a), and advanced material flow costing in terms of 'wasted material purchase value' (quite different from environmental costing in the SEEA) (United Nations, 2001a). The Environmental Management Accounting Research Center provides a web site on the US EPA Environmental Accounting Project and offers links to international activities and networks (http://www.emawebsite.org/).

The Environmental Management Accounting Network (EMAN), an EU-sponsored forum for sharing information about EMA, intends to focus on 'sustainability accounting' in its future publications (http://www.eman-eu.net/).

The British Association of Chartered Certified Accountants (ACCA) published a comprehensive study on 'full cost accounting' (Bebbington et al., 2001), following a call for environmental cost internalization by the EU's Fifth Environmental Action Programme (http://europa.eu.int/comm/environment/actionpr.htm). The 2004 ACCA report (http://www.accaglobal.com/pdfs/environment/tech-ea2-001b) seems to be more pessimistic about implementing the 'holy grail' of full-cost accounting; it still sees an opportunity for liability accounting in the context of the EU's emission trading scheme (Casamento, 2004 in ch. 4). The Canadian Institute of Chartered Accountants (1997) and the Center for Waste Reduction Technologies (1999) advanced similar proposals.

FR 9.3 Environmental Management and Reporting

The following web sites present the two main international environmental management guidelines:

ISO 14000: http://www.iso.org/iso/en/prods-services/otherpubs/iso14000/index.html and the EU Environmental Management and Audit Scheme (EMAS): http://europa.eu.int/comm/environment/emas/index_en.htm. The ISO guidelines (ISO 14040-43) incorporate life cycle analysis (LCA). UNEP promotes LCA in its life cycle 'assessment' and 'initiative' (http://www.uneptie.org/pc/pc/tools/lca.htm). The World Resources Institute provides a concise overview of LCA: http://www.gdrc.org/uem/lca/life-cycle.html.

The Global Reporting Initiative (GRI) (http://www.globalreporting.org/about/brief.asp) could be seen as a direct application of the communication module of environmental management (Fig. 9.1). There are no explicit links, however, to ISO 14000 and EMAS. Part C of the GRI's 'Sustainability Reporting Guidelines 2002' contains a detailed description of sustainability performance indicators: http://www.globalreporting.org/guidelines/2002/contents.asp.

Review and Exploration

- Should corporations get involved in improving the social and environmental conditions of their neighbourhood communities?
- Why should business account for external effects of its activities?
- Describe the benefits of the micro-macro link in green accounting.
- Compare the scope, coverage and contents of ISO 14000 and EMAS II.
- Do environmental accounting and management improve the bottom line (profits) of corporations?

Part IV
Analysis – Modelling Sustainability

Applied mathematical models can combine *eco*–nomic theory, sketched out in Chapter 2, with suitable measurement as presented in the green accounting systems of parts II and III. As a result, applied models could

- Explain the complex environment-economy interaction transparently, rather than intuitively, and
- Predict environmental impacts for formulating policy options.

Inevitably, modelling entails some abstraction from real-world complexities. In order to minimize this information loss, part IV focuses on those models and techniques that are closely related to the accounting systems, i.e. input-output analysis.

Computerized models can handle vast amounts of economic and environmental variables and their complex interrelationships. Measurability and data availability pose limits, however, to representing reality with reasonable accuracy. Several models in this part take, in fact, CO_2 emission as a convenient surrogate for environmental impacts. Green accounting case studies do indicate a heavy burden from, and considerable mitigation cost of, this greenhouse gas.[1] However, as discussed in section 4.3, such a reductionist view carries the risk of distorting the significance of environmental concerns themselves and their role in sustainability analysis. The presentation of CO_2-focused models in this part serves, therefore, mostly illustrative purposes; it also points to the need for better coverage of environmental impacts.

Chapter 10 reviews first the results of sustainability measurement obtained from the physical and monetary accounts. It enters 'analysis' by transforming the supply and use accounts of the national accounts into input-output tables. Input-output and related techniques permit tracing the full, direct and indirect, environmental impacts of different economic activities and identifying the main driving forces behind these impacts. Chapter 11 moves from descriptive to predictive analysis.

---

[1] For instance, hybrid accounts in the Netherlands showed the weight of CO_2 emission to exceed the weight of all other pollutants by several orders of magnitude (Section 7.3, Table 7.2). In Germany, half of the pollution cost, which makes up the bulk of environmental cost, stems from CO_2 emission (at a 25% reduction standard: see Annex III).

The chapter also explores econometric and simulation techniques in two applications that test the connection between economic growth and environment at national and global levels. Chapter 12 turns then to more prescriptive models, which seek to show how sustainability and optimality can be reconciled in economic policy analysis.

Chapter 10
Diagnosis: *Has* the Economy Behaved Sustainably?

The title question of this chapter cannot be answered unequivocally. Economic welfare measures may refer to the ultimate goal of economic activity. However, they suffer from problems of measuring the utility of economic benefits and the disutility (damage) from environmental impacts. Material flow indicators are more specific: they indicate that the relatively strong sustainability concept of dematerialization is still an elusive goal, nationally and globally. Green accounting case studies show weak sustainability for most economies, with some exceptions, notably of African countries which appear to live off their produced and natural capital base.

Tracing the total, direct and indirect, environmental impacts of economic activities and their driving forces is the task of input-output and decomposition analyses. To date, such studies are still isolated efforts, dealing with selected pollutants or the usual environmental placeholder of CO_2 emission.

10.1 Welfare Secured? Dematerialized? Capital Maintained?

10.1.1 Welfare Indices: Confirming the Threshold Hypothesis?

The closest economists have come to measuring welfare is by adding or deducting selected (quantifiable) effects on human well-being to/from utility-generating personal consumption or income. Time series of indices such as the Measure of Economic Welfare (MEW), the Index of Sustainable Economic Welfare (ISEW) or the Genuine Progress Indicator (GPI) supposedly indicate past and, by extrapolation, future trends of economic welfare generated by production and consumption (Section 7.1.1). A persisting decline of national welfare would indicate the non-sustainability, at least for the period covered, of the outcomes of economic activity.

An opening-scissor trend of GDP and the welfare indices provides the main evidence – at least for ecological economists[2] – for non-sustainable economic growth

[2] For example, Friends of the Earth: http://www.foe.co.uk/progress/java/UserDataServlet; see also Costanza et al. (1997a), Sachs et al. (1998), and Daly and Farley (2004).

P. Bartelmus, *Quantitative Eco-nomics,*

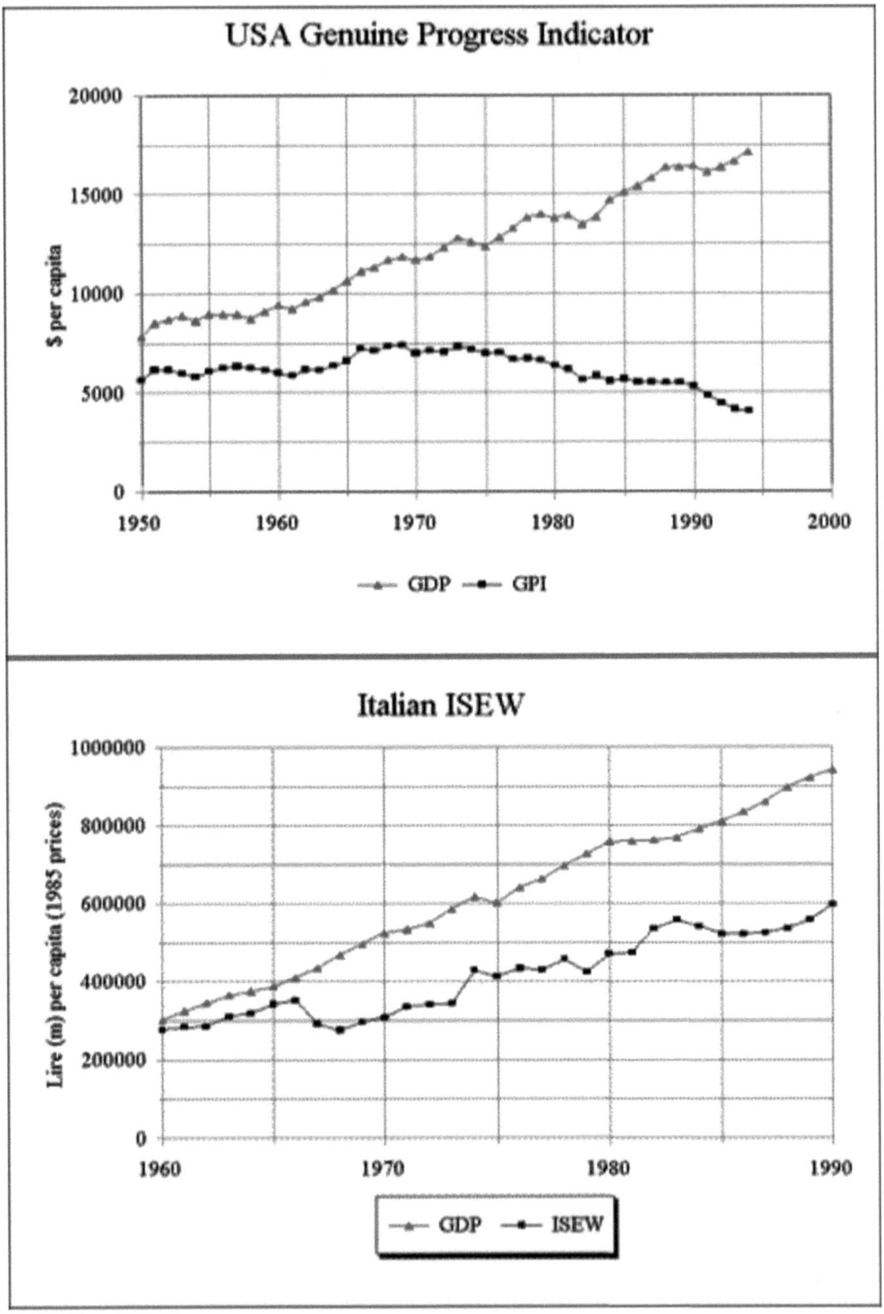

Fig. 10.1 ISEW and GDP: No threshold in Italy?
Source: http://www.foe.co.uk/campaigns/sustainable_development/progress/international.html.

that undermines the human quality of life. Methodological and data problems render the validity of this evidence questionable. Moreover, actual index compilations (cf. Fig. 7.2) do not generally support the 'threshold hypothesis' (Max-Neef, 1995) of gaping trends of welfare and economic growth, once a certain level of growth is reached. Contradictory interpretations of these indices might stem from short time series available after the presumed turning points, preventing any meaningful extrapolation of trends. Figure 10.1 exemplifies for Italy that the ISEW calculation does not provide the distinct scissor movement observed in the USA.

Replacing monetary valuation by averaging physical indicators in indices of the quality of life, well-being or sustainable development blurs the meaning of the indices by weighting equally unequal concerns; it also loses comparability with measures of economic performance (Section 5.3). As a consequence, these indices do not attempt to assess the sustainability of economic growth. Rather, they compare relative 'sustainability' in country rankings or show well-being scores only. The 'sustainability barometer' is deemed to be an indicator of such well-being; it sets a quite arbitrary sustainability level of 80 points (out of 100) and claims that no country has achieved this level (Section 5.2).

All in all, it does not seem to be possible to assess the (non)sustainability of economic growth with opaque measures of economic well-being (welfare) or the human quality of life.

10.1.2 Dematerialization: Delinkage of Economic Output and Material Input

Material flow accounts have distinct systemic advantages over ad hoc attempts at aggregating indicators by averaging or other types of weighting. The reason is that they base the measurement of material and energy flows on thermodynamic theory. Still, besides equal weighting of unequal environmental pressures, physical material aggregates are not directly comparable with economic indicators. The comparison of material flows and economic indicators resorts therefore to comparing their speed rather than their levels.

The result is the assessment of sustainability as a matter of decoupling the material indicators, notably total material input, from GDP. The questions are then: how much decoupling do we need, and for how long? On their own, material inputs may capture actual and potential pressures on national or global carrying capacities. Assessing sustainability has to go farther, however, by setting a standard for the maximum permissible pressure on environmental source and sink functions. Measuring the ecological sustainability of economic growth becomes thus distinctly normative.

The popular Factor 4 standard uses the relatively opaque notion of available 'environmental space' to defend its call for halving material inputs into the planet's economies. A more cautious but at the same time fuzzier approach reduces the Factor 4 or 10 standards to safe guardrails guiding development, rather than prescribing precise targets (cf. Section 2.4.2).

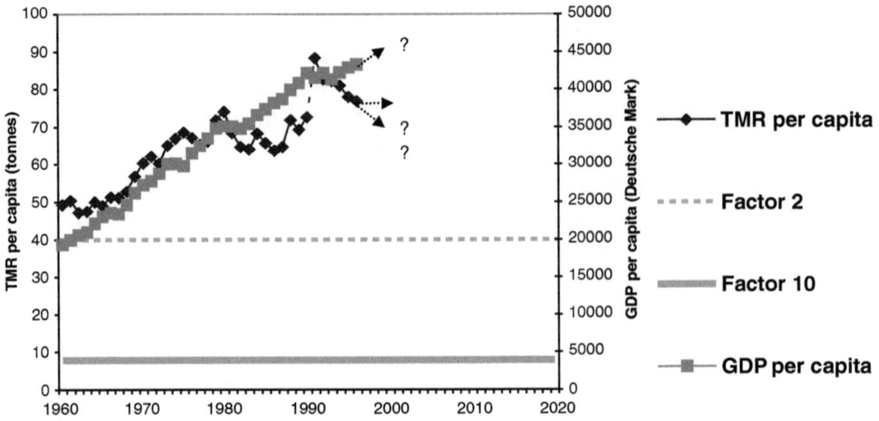

Fig. 10.2 Is Germany's economy sustainable?

Notes: 1960–1990 Federal Republic of Germany (West), 1991–1996 Federal Republic of Germany; TMR compiled by the Wuppertal Institute for Climate, Environment and Energy; GDP in 1996 prices.

Figure 10.2 indicates that, at least in Germany, economic performance is still far away from halving material inputs (from levels in the 1990s), and the outlook for getting there does not look good. Clearly, from an ecological sustainability point of view, this economy, just as those of most other industrialized countries (see Section 6.3.2), has not behaved sustainably over the last 40–50 years. The EU strategy on the sustainable use of natural resources comes to the same conclusion, pointing out that material consumption has remained constant over the last 20 years. At the same time, the strategy rejects setting quantitative targets due to lack of knowledge and indicators (Commission of the European Communities, 2005).

10.1.3 Capital Maintenance: Has Economic Growth Been Sustainable?

Greening the national accounts achieves comparability of environmental impacts with economic indicators by costing the impacts as natural capital consumption. The economic sustainability notion of capital maintenance calls for reinvesting the cost allowance for capital maintenance (Sections 2.3.1, 8.2.1). Industrialized countries and many developing ones increased their capital base through truly net capital formation – net of produced *and* natural capital consumption (Section 8.3). However, a significant number of nations, mostly from Africa, seemed to have lived off their produced and natural capital base, if the rough World Bank estimates can be trusted

Table 10.1 Environmental depletion and degradation cost in selected countries (% of NDP)

Developing countries		Industrialized countries	
China	6	Germany	3–4
Costa Rica	4–11[a]	Japan	2–3
Ghana	17–25	Korea, Republic of	2–4
Indonesia	13–31[a]	UK	0–5[b]
Papua New Guinea	3–10	USA	0.4–1.5[c]
Philippines	0.5–4[a]		

Notes: [a]Depletion only. [b]Oil and gas depletion only. [c]Subsoil resources only.
Source: Table 8.1.

(see Table 8.2). Of course, maintaining the total value of capital is only a necessary condition for weak sustainability of economic growth, ignoring critical natural capital and the role of other human and social capital categories.

Conventional economic net indicators of value added, domestic product and capital formation overstate economic performance with regard to the social (environmental) costs generated during the accounting period. A more accurate reckoning of these costs reveals the necessary economic effort that should and could have been made for replacing, avoiding or reducing the natural capital loss during the accounting period. Amounting to a few percentage points of NDP (Table 10.1), these environmental costs are well within the reach of industrialized countries. Developing countries, on the other hand, seem to face relatively high costs of natural resource depletion. At the same time, many developing countries are endowed with significant natural resources. Rent (profit) absorption and reinvestment by government might be the crucial way of fostering their economic development, rather than relying on fickle aid and debt relief (Section 13.3.2).

Asset accounts, including environmental assets, are a more forward-looking tool of assessing capital maintenance. The availability of produced and natural capital indicates economic growth *potential*. However, measurement and valuation problems of different types of produced capital stocks, natural resource deposits (ranging from speculative to proven reserves) and a large variety of environmental sinks have prevented so far the regular compilation of these stocks in the national and environmental accounts. Section 7.1.2 showed the flaws of an attempt at comprehensive wealth measurement; it also stressed the importance of wealth for future economic growth and development.

In principle, asset accounts include the 'other asset changes' of natural and political disasters, discoveries, regrowth and revaluation. Contrary to exhaustible natural capital that is lost in destructive disasters, produced capital can be reproduced. The write-off of disastrous capital loss as economic disappearance under other asset changes does not affect national product and income, but a remedial increase in public and private capital formation does. As a consequence, the full (social and economic) cost of wars and natural disasters are generally underestimated overstating the net economic 'stimulus effect' of such events.

To answer the question of this chapter: most nations show weakly sustainable growth, with notable exceptions in the poorest countries. These countries did not have the means to reserve enough resources for capital maintenance, unable even to replace the wear and tear of produced capital. On the other hand, if we accept the normative Factor 4/10 targets, we may safely conclude that nearly all countries are still far away from relatively strong sustainability of economic growth.

Further analysis is needed to predict whether there is a good chance of reaching these targets within the next few decades as proclaimed by the Factor 4 stipulation. The following sections discuss first the analytic techniques that can reveal the full (direct and indirect) environmental impacts of different economic activities, as well as the driving forces behind these impacts. The next chapters will then extend these tools into examining future trends and policy scenarios.

10.2 What Are the Causes? Structural Analysis of Environmental Impact

Figure 10.2 connects the aggregate analyses of the physical MFA and the monetary SNA accounts. By plotting the time series of TMR and GDP next to each other the figure represents the overall outcome of hybrid accounting. Extrapolation of these indicators might or might not show (as indicated by different arrows in the figure) a sufficient dematerialization of future economic growth, and hence its potential ecological sustainability.

This section turns from the bird's-eye view of the economy and environment to the ground of structural analysis. The objective is to find which sectors and driving forces are responsible for environmental impacts. Three basic approaches can be distinguished:

- Comparison of sectoral economic performance with direct environmental impacts in environmental-economic profiles
- Modelling of direct and indirect impacts from economic activities by means of input-output analysis
- Time-series analysis of the driving forces behind environmental impacts by means of structural decomposition.

The value of such analyses depends crucially on meaningful aggregation and disaggregation of environmental impacts. However, weighting and valuation problems render most structural analyses of environmental concerns highly selective: typically they deal with one (notably CO_2) or selected pollutants only. The revised SEEA defends the 'legitimacy' of selecting 'the most urgent environmental concerns' in hybrid accounts as building 'a bridge between (aggregate) policy assessment and (underlying) policy research' (United Nations et al., in prep.). The structural flaw of this bridge is that selectivity adds a further assumption of 'representativity' for total environmental impact to the difficulties of comparing physical and economic indicators. Anticipating the building of safer bridges, which can carry the full load of environmental impacts, this section reviews briefly the use of hybrid accounts and input-output analysis in assessing selected pollutants from economic sectors.

10.2.1 Environmental-Economic Profiles

Hybrid accounts are a good starting point for generating 'environmental-economic profiles' (United Nations et al., in prep.). The profiles compare directly the sectoral contributions to GDP with their share of natural resource inputs and residual outputs. They also give a first indication of the structural causes for environmental impacts and of possible trade-offs between economic benefits and environmental deterioration.

The aggregated Table 10.2 describes in the first column Germany's post-industrial economy, where the service sector accounts for over two thirds of the value added generated in all production sectors. At the same time, the industrial sector of mining, manufacturing, construction and utilities is responsible for the bulk of environmental deterioration. As usual, CO_2 emission in column 2 may stand for environmental degradation. A further breakdown by industries reveals a similar pattern, with the energy sector contributing 2% to GDP but accounting for 40% of CO_2 emissions.

Table 10.2 presents energy and pollution intensity, abstracting from the level of economic activity by showing the environmental effects of production and consumption patterns. Generally, CO_2 emission intensity has declined since 1991. On the other hand, the large variation of both energy and pollution intensities among economic sectors indicates that environmental impacts are not only a matter of scale but also of structure. Section 10.2.3 attempts to quantify and compare these influences of structure and level. But first, let us explore the impact side of economic activity in greater detail.

Table 10.2 Environmental-economic profiles: Energy and CO_2 intensities, Germany 2000 (1991)

	Gross value added[a] (%) (1)	CO_2 emission (%) (2)	Energy consumption per gross value added[a] (Mj/€) (3)	CO_2 per gross value added[a] (kg/'000€) 2000 (4)	1991 (5)
Agriculture, forestry, fishery	1.3	1.0	5.8	350	581
Mining, manufacturing, construction, utilities	29.6	62.8	15.3	1020	1,194
Services	69.0	13.2	1.7	92	106
Total industries	100	(76.9)	5.8	370	489
Domestic private consumption		23.1	3.6	188[b]	240[b]
Grand Total		100			

Notes: [a]1995 prices. [b]CO_2 emissions per private household consumption in constant prices.
Source: Statistisches Bundesamt (2002, data from Annex Tables 18, 26, 34, 37).

10.2.2 Direct and Indirect Impacts: From Accounting to Modelling

The basic assumption of input-output analysis [FR 10.1] is, quite realistically, interdependence of different industries. Each industry may thus provide, in principle, inputs to all other industries. In this case it is not sufficient to measure only the immediate environmental impact from using a specific set of inputs in the production of a particular product. Rather, for an assessment of the total impact of a product, one would have to assess all the impacts resulting from the full chain of different inputs used – not only in the last-stage production process but also in all 'antecedent' industries.

Classic input-output analysis determines the total amount of an output x_i required for delivering final goods and services in an inter-industry exchange system. A 'squared' input-output table with fixed-coefficients linear production functions facilitates standard input-output analysis.[3] The inclusion of environmental impact generation activities, which use economic products as the 'inputs' into the impact process, allows then determining the full – directly and indirectly – generated impact per unit of a particular output, sector or the economy.

For example, the set of Equations (10.1) presents an n-product x_i (i = 1,2…n) and one-pollutant p production system, with given final demand y_i, a set of fixed input a_{ij} and pollution a_{pj} coefficients (j = 1,2…n), and the – unknown – total amount of pollution y_p generated by this system. Equation (10.2) is the solution of this system for all outputs $\boldsymbol{x}$, with $(\boldsymbol{I} - \boldsymbol{A})^{-1}$ (the inverse of the direct coefficients matrix $\mathbf{A}$) representing the matrix of total (direct and indirect) production and pollution coefficients:

$$\begin{aligned}
&(1-a_{11})x_1 - a_{12}x_2 - \ldots - a_{1n}x_n = y_1 \\
&-a_{21}x_1 + (1-a_{22})x_2 - \ldots - a_{2n}x_n = y_2 \\
&\qquad\vdots \qquad\qquad\qquad\qquad \equiv \boldsymbol{x} - \boldsymbol{Ax} = \boldsymbol{y} \qquad (10.1) \\
&-a_{n1}x_1 - a_{n2}x_2 - \ldots + (1-a_{nn})x_n = y_n \\
&-a_{p1}x_1 + a_{p2}x_2 + \ldots + a_{pn}x_n = y_p
\end{aligned}$$

$$\boldsymbol{x} = (\boldsymbol{I} - \boldsymbol{A})^{-1}\boldsymbol{y} \qquad (10.2)$$

[3] Solving the equations of an input-model model by calculating the inverse matrix of its input-coefficients requires a squared input-output table. Typically, squaring needs to be done when using the supply and use tables (SUT) of the national accounts as the database for input-output tabulations. The SUT are usually rectangular, since they combine an unequal number of industries and products. Converting industries into products or bundling products into industries to equalize the numbers and rows of the input-output system requires considerable estimation and data 'manipulation' (United Nations et al., in prep.).

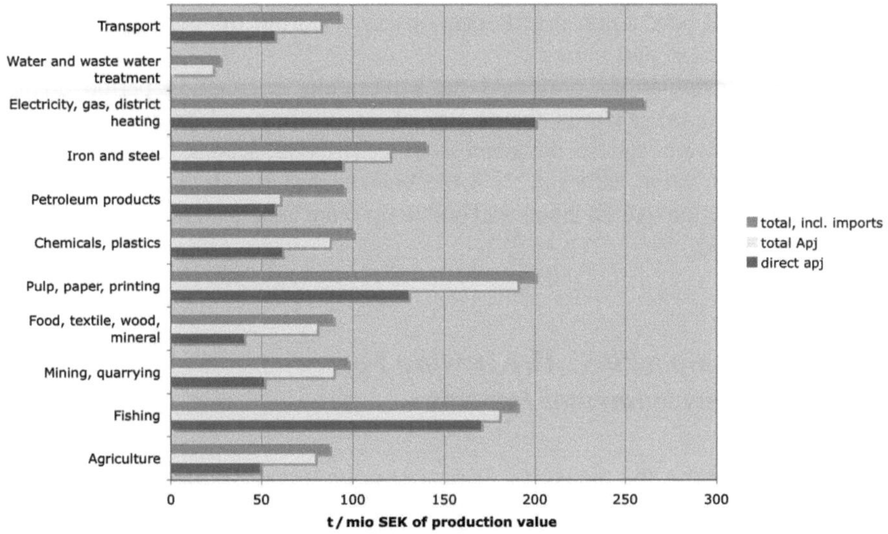

Fig. 10.3 Direct and total CO_2 emission coefficients, Sweden 1991
Source: Hellsten et al. (1999), table 9, p. 63; with permission by the copyright holder, Elsevier.

Figure 10.3 compares the inverted (total) pollution coefficients A_{pj} with the direct pollution coefficients a_{pj} for selected industries and one particular pollutant, p (CO_2). Based on a hybrid input-output table, the coefficients show CO_2 emission per million Swedish krona (SEK) of output. The differences between direct and total emissions are particularly high in transport, energy, paper and water industries. Note that for water supply and treatment there are no direct emissions, and the total stems thus from emissions embodied in the traced-back input chain. The A_{pj} are calculated for a closed economy; they represent therefore the emissions from domestic production and final use only. For an open economy such as Sweden's one should calculate the additional emissions generated by the production chain of imported goods in order to reflect the responsibility of domestic final demand for emissions generated not only at home but also abroad. Figure 10.3 indicates that the inclusion of imported petroleum products is responsible for a particularly high increase in CO_2 emissions from domestic demand for foreign products.

Assuming a fixed-coefficient homogeneous production function with constant returns to scale for each industry (the classic Leontief assumptions) is the smallest but nonetheless definite transition from descriptive accounting to modelling. The transition is small because it retains the observed technical coefficients as model parameters (unless obtained from data 'manipulation' when squaring the supply/use table). Emissions, generated in a 'whirlpool' (Dorfman, Samuelson & Solow, 1958) of preceding production processes, might lead back into past accounting periods when inputs were actually produced and used. It is far from obvious that the emission and production coefficients observed during the current accounting period hold for these past periods.

In general, calculations presented for a particular accounting period do not reveal this fact. In other words, the display of total (direct and indirect) emissions for a particular year and in the context of descriptive accounting (e.g. United Nations et al., in prep., tables 4.15, 4.16; Statistisches Bundesamt, 2002, p. 16) leaves the impression that all inputs and pollutants were generated during a particular accounting period.

10.2.3 Decomposition: The Driving Forces of Environmental Impacts

One way of making the measures of past environmental impacts more policy-relevant is to trace the driving forces that increased or decreased these impacts over an extended period of time. Structural decomposition analysis (SDA), applied to time series of input-output tables, is an analytical tool of teasing out the main causes of impacts as changes in the parameters and variables of these tables [FR 10.2].

The first step in applying decomposition analysis is to 'explain' the variable under scrutiny as the mathematical product of predetermined influences. The next step is to apply the product rule of differentiation. This obtains a difference equation, which explains the change of environmental impacts between two points in time as the sum of weighted changes in its driving forces. Depending on the weights, which can be taken from the base or end period, or can be combined, one can formulate alternative, equally valid decomposition forms.

De Haan (2001) applied SDA to an input-output table, which generalizes the above model (10.1) to include p_k (k = 1,2,...m) pollutants. The model can then be formulated as

$$\boldsymbol{p} = \boldsymbol{E}\boldsymbol{x}$$

$$\boldsymbol{x} - \boldsymbol{A}\boldsymbol{x} = \boldsymbol{y} \qquad (10.3)$$

with $\boldsymbol{E}$ denoting the diagonal matrix containing the emission coefficients e_{kj} of pollutant k per unit of output j.

In order to obtain the driving forces of structural change in final demand total demand, $\boldsymbol{y}$ can be expressed as the product of 'bridge coefficients' $\boldsymbol{B}$ and final demand $\boldsymbol{y}$ (Dietzenbacher & Los, 1998):

$$\boldsymbol{y} = \boldsymbol{B}\boldsymbol{y} \qquad (10.4)$$

The bridge coefficients b_{li} of matrix $\boldsymbol{B}$ measure the fraction of final demand in category l, which is spent on outputs from sector i.

Calculating the Leontief inverse (10.2) and denoting $(\boldsymbol{I} - \boldsymbol{A})^{-1}$ as $\boldsymbol{S}$, solves the model as

$$\boldsymbol{p} = \boldsymbol{ESBy} \tag{10.5}$$

Decomposition is applied to this basic multiplicative relationship, resulting, for instance, in the following decomposition form:

$$\begin{aligned}\Delta\boldsymbol{p} = \boldsymbol{p}(1) - \boldsymbol{p}(0) = \Delta\boldsymbol{E}\,\boldsymbol{S}(1)\boldsymbol{B}(1)\boldsymbol{y}(1) + \boldsymbol{E}(0)\Delta\boldsymbol{S}\,\boldsymbol{B}(1)\boldsymbol{y}(1)\\ + \boldsymbol{E}(0)\boldsymbol{S}(0)\Delta\boldsymbol{B}\,\boldsymbol{y}(1) + \boldsymbol{E}(0)\boldsymbol{S}(0)\boldsymbol{B}(0)\Delta\boldsymbol{y}\end{aligned} \tag{10.6}$$

The driving forces or determinants (Δ terms), which cause a change in the emission of pollutants, add up in 'exact' solutions (op. cit.) to 100% of the total emission change $\Delta\boldsymbol{p}$; they are

- $\Delta\boldsymbol{E}$, the change in eco-efficiency of production as a change in total emission coefficients (per unit of output)
- $\Delta\boldsymbol{S}$, the change in production technology as the change in the total input coefficients (per unit of output)
- $\Delta\boldsymbol{B}$, the change in the structure of final demand, notably in final consumption patterns
- $\Delta\boldsymbol{y}$, the change in the volume of final demand.

Equation (10.6) can be reformulated in 4! = 24 different decomposition forms, which are all equally valid on theoretical grounds. The problem is to choose from these forms. The apparent lack of a unique way of decomposing a time series into its causal determinants is a problem similar to (but aggravated by the number of determinants) the weighting of price indices: the production or consumption baskets of these indices may refer either to the base or the end period, or could be calculated as a mean of the two baskets. De Haan (2001) takes the averaging approach, creating a basket of weights, which is obviously difficult to interpret in an economic or environmental sense. However, this average seems to generate a relatively low standard error in the variation of the determinants (for the different decomposition forms).

Combining the two types of structural change in production and final demand, $\Delta\boldsymbol{S}$ and $\Delta\boldsymbol{B}$, Fig. 10.4 shows the results of decomposing the increase of total CO_2 emission (bold line) in the Netherlands over an 11-year period. The main contributor to this increase was economic growth, represented by the total-volume (of final demand) effect $\Delta\boldsymbol{y}$. Gains in eco-efficiency $\Delta\boldsymbol{E}$ offset some of this driving force. The structural effect had little influence.

The full SDA case study also produced results for the different economic sectors (de Haan, 2001, table 2). For some industries, structural effects can play a bigger role as, for instance, in the utility sector (emission decreasing influence) and air transport (emission increasing influence). Still, the volume of final demand maintains its dominating role. Eco-efficiency, on the other hand, decreased emissions especially in the chemical and air transport industries.

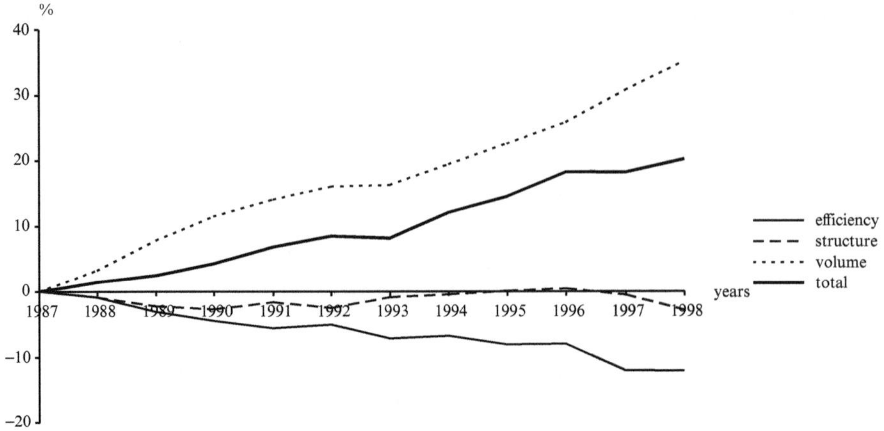

Fig. 10.4 Decomposition of annual changes in carbon dioxide emission, the Netherlands, 1987–1998

Source: de Haan (2001), fig.2; with permission by the copyright holder, Taylor & Francis: http://www.informaworld.com.

The calculation of direct and indirect environmental impacts and the decomposition of their annual changes into driving forces provide information about the origin and causes of pollution and natural resource use. Such information is useful for setting priorities for environmental protection and natural resource management. It also permits to relate environmental impacts and their changes to the results of economic activities responsible for these impacts. However, any assessment of sustainability by showing the delinkage of economic growth from environmental concerns requires comprehensive analysis of all significant environmental impacts. So far almost all input-output studies deal only with selected pollutants and a few energy sources. The reasons are lack of data and time/cost constraints for compiling or estimating the data. An aggravating factor is the notorious reluctance of corporations to provide information about their production processes and environmental impacts.

On the other hand, the policy relevance of input-output analyses and the closeness of input-output tables to the supply/use accounts seem to have seduced national accountants into embracing input-output modelling for greening their physical accounts. The revised SEEA-2003 presents thus the backward modelling ('backcasting') of direct and indirect environmental impacts and their decomposition as part of hybrid accounting, without discussing the underlying model assumptions (United Nations et al. in prep., ch. 4). There is perhaps some realization that simply juxtaposing environmental statistics and economic indicators in hybrid accounts does not really green the accounts themselves.

It is important, however, to clearly distinguish between more or less objectively observed statistics and modelled information. Even backcasting represents a distinct step away from reality into a hypothetical situation. An even greater step is modelling that reaches into the unknown future. The next chapters will make this step as transparent as possible by remaining closely linked to the national accounts and their input-output tables.

Further Reading

FR 10.1 Input-Output Tables and Analysis

Leontief (1951) developed input-output analysis for describing and explaining the structure of a market economy. He was also one of the first to introduce environmental pollution and its control into the input-output model (Leontief, 1970). The United Nations System of National Accounts (SNA) (United Nations et al., 1993) devotes one chapter (ch. XV) to the relationships between the supply-and-use table of the national accounts and input-output tabulations. A handbook of the SNA (United Nations, 1999) elaborates on these relationships (including squaring techniques). The handbook also presents a clear, practical review of the concepts, methods and compilation of input-output tables, including the calculation of a greened GDP (see Section 12.1).

Many economic and mathematical textbooks include descriptions of methods and uses of input-output analysis. Lahr and Dietzenbacher (2001) present extensions of input-output models into regional and environmental analysis. The European Network of Environmental Input-Output Analysis focuses on the use of input-output techniques for life cycle analysis: http://www.leidenuniv.nl/cml/ssp/projects/envioa/proceeding1.pdf.

FR 10.2 Decomposition Analysis

Decomposition techniques, applied to input-output tables, have become known as structural decomposition analysis (SDA). SDA can in fact be traced back to early work of Leontief: see for a brief history and methodological review Rose and Casler (1996). Typically, SDA explains changes in output and employment, but more recently the focus has been on natural resource use (especially energy) and pollution. Dietzenbacher and Los (1998) describe the techniques applied in the illustrative example of pollution in the Netherlands (de Haan, 2001). Rørmose and Olsen (2005) apply SDA to CO_2, NO_x and SO_2 emissions in Denmark; they also give a detailed description of the applied input-output and decomposition methods.

Review and Exploration

- What do the data tell us: has economic growth been sustainable? Evaluate the results of green (physical and monetary) accounts for the assessment of sustainability.
- What is the purpose of structural decomposition analysis of environmental impacts?
- Why is it important to distinguish between accounting and modelling? Is the calculation of total (direct and indirect) emissions from an input-output table a modelling or an accounting exercise?
- Explain the difference between *ex post* (descriptive) and *ex ante* (predictive) modelling.

Chapter 11
Prediction: *Will* Economic Growth Be Sustainable?

Looking back in Ch. 10 at the causes for past environmental impacts makes it possible to stay close to the observed data while using the powerful tools of input-output analysis. The real challenge for sustainability policies is however to predict future trends and anticipate the success or failure of policy options. The trade-off of taking the analysis to the future is the need for additional assumptions that remove the models further from reality. The focus of this book on quantitative assessment justifies concentrating on those analyses that build upon empirical data.

Econometrics, as its name implies, is modelling that remains closest to measurement (of parameters and trends). The well-known and fiercely contested *Environmental Kuznets Curve hypothesis* about the correlation of economic growth and improvement of the environment provides a good example for econometric tests. The testing of this hypothesis is mostly applied at the national level. At the global level, the popular *Limits-to-Growth model* also uses some econometric parameter testing. It is a dynamic simulation model, criticized, however, for mostly relying on untested feedback loops and exponential growth assumptions. The model has become known for its rather pessimistic baseline scenario of 'overshooting' environmental limits and consequential social collapse. Chapter 12 analyses how these limits can be addressed while seeking optimality and sustainability of economic activity.

11.1 Econometrics: The Environmental Kuznets Curve Hypothesis

Do we need 'economic growth that is ... socially and environmentally sustainable' (WCED, 1987), or is it 'qualitative' development (Daly, 1996)? The first statement pleads for sustainable development with, the second without, 'quantitative' economic growth. As discussed in Ch. 2, the two views reflect the prevailing dichotomy between ecological economists, who see economic growth as the cause of environmental degradation, and environmental economists, who see it – with some modification – as the solution. The continuing discussion of the Environmental

Kuznets Curve (EKC) hypothesis (cf. Section 2.2.2) reflects the search for evidence in support of either view. The protagonists of the hypothesis claim that there is 'no evidence that environmental quality deteriorates steadily with economic growth' (Grossman & Krueger, 1995). Rather, they find an inverted-U relationship where environmental impact increases at low levels of national income and decreases at higher ones.

This section reviews critically the testing of the EKC hypothesis since it is implicitly, and sometimes explicitly, at the heart of the environmental-economic dispute. The relatively simple and transparent 'metric' analysis points to the crucial role of empirical data; it reveals also the limits of correlational methods for explaining the causes of environmental trends.

11.1.1 Regression Analysis: Testing the Hypothesis

Figure 10.2 puts question marks behind the decoupling of primary materials use from economic growth. Regression analysis is the tool of assessing empirically the potential linkage or delinkage of economic growth and environmental quality. For testing the EKC hypothesis, cross-country data typically establish the parameters of the regression function. The next step is to apply the parameters to time series of the explanatory variable of GDP (or national income) per capita. The emissions or concentrations of different pollutants E_i represent usually the environmental impacts during a year of income (Y = GDP p.c.) generation. A quadratic equation then specifies the EKC hypothesis as

$$E_i = \beta_1 + \beta_2 Y + \beta_3 Y^2 + \varepsilon_i \tag{11.1}$$

As in most regression analyses the error term ε represents all those influences (in this case on changes in environmental quality from pollution), which are excluded from the analysis as unknown, peripheral or due to data deficiencies. This simplification, inherent in a 'reduced-form' equation of one dependent and one explanatory variable, has been the focus of critique and rejection of the EKC.

Adding a further cubic term for Y allows to test the case of relinkage, introducing a possible second turning point after an initial EKC phase:

$$E_i = \beta_1 + \beta_2 Y + \beta_3 Y^2 + \beta_4 Y^3 + \varepsilon_i \tag{11.2}$$

Figure 11.1 illustrates the cases of EKC confirmation (the inverted U) and rejection (because of relinkage). Assuming ε to be constant, we

- Obtain, for $\beta_4 = 0$, $\beta_2 > 0$ and $\beta_3 < 0$, the EKC of graph A
- May obtain, for β_2, β_3 and $\beta_4 \neq 0$, the so-called N-curve of relinkage (graph B)
- Reject the EKC hypothesis for $\beta_2 \neq 0$ and $\beta_3, \beta_4 = 0$, owing to a linear positive or negative association of environmental impact and economic growth.

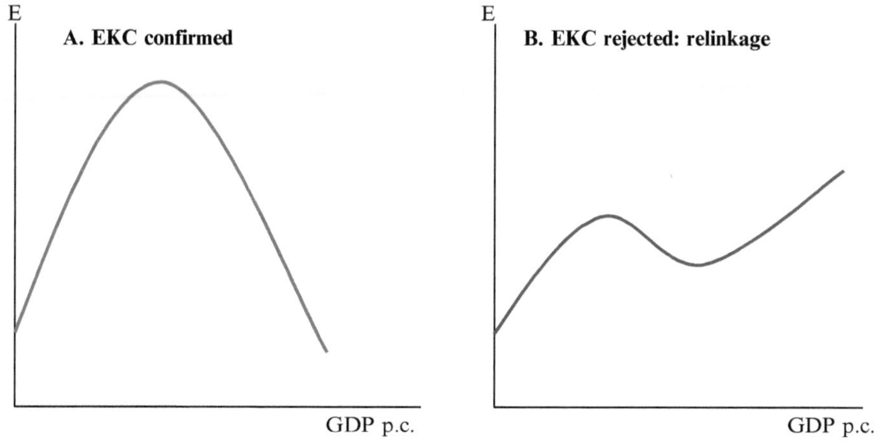

Fig. 11.1 EKC hypothesis – confirmed and rejected

11.1.2 Results: Rejecting the Hypothesis?

Numerous authors set out to refute the EKC hypothesis – and with it the belief that economic growth and its market forces could be beneficial for the environment [FR 11.1]. The defence, on the other hand, seems to be less pronounced since economists would rather deal with environmental impacts as a distortion of short-term market equilibrium (see Ch. 12).

The classic study by Grossman and Krueger (1995) finds an EKC relationship for urban air and river quality. Based on panel data from UNEP's Global Environmental Monitoring System, the study includes ambient concentrations of SO_2, smoke, heavy particles, heavy metals, nitrates, BOD, COD (biological, chemical oxygen demand) and faecal coliform. The authors reject the hypothesis for faecal coliform only. In all other cases they find an inverted-U relationship with a turning point at less than $8,000 per capita GDP (in 1985 US$). Relinkage for some of the variables at very high income levels is explained away as stemming from poor data at these levels.

The authors also refer to other studies that confirm their findings and in fact extend it to deforestation, CO, and access to safe water. They cite, in particular, a World Bank (1992) report, which establishes an EKC relationship for SO_2 and suspended particulates for high-, middle- and low-income cities, but rejects the hypothesis for CO_2 and municipal waste. Summing up, Grossman and Krueger (1995) assert that while their study covers 'relatively few dimensions of environmental quality' it is 'the most comprehensive possible, given the limited availability of comparable data from different countries'.

Most other studies find the EKC relationship to hold only for local air pollutants with low pollution control cost, rejecting the hypothesis for global pollutants such as CO_2. Turning points of per capita income also vary widely. For instance, SO_2 reaches the maximum at values (at constant 1985 prices) between $823 and $10,800 (Barbier, 1997). [FR 11.1]

Where countries did achieve a turning point or slow-down in environmental degradation, one could imagine using their experience for 'tunnelling through' the actual or incipient EKC (Fig. 11.2). For low-income countries such tunnelling would require the transfer of environmental knowledge and technologies – a common refrain in nearly all international environmental conferences.[1] Unfortunately such transfer – 'on concessional and preferential terms' (United Nations, 2003) – has in general not come true. It remains to be seen if developing countries can overcome the obstacles (patents) to the transfer of technology and/or create their own 'eco-techniques' (cf. Section 3.2.3).

The use of an Index of Material and Energy Intensity (IMEI), which averages consumption of selected materials per GDP, yields the results of Fig. 11.2, shown in a stylized format. The index pictures a possible delinkage of economic growth and environmental pressure for industrialized countries (Japan, UK, USA) and linked growth and environmental pressure for newly industrializing countries (NICs: Indonesia, Mexico, Republic of Korea, Thailand).[2] The trend for the developing

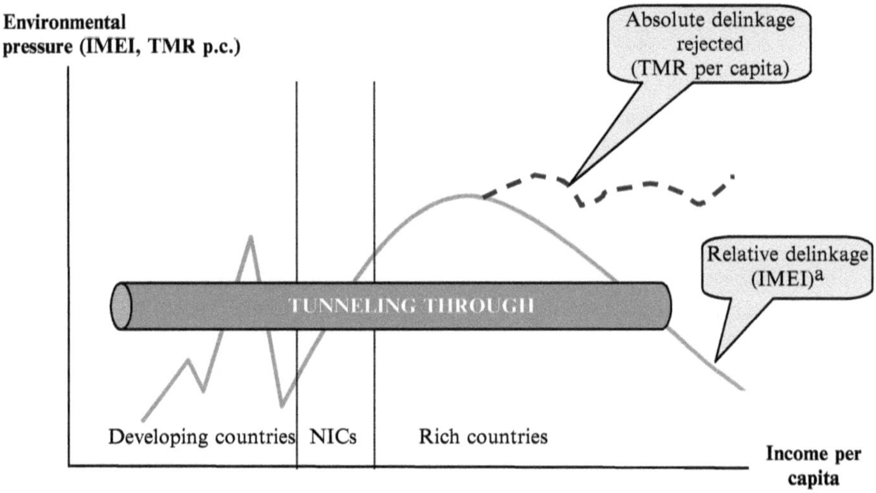

Fig. 11.2 Inconclusive evidence for the EKC hypothesis
Note: [a]Relative delinkage is based on an Index of Materials and Energy Intensity (IMEI) (Bartelmus, 1997b, pp. 333–335); materials included are cement, steel, freight and energy carriers (source of data: United Nations Statistics Division).
Source: Bartelmus (2000), fig. 4B.

[1] Cf. chs. 34 and 37 of Agenda 21 of the Earth Summit (United Nations, 1994) and throughout the Plan of Implementation of the Johannesburg Summit (United Nations, 2003).

[2] The reason might be fear of the NICs to lose their competitive edge by complying with higher environmental standards (cf. Section 14.1). Accelerated economic growth in emerging economies such as China and some transition (to market systems) countries make the NIC classification somewhat redundant. Figure 11.2 serves therefore just as an illustration for an international comparison of de- and (re)linkage of environmental pressure from/with economic growth and potentials for some technological leap-frogging of developing countries.

countries of the Philippines and Costa Rica appears to be erratic. At first sight, Fig. 11.2 seems thus to confirm an EKC effect for NICs and industrialized countries. However, when taking absolute levels of TMR (per capita) for industrialized countries, rather than material intensity, into account, the EKC effect disappears (dashed line). The result is a 'relative delinkage' only of TMR from economic growth (cf. Section 6.3.2).

11.1.3 Critique

The critique of the EKC addresses the econometric analysis and its database, on the one hand, and the interpretation of a statistical regression as a cause-effect relation, on the other hand. In particular, it includes the

- Application of cross-country parameter estimates to time series analyses of individual countries
- Reduced-form analysis of associating environmental quality solely with economic growth, ignoring underlying exogenous influences
- Selectivity in choosing the environmental impact variables, covering only a few pollutants
- Cause-effect interpretation, assuming economic growth to be the cause of environmental quality change.

The applicability of cross-panel estimates to time trends is a general problem of regression analysis. Switching from cross-country analysis, comparing countries at different income levels, to modelling the time paths of economic growth and environmental quality in individual countries ignores a time trend brought about by exogenous forces. This trend may shift the EKC in different directions and may turn a national EKC into a different shape such as an N-curve (de Bruyn, van den Bergh & Opschoor, 1998).

The reduced form of explaining the dependent variable, environmental quality E, by one explanatory variable, income or GDP (Y), in Equations (11.1, 11.2) distorts the relationship $E = f(Y)$ when exogenous forces affect environmental quality. Such forces include structural changes in the composition of inputs, outputs and pollution, technological innovation, high income elasticity of household demand for environmental quality, foreign trade in natural resources, and autonomous economic and environmental policies.

Section 10.2.3 described the technique of decomposition to identify separately the significance of these driving forces. The critique of the EKC frequently cites the results of decomposition studies (e.g. de Bruyn, 1997; Komen et al., 1997), when confirming the relevance of technological progress, and scale changes of economic activity. However, in the context of testing the EKC, income-induced policy responses to environmental impacts emerged as the dominant force behind the EKC hypothesis.

Mainstream economists view environmental policy response as a result of economic growth: 'As incomes rise, the demand for improvements in environmental quality will increase, as will the resources available for investment' (World Bank,

1992). Others contest this view, pointing out that *laissez-faire* induced economic growth does not necessarily bring about the required policy reform. In fact, studies in the Netherlands find that 'reductions in emission in developed economies, as forecasted by the EKC, may ... be realized, not due to, but rather in spite of economic growth' (de Bruyn et al., 1998).

As with many environmental assessments, selecting only a few or one placeholder variable(s) cannot represent comprehensively environmental quality, natural resource depletion, and their health and welfare effects. To overcome this selectivity some EKC tests use material flow aggregates as comprehensive measures of environmental pressure. Mostly, they reject the hypothesis (e.g. Seppälä et al., 2000). Physical aggregates do not measure, however, actual environmental impacts; they can take account, though, of the import or export of sustainability from/to pollution and resource 'havens'. Alternatively, one could imagine using overall environmental cost estimates such as those of the SEEA.

In summary, all these studies render the results of EKC testing quite inconclusive. Some EKC relationships seem to exist for selected pollutants. Most authors also conclude that out of the above-discussed driving forces, income-induced environmental policy is the dominant force for any environmental improvement. This does not necessarily mean automaticity in policy response, and active policy intervention would still be called for (Barbier, 1997). Chapter 12 will evaluate the potential outcomes of this policy advice by means of prescriptive modelling. But let us look first at another prediction, which claims to provide evidence for the transgression of *global* environmental limits.

11.2 Simulation of Non-sustainability: The Limits-to-Growth Model

11.2.1 Model Features and Results

The first edition of the *Limits to Growth* (LTG) report in 1972 created uproar by the clash of environmentalist acclamation and mainstream economists' rejection [FR 11.2]. Disdaining economic theory as a puzzle-solving marginalization of the environment (cf. Section 2.1), environmentalists rely on metaphors such as the metabolism of society and on symptomatic evidence for reaching the limits of carrying capacities. Lacking a theoretical system like market equilibrium, they embraced the LTG model as a systemic confirmation of their conviction about transgressions of environmental limits. Commenting on the second (1992) edition of the LTG model, von Weizsäcker et al. (1997) assert: 'the Meadowses may be right' ('are right' in the German original!). Economists were quick to reject the Malthusian 'model of doom' (Cole et al., 1973) because of its neglect of counteracting market forces and technological innovation.

This section does not take a definitive stand on who is really 'right' but leaves it to the reader (and further reading) to evaluate the critique of the model. The reason is that many of the model assumptions are quite normative in nature. Still, the model is discussed here because it

- Serves as a main underpinning, if not *raison d'être*, of ecological economics, which otherwise relies on sets of diverse indicators of environmental impacts
- Can be seen as an attempt to overcome the hardly quantifiable complexities of optimal and sustainable growth analysis (see Section 12.3.2) by means of simulating 'plausible' feedback loops[3]
- Raises basic questions about the use of sustainability modelling in policy analysis, further discussed in Chs. 12 and 13.

The model is a computer-based dynamic simulation of the interaction among population, industrial and agricultural production, natural resource use, and pollution. Figure 11.3 describes the interactions as direct influences and feedbacks among stocks

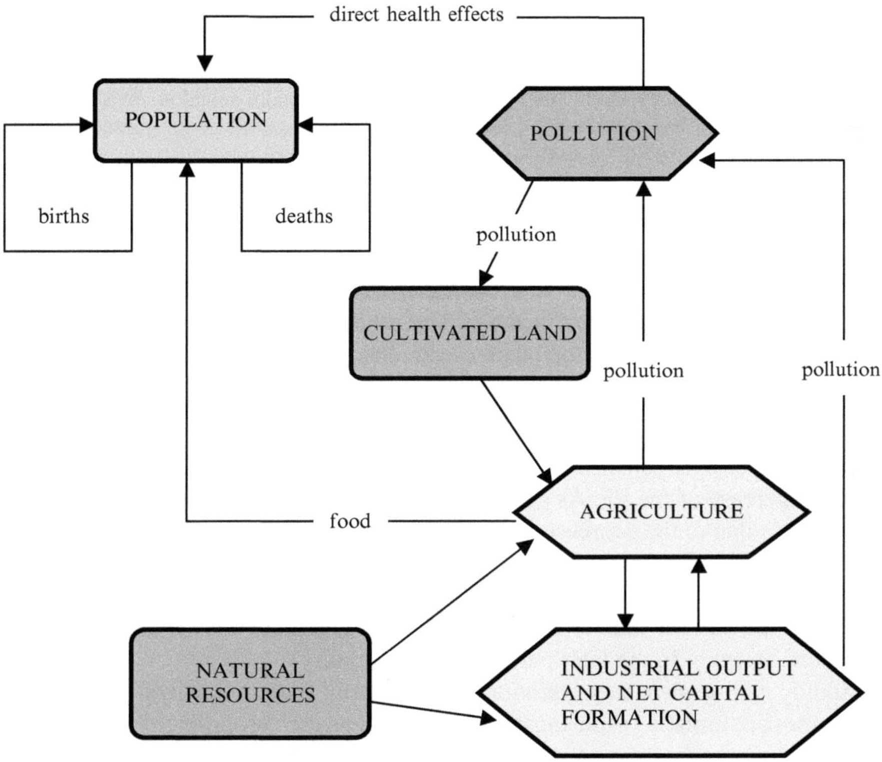

Fig. 11.3 Limits to growth model – components and interactions
Source: Meadows et al. (2004), figs. 4 and 5 and 4–6, modified for a better presentation of stocks and flows.

[3] Unless otherwise stated quotes refer to the third edition of the LTG model (Meadows et al., 2004). Note that there is little change in the model structure over the three editions (op. cit., appendix 1).

of population, natural resources, land, and flows of outputs, capital formation and consumption, and pollution.[4] In mathematical terms the model consists of differential and difference equations, which specify the time paths and interactions of the variables.

In response to the doomsday interpretation of the model, the authors stress that the purpose is 'not to make point predictions, but rather to understand the broad sweeps, the behavioural tendencies of the system'. Their objective is to present scenarios of alternative social responses, although some of the (more optimistic) responses are characterized as 'unrealistic' or 'wishful thinking'.

A closer look at the overall results in the main scenarios may shed light on the question of prediction vs. scenario building. For an aggregative analysis the model employs the above-described (Sections 3.1.2, 5.2) indices of human development (HDI) and the ecological footprint (EF) as measures of human welfare and overall environmental impact, respectively. Figure 11.4 shows the development of welfare and environmental impact from 1900 to 2100 for key scenarios:

- The *business-as-usual* scenario (scenario 1) leads to the collapse of society and economy, famously described in the first LTG report (Meadows et al., 1972):

  > If the present growth trends in world population, industrialization, pollution, food production, and resource depletion continue unchanged, the limits to growth on this planet will be reached sometime within the next one hundred years. The most probable result will be a rather sudden and uncontrollable decline in both population and industrial capacity.

The figure shows that continuing exponential growth in the model variables will bring about decline of our standards of living (welfare) within the next 20–30 years, and down to year-1900 levels in 2100. Environmental impact declines correspondingly with reduced economic activity. The other scenarios show the progressive introduction of positive factors that could delay or prevent the socioeconomic collapse:

- The *doubling-of-resource-availability* scenario (scenario 2) delays collapse by a decade or two. It generates, however, a 'global pollution crisis' reflected in the steep increase of the EF. As a consequence, land fertility declines and mortality increases due to the decrease in food production.
- The further introduction of *resource-saving and environmental protection technology* (scenario 6) generates a more positive picture of human welfare oscillating around current standards of living. Ultimately, however, the authors believe decline to begin in the later part of the 21st century.
- Adding *zero-growth in population and industrial output* (scenario 9) generates the most optimistic result (apart from an unrealistic 'no-problem' scenario). It leads to a steady-state economy with welfare slightly above year-2000 levels and no indication of decline. Environmental impact stabilizes at a low level.

[4] The figure is a simplification of the many feedback loops used in the model; it also distinguishes more clearly between stocks (rounded rectangles) and flows (hexagons) and omits non-material 'knowledge' flows. This should help understand the model's stock and flow variables in more conventional economic terms.

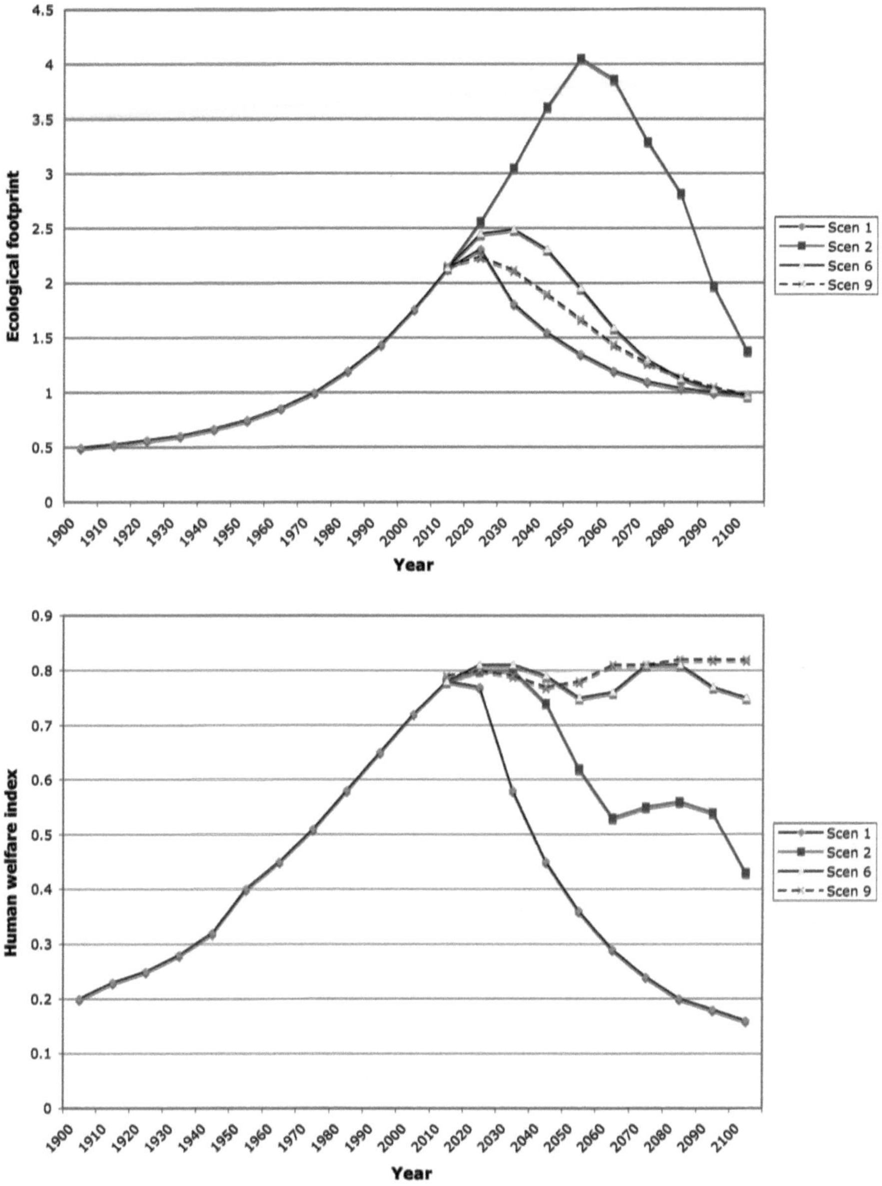

Fig. 11.4 Selected scenarios of the LTG model
Source: Meadows et al. (2004); with permission by the copyright holder, Dennis Meadows.

11.2.2 Critique and Counter-critique

As discussed in Ch. 5, we can question the capability of the HDI and EF to capture human welfare and environmental impact. A closer examination and interpretation would have to look at the underlying trend variables of population, food consumption, output, life expectancy and pollution. The focus here is, however, on the fundamental critique, in particular from economists, who see most of their own analyses ignored by modelling 'engineers and scientists' (Nordhaus, 1973).

Table 11.1 summarizes the main criticisms against the model. They refer to the original MIT model and its application by Meadows et al. (1972). Only limited changes were introduced since then, and the main critique appears to hold for the 2004 update, including the

- Lack of empirical validation of functional relationships, famously dubbed as 'measurement without data' (Nordhaus, 1973)

Table 11.1 Assumptions, purpose and critique of the LTG model

LTG Assumptions and Purpose	Critique
1. The basic assumption: exponential growth in a limited world	
- Exponential growth of population and the material economy	- Declining population growth rates (from falling birth rates and AIDS) in the second half of the 21st century (United Nations projections); transition to a (non-material) service economy of rich countries (OECD business and policy forum)
- Limits: natural resource stocks and increasing exploitation costs	- Falling resource prices indicate decreasing resource scarcity due to discovery and substitution (Simon: see FR 11.2)
- Limits: source and sink functions (carrying capacities) already exceeded (in terms of the EF)	- Questionable concept and estimates of the EF (see Section 5.3)
2. Purpose: prediction or warning?	
- No prediction of the future; rather: range of scenarios	- 'Unrealistic' (optimistic) scenarios ruled out (Meadows et al., 2004); 'most likely' model outcome is 'collapse' (Meadows et al., 2004)
- System dynamics for assessing the long-term causes and consequences of growth	- Causes and consequences predetermined by mental model (Meadows et al., 2004) of the modellers
3. Database: modelling without data?	
- World system dynamics 'can be understood and discussed independently of the precise numerical assumptions of any model' (response to Cole et al., 1973)	- 'Not a single relationship or variable is drawn from actual data or empirical studies'; 'subjective plausibility rather than 'empirical validation' (Nordhaus, 1973)
- Best data available used; improved data base for LTG update	- Cross-sectional data cannot represent time series (Cole et al., 1973)
4. Technology – the saviour?	
- Man is not 'omnipotent' (response to Cole et al., 1973); negative impacts of technology	- Main general critique: ignoring the power of technological advance, prompted by market forces (see pt. 5)
- Adaptive technologies on their own are too late and not enough	- Inclusion of 'continuous technological change' postpones catastrophe 'indefinitely' (Cole et al., 1973)

(continued)

Table 11.1 (continued)

5. Economics: the force of market behaviour	
- 'Supplement cleverness [of *homo oeconomicus*] with wisdom' [of the modellers?] (Meadows et al., 2004)	- 'Insentient man' (Nordhaus, 1973) of the model ignores the economic man (capable of optimization, adaptation and control)
- No explicit price mechanism due to the intermediary role of short-term price effects; price omission avoids 'delays and inaccuracies ... in real market systems' (Meadows et al., 2004); prices and costs implicitly included in the feedback loops of technology impacts	- No price mechanism for adjusting to relative scarcities by means of price-induced substitution, discovery and technological advance (e.g. Nordhaus, 1973)
- No 'wonderful' Cobb-Douglas production function in the 'real world' (Meadows et al., 2004)	- No concepts of output and production function (Nordhaus, 1973)
6. Social response: policy intervention and value change	
- Adaptive change in social values and other social factors cannot be represented in the model; the model lesson is the need for deliberate social choices to limit growth	- Technocratic (mechanistic, computerized) system ignores adaptability of individuals and the public system through changes in goals, values and norms

- Denial of adaptive behaviour by economic agents and governments, reacting to natural resource scarcities and pollution through innovation, substitution, legislation and regulation.

In response, the authors point out that their intention is not to predict any precise catastrophic development. Rather, they want to show potential scenarios that may occur if there are no radical changes in social values and choices. Nevertheless, they deem the more optimistic scenarios unlikely and reject faith in the ability of markets and human ingenuity to prevent the collapse of society. The 'most likely' mode of behaviour of the model is thus 'overshoot and collapse'. The authors apparently believe that we are already beyond the limits – a point that is pressed in the descriptions of environmental impacts. Consequently, the model seems indeed to be less prediction than warning about imminent disaster.

Taking the model's warnings seriously would require restricting the growth of both, population and the 'physical' economy. The 'ideal' would be 'to increase "human welfare" while ensuring that the "ecological footprint" ... stays below the global carrying capacity'. The most optimistic model scenario seems to be content, though, with attaining long-term *equilibrium* in the sense of constant levels of the trend variables (scenario 9). The authors hasten to stress that this does not mean zero growth, as widely perceived, but different 'kinds of growth and purposes for growth', i.e. 'qualitative development'. Still, 'growth in population and capital must be slowed and eventually stopped'.

All in all, the LTG model does not succeed in bridging the environmental-economic dichotomy. Rather it seems to have added fuel to the sustainability debate [FR 11.3]. The next chapter discusses the extent to which generic proclamations on growth or no-growth can be assessed in dynamic models, which incorporate goals of sustainability *and* optimality.

Further Reading

FR 11.1 EKC hypothesis

Two journals happen to devote a special issue to the topic: *Environment and Development Economics* 2 (1997) and *Ecological Economics* 25/2 (1998). *Environment and Development Economics* aims at assessing the underlying factors of the growth-environment relationships. *Ecological Economics* reads more like an assembly of arguments for rejecting the hypothesis. Much of the critique of Section 11.1.3 is taken from this issue. Barbier (1997) and Perrings (1998) argue that the turning points of most studies are beyond current national income levels. Barbier (1997) gives a concise review of evidence for and against the hypothesis. Beckerman (1992) is usually cited as an advocate of economic growth, which eventually generates enough prosperity for funding environmental protection. Stern reviews the literature and case studies, finding little evidence for the existence of the EKC: http://www.ecoeco.org/publica/encyc_entries/Stern.pdf.

FR 11.2 The Limits-to-Growth Model

There are now three editions of *Limits to Growth* (LTG): (1) the original publication with its famous prediction of Malthusian collapse (Meadows et al., 1972), (2) the 20-year update, which confirmed the prediction and in fact argued that we have already transgressed the limits (Meadows et al., 1992), and (3) the latest 30-year update, which 'is still making basically the same points as the two previous books', but in a 'more understandable' manner (Meadows et al., 2004).

A research unit of the University of Sussex provided the main critique against the 1972 version, with a response by the LTG authors (Cole et al., 1973). Nordhaus (1973) translated the LTG model into a simplified – economic – model, in order to test the model assumptions in more established economic terms and variables; he also demonstrated alternative (non-disastrous) growth options. Nordhaus and others (e.g. Beckerman 1992) criticized, in particular, the denial of market solutions responding to environmental scarcities and damages.

FR 11.3 Doomsayers and Doomslayers – A Bet

The debate between the neo-Malthusian 'doomsayers' and 'doomslayers' was trivialized in the famous bet on commodity prices, decreasing with reduced demand (Julian Simon) vs. increasing prices with depletion and greater scarcity (Paul Ehrlich). Ehrlich lost the bet. Ed Regis applauds unabashedly Simon's writings, but in a quite amusing presentation: http://www.wired.com/wired/ archive/5.02/ffsimon.html. A more concise description of the Simon-Ehrlich wager is: http://en.wikipedia.org/wiki/

Ehrlich-Simon_bet. Note that Simon's writing also inspired Lomborg's critical review of Malthusian trends, including those of the LTG model (Lomborg, 2001).

Review and Exploration

- Is economic growth a remedy for environmental deterioration? What is the evidence for the EKC hypothesis?
- Is decomposition analysis an alternative for testing the EKC hypothesis?
- Does the EKC hypothesis take account of policy response to environmental impacts?
- What are the advantages and chances of developing countries for tunnelling through the EKC?
- Compare the EKC hypothesis with the findings of the LTG model.
- Which LTG scenario is most likely to occur? Why?
- Compare and assess the critique of the LTG model. Who is 'right'?

Chapter 12
Policy Analysis: *Can* We Make Growth Sustainable?

Computable general equilibrium (CGE) models promise quantification of the effects of environmental policy on economic behaviour and on reaching a new general equilibrium. Introducing environmental costs or constraints into input-output analysis permits calculating a greened (more sustainable) GDP, which could be attained by environmentally sound production processes. Maximizing GDP under environmental and other constraints in linear programming models is more ambitious and also less realistic. Even more removed from reality is finding the time path for moving from one state of equilibrium to another and selecting the optimal state of equilibrium in dynamic and optimal growth analysis. However, abstract models enrich the discussion of new concepts and paradigms – notably of sustainability – and help define them with greater precision.

12.1 Environmental Policy Measures in General Equilibrium and Input-Output Analysis

Under perfect market conditions, individual optimal behaviour of utility and profit maximization brings about Pareto-efficient market equilibrium. As any textbook of economics will explain, in such equilibrium no person's economic well-being can be improved without impairing the well-being of someone else. Unfortunately, one of the main market failures, which upset perfect conditions, is the generation of environmental externalities. As discussed in Section 2.3.2 and Annex I, re-attaining Pareto-efficiency would require the internalization of these social costs into the planning and budgeting of those who generated them.

Green accounting facilitates measuring environmental social costs and allocating them to responsible economic agents. Accounting for past economic activities and their impacts cannot explain or predict, however, the behaviour of economic agents after being prompted into cost internalization. This requires modelling the extent to which households and enterprises absorb, shift or avoid the new cost burden by full or partial cost absorption and changes of the scale and pattern of production and consumption. Price changes, induced by these responses, might then bring about a new market equilibrium.

CGE models typically consider induced price changes as the result ('shock') of an exogenous policy measure. The comparative-static approach to CGE modelling assesses only the results of the new market equilibrium – rather than the time path towards it. It compares the new equilibrium with the original one in terms of outputs, income and consumption. In contrast, dynamic analysis (Section 12.3) introduces inter-temporal optimizing behaviour, usually for capital accumulation by producers who anticipate future cost and price changes [FR 12.1].

Focusing on the quantifiability of the environment-economy interaction justifies seeking out those models that make direct and transparent use of observable data. The fixed input-output relationships and the condition of market clearance of the basic input-output model represent a simple, if not the simplest, CGE model. While other models also carry the CGE label, the closeness of input-output tables and input-output analysis has made this approach a favourite of environmental-economic analysis. The Leontief prototype can also explain important extensions into linear programming and dynamic optimal growth analysis.

Basically there are two options for incorporating environmental concerns into input-output analysis: (1) introducing environmental impacts as joint outputs of production and costing the tools of mitigating the impacts, and (2) introducing, a priori, limited environmental capacities for natural resource supply and waste disposal to/from economic activity. The first option adds a further environmental protection industry to the basic Leontief model. The second option introduces constraints for economic activities. Linear programming extends the second option by combining restrictions and optimization.

12.1.1 Environmental Cost Internalization in the Static Leontief Model

The simple Leontief model of Ch. 10 is the starting point for showing how environmental protection and its cost may affect sectoral output and national income or GDP in a new market equilibrium. Leontief (1970) himself laid the foundation for such analysis by introducing, besides the generation of pollutants, an 'anti-pollution industry' and by pricing its activity in terms of additional labour cost.

Before discussing this approach let us go back to the physical input-output model (Equations 10.1, 10.2). Extending the model to p_k (k = 1, 2 … m) pollutants (without any pollution from final demand) and distinguishing between the generation of n conventional outputs $\boldsymbol{x}$ and m pollution abatement activities $\boldsymbol{x}_p$ obtains a new model

$$\begin{aligned} \boldsymbol{x} &= \boldsymbol{A}\boldsymbol{x} + \boldsymbol{A}_p\boldsymbol{x}_p + \boldsymbol{y} \\ \boldsymbol{x}_p &= \boldsymbol{E}_1\boldsymbol{x} + \boldsymbol{E}_2\boldsymbol{x}_p - \hat{\boldsymbol{y}}_p \end{aligned} \tag{12.1}$$

where $\boldsymbol{A}_p$ is the direct input-coefficient matrix of pollution abatement activities, $\boldsymbol{x}_p$ is the vector of eliminated or reduced pollutants, $\boldsymbol{E}_1$ and $\boldsymbol{E}_2$ are the direct pollution coefficients for pollutants from production and abatement activities, and $\hat{y}_p \in y_p$ is the 'uneliminated' or 'tolerated' amount of pollution, rather than a 'demanded' one (Leontief, 1970). In actual model applications this amount would have to be specified in accordance with national pollution standards.

It is easy to show that – with given total outputs $\boldsymbol{x}$ – the fixed technical coefficient matrix reduces the final use of products $\boldsymbol{y}$ by the amount of products diverted to intermediate consumption in abatement activities (United Nations, 1999). In a hybrid input-output model (with all product flows measured in monetary units, and only pollutants measured in physical ones) the original GDP (without pollution abatement: $\boldsymbol{y}_0 = \boldsymbol{x} - \boldsymbol{Ax}$) is higher than the 'greened-economy' GDP ($\boldsymbol{y}$), which includes pollution abatement. Substituting $\boldsymbol{x}$ from the first equation in (12.1) into the second one and solving for $\boldsymbol{y}$ obtains

$$\boldsymbol{y} = \boldsymbol{x} - \boldsymbol{Ax} - \boldsymbol{A}_p (\boldsymbol{I} - \boldsymbol{E}_2)^{-1} \boldsymbol{E}_1 \boldsymbol{x} < \boldsymbol{y}_0 = \boldsymbol{x} - \boldsymbol{Ax} \tag{12.2}$$

The original $\boldsymbol{y}_0$ is lessened by $\boldsymbol{A}_p(\boldsymbol{I} - \boldsymbol{E}_2)^{-1}\boldsymbol{E}_1\boldsymbol{x}$, which is the amount of the total (direct and indirect) inputs used in abatement activities.

The basic physical and hybrid models do not incorporate price formation affected by environmental cost internalization. An input-output model, which simply adds pollution abatement as one or more additional production activity(ies), is thus not really presenting a new market equilibrium. It reflects technological capacities of dealing efficiently, i.e. without product shortages and surpluses, with an additional production activity. In such a system no price/cost-induced changes in supply and demand take place. In order to introduce market behaviour of economic agents, one has to introduce pricing and costing of economic activities, including pollution abatement. According to the polluter-pays principle the necessary abatement cost should be allocated to pollution-causing activities as a proxy for the environmental damages they cause (Sections 8.1.2, 13.3.2).

Environmental costs can be either introduced exogenously, for instance as the 'historic' maintenance cost compiled by the SEEA, or modelled internally ('shadow-priced') with regard to complying with desirable environmental standards. In both cases, the basic Leontief ruling is that total costs must equal total income generated. Such ruling can be justified in perfect competition and a fixed-coefficients linear production system where average costs equal marginal ones, which in turn equal equilibrium prices. Leontief (1970) thus introduces a primary production factor, labour L, whose remuneration, at a particular wage rate (put at 1$ in Equation 12.3), represents national income and covers value added v_i of the n production sectors[1]:

$$a_{l1}x_1 + a_{l2}x_2 + \ldots + a_{ln}x_n = L = \sum v_i \tag{12.3}$$

[1] This resembles the basic accounting identity of factor income and net output/value added. However, in a perfect-competition equilibrium no profits are made, distinguishing the model outcome from national accounting where an operating surplus balances the accounting results. Note also that, given the relative novelty of green accounting, the effects of introducing green accounting costs into the input-output system have not yet been modelled and compared to the standard-costing analysis.

This monetization of the physical model applies also to the abatement activities $\boldsymbol{x}_p$. In the one-pollutant case, the price p_p for eliminating one unit of the pollutant x_p has to be high enough to pay the primary factor input used v_p, after paying for the purchases of other inputs needed for the abatement activity:

$$p_p - a_{1p} p_1 - a_{2p} p_2 - \ldots - a_{np} p_n = v_p \qquad (12.4)$$

The introduction of an abatement activity thus increases value added and GDP by the product of the total amount of the pollutant abated x_p times the unit value added generated by the abatement activity v_p. This result is quite different from the no-price model (12.2). The reason is a different model assumption about the exogenous variables. Rather than assuming a set of given outputs $\boldsymbol{x}$, Leontief sets out from a given bill of final demand $\boldsymbol{y}$, which should not be reduced by an additional abatement activity. In order to meet this condition one has to assume freely available inputs for the abatement industry. One could of course doubt whether such an increase, generated by a defensive expenditure, increases social welfare (see Section 8.2.3). On the other hand, the model treats economic performance and net output correctly according to national accounts conventions.

One lesson of these apparently contradictory results is the significance of model assumptions about the setting of certain variables exogenously, and others to play out the internal model rules. Obviously, there is a risk of manipulating model results, especially when hiding the assumptions behind complex mathematics, or by describing results only.

12.1.2 Two Case Studies: Greened GDP in Input-Output and CGE Models

Specifying different policy options in terms of exogenous variables, such as the rate of economic growth and the level of 'tolerated' pollutants, together with widely differing modelling methods and assumptions, generated a large variety of models. It is therefore hardly possible to generalize about the results and uses of CGE models and input-output analyses. Instead, this section sketches the approach and outcomes of two models with regard to the calculation of a greened GDP, or 'greened-economy GDP' (in SEEA terminology: United Nations et al., in prep.).[2] The two case studies also illustrate the difficulty of comparing results because of different model assumptions.

[2] Note that, in contrast to the prediction of environmentally sustainable economic growth by a green*ed* GDP or NDP, the green accounting indicators of Section 8.2 assess the past sustainability performance of the economy.

12.1.2.1 Comparative-Static CGE Model: Sweden

In perfect market equilibrium, relative prices of goods and services are equal to the marginal utilities of the goods and services consumed, which themselves are equal to the marginal costs of their production. These fundamental equalities can be upset by an exogenous policy impact such as an eco-tax or environmental standard. CGE models translate these impacts into relative price changes and determine a new general equilibrium with different production and consumption patterns and a new overall level of economic activity.

Bergman's (1990) CGE model is a straightforward and transparent application of such comparative-static general equilibrium analysis. The objective of the model is to assess environmental policy impacts on gross national product (GNP) growth. A sequence of annual solutions compares the baseline situation with the effects of environmental standards and their enforcement. The baseline scenario of GNP and emissions of SO_x, NO_x and CO_2 is modelled first, using average growth rates for the exogenous variables from their benchmark values in 1985 to the year 2000. There are no emission constraints in the baseline analysis. The comparative environmental policy case introduces Swedish standards of emission reduction: 80% and 30% for SO_x and NO_x, respectively, during 1980 and 1993, and constant CO_2 emission at 1980 level.

Although based on a highly aggregated input-output table, the model differs from the simple input-output model described in Section 12.1.1 by giving up a good deal of its linearity in the use of labour and capital (including natural capital). Constant elasticities of input substitution affect thus prices and costs, which in turn are determined by the usual marginal cost-price identity in perfectly competitive markets. Rather than assuming direct pollution control the model introduces a cap-and-trade policy for emissions with permits set to prevent exceeding the limits of the environmental standards.

Table 12.1 indicates that compliance with environmental standards would produce a greened GDP in 2000, which is 4% lower than the conventional one (expected to grow by 2% annually in the baseline case). On the other hand, emissions are reduced in line with the 1980/1993 standards. Note that 588×10^5 tons of CO_2 emissions

Table 12.1 Economic growth and effects of environmental standards, Sweden, 1985/2000

	Baseline scenario: average annual growth rates (1985–2000) (%)	Environmental scenario[c]: change in key variables, compared to baseline values (%)
GNP[a]	2.0	–4.2
SO_x[b]	1.8	–52.1
NO_x[b]	2.1	–35.1
CO_2[b]	3.2	–38.5
Electricity consumption[a]	0.9	–12.5
Fuel import[a]	2.8	–28.6

Notes: [a] 1985 prices; [b] tons; [c] SO_x = 35% of 1980, NO_x = 85% of 1980.
Source: Bergman (1990), data from tables 2–6.

in 1980 (not shown in the table) are about the same as for 2000 (593×10^5 tons) in the environmental policy scenario. Meeting the SO_x and NO_x standards reduced electricity consumption and fuel imports by 13% and 29% respectively; this reduction achieved the CO_2 standard of zero emission growth as a complementary effect.

12.1.2.2 Econometric Input-Output Model: Germany

The German Panta Rhei model (Meyer, 2005) applies econometric trend and parameter estimation to an input-output model with 59 sectors and emissions from the use of 30 energy carriers. The focus is on energy consumption, concomitant air pollutants and the use of an eco-tax to reduce CO_2 emissions.

The model differs from conventional, fixed-coefficient input-output analysis by introducing variable, input price and trend-dependent coefficients. The coefficients are deemed to reflect technological change. However, in the absence of production functions, the model largely ignores substitution and innovation among the (separately modelled) economic sectors. Non-linear relationships can also be found in dynamic behavioural equations. At the same time, the model builders reject the optimizing behaviour of economic agents in perfect-market CGE models as unrealistic since these models generally lack 'empirical validation'. Instead, they opt for 'oligopolistic mark-up' (of unit costs) pricing. Prices do not reflect, therefore, equilibrium of supply and demand but are determined by unit costs, and trade and transport margins. A separate (world) model provides sectoral import and export prices.

'Bottom-up modeling and full integration' seeks to achieve systemic consistency. Input-output analysis generates sectoral results, which are aggregated into the macro-indicators of the national accounts. The combination of econometric with input-output analysis in a national accounts framework maintains indeed a close link to the databases of the official statistical system. The model loses, however, some of its transparency through simulations of relative price changes and their relationships with about 50,000 different micro- and macroeconomic variables.

Figure 12.1 presents the basic structure of the model. For illustrative purposes the figure stays with the original (1999) version, which integrates the emission of air pollutants directly into the economic system. Emissions (bold black arrows) flow from households (final demand) and producers – in part – to abatement activities of the institutional sectors. Unabated emissions create a cost burden as the basis for an eco-tax, affecting the cost of production and – through cost incidence – prices (grey arrows).

Figure 12.2 presents the effects of the eco-tax in Germany. It indicates growing abatement costs from 10 DM per ton of CO_2 in 1999 to 277 DM in 2010. The particular scenario shown assumes that all eco-tax revenues are used to reduce employers' (social) labour costs – the so-called ecological tax reform (see Section 13.3.3 below). The greened, eco-taxed, GDP is nearly 6% lower than what it would have been without eco-taxation and labour cost reduction. A greened GDP reflects however only

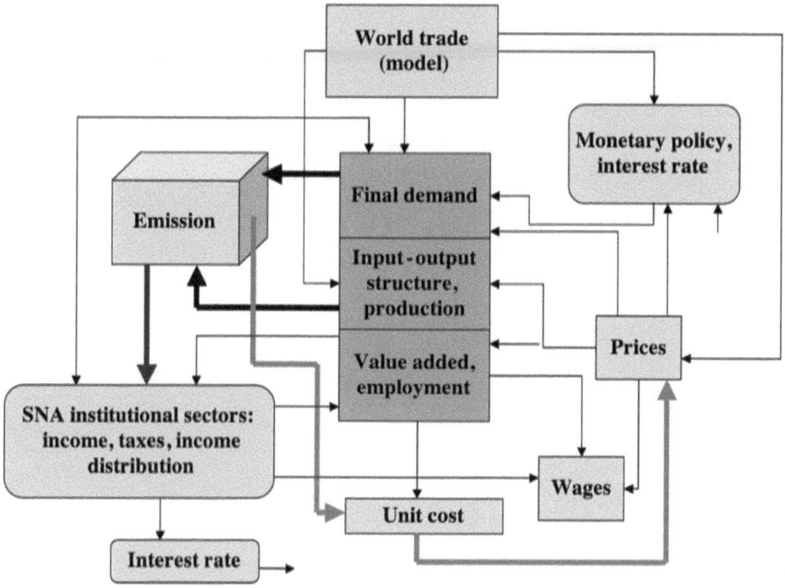

Fig. 12.1 Econometric input-output model (Panta Rhei)
Source: Meyer (1999), fig. 1, simplified; authorized copyright permission: European Communities.

environmental concerns; it ignores produced capital maintenance cost or assumes a constant share of fixed capital consumption in GDP. GDP-based models thus assess the potential economic cost of environmental policy, rather than the sustainability of economic growth (cf. Section 8.3).

Figure 12.2 shows a decrease of CO_2 by 17% since the introduction of the eco-tax in 1999. Since 1991, the total reduction amounts to about 25% – in line with governmental targets at the time. The figure also presents a revised baseline scenario, reflecting the policy situation in 2004. This scenario assumes, among others, the introduction of EU-wide trading of capped pollution permits. As a result, Germany should be below the year 2020 target of 800 million tons of emission, set for the country by the Kyoto protocol.

One of the modules added in the latest version is the material flow account. Based on export-driven demand for capital goods and diminishing effects of the unification-caused decrease of lignite production (cf. Section 6.3.2), the model predicts a relinkage of TMR with GDP increase. For the period 1991–2020 we might thus see an inverted Kuznets curve, i.e. initially falling and later increasing environmental pressure with continuing economic growth. In Factor-4 terms, the sustainability gap shown in Figure 10.2 would be widening.

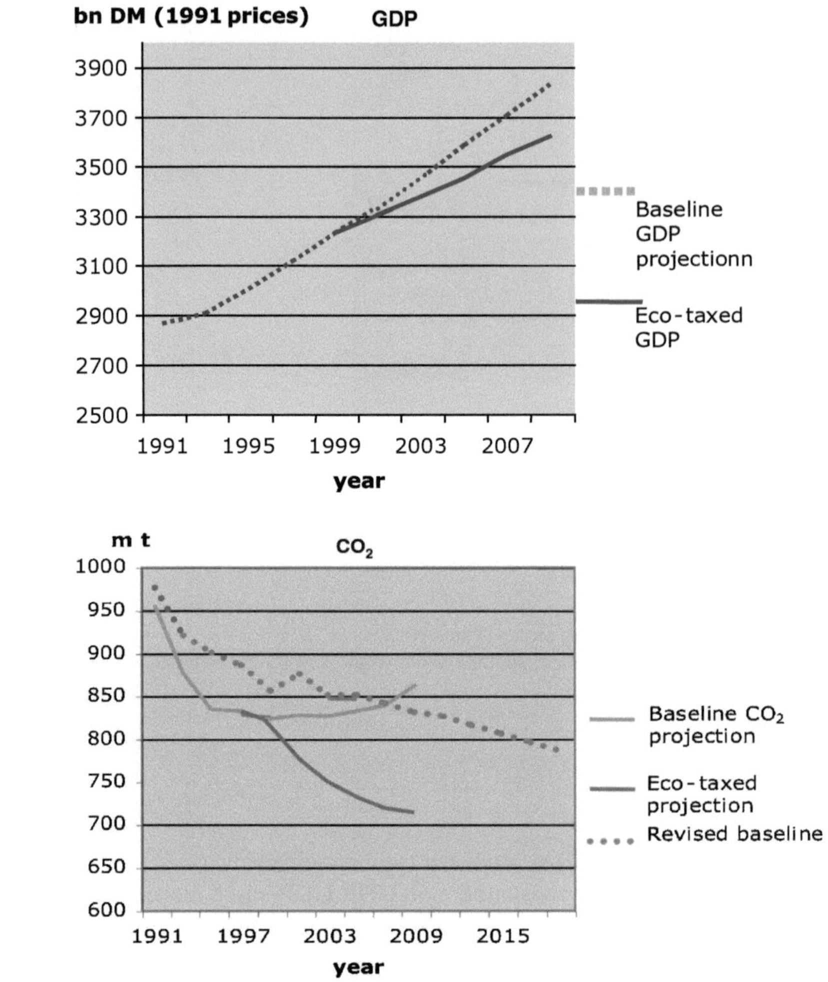

Fig. 12.2 Panta Rhei projections of GDP and CO_2 emissions, Germany 1991–2007/2015
Source: Meyer (1999, 2005); authorized copyright permission: European Communitites.

12.2 Environmental Constraints and Optimality: A Linear Programming Approach

The basic input-output model does not leave anything to choice and hence to *optimal*, cost minimizing or output maximizing, behaviour. As indicated (Section 12.1.1), the introduction of pollution control cost is bound by the (shadow-priced) equality between income and cost. Optimal behaviour is thus 'locked' (Dorfman et al., 1958) in the fixed-technology model, where the equality sign of Equation (10.1) ensures that output $\boldsymbol{x}$ is just enough to produce the given bill of final demand $\boldsymbol{y}$.

Relaxing this built-in condition, allows production of more outputs than necessary for predetermined $\boldsymbol{y}$. This invites inefficiency and at the same time, opens the door to the possible increases of **y**, i.e. higher standards of living – indeed a more realistic assumption. To stem the risk of 'going wild' (Chiang, 1984) with (unlimited) final demand maximization one would have to introduce production constraints from limited availability of primary production factors such as labour and/or environmental source and sink capacities. This converts the basic input-output analysis into optimization under constraints, i.e. into a linear programming problem [FR 12.2].

Figure 12.3 illustrates the introduction of social and environmental constraints into the model of interdependent economic activities. Two industries of food x_1 and shelter x_2 production face minimum requirements for food $\bar{c}_1$ and shelter $\bar{c}_2$, and maximum environmental limits for the emission of a pollutant $\bar{x}_p$ and the availability of a natural resource $\bar{x}_r$. Leaving out for now the optimizing function, these limits can be expressed as constraints in a linear programming model:

$$\begin{aligned} (1-a_{11})x_1 \quad & -a_{12}x_2 \geq \bar{c}_1 \\ -a_{21}x_1 + & (1-a_{22})x_2 \geq \bar{c}_2 \\ a_{r1}x_1 \quad & +a_{r2}x_2 \leq \bar{x}_r \qquad \equiv \mathbf{x} - \mathbf{Ax} \geq \bar{\mathbf{c}}, \leq \bar{\mathbf{x}} \\ a_{p1}x_1 \quad & +a_{p2}x_2 \leq \bar{x}_p \\ & x_1, x_2 \geq 0 \end{aligned} \tag{12.5}$$

The restrictions delimit a feasibility space (shown in highlighted boundaries in Fig. 12.3) for different production levels and product combinations. Note that labour is not considered a limitation in this particular model. Introducing new environmentally sound technologies would change the pollution and resource use coefficients, turning $\bar{x}_p$ and $\bar{x}_r$ further outward. The feasibility space would increase, facilitating a greater scope and level of sustainable economic activity.

We can interpret the minimum requirements for food and shelter as basic human needs of development. At the same time development is constrained by environmental

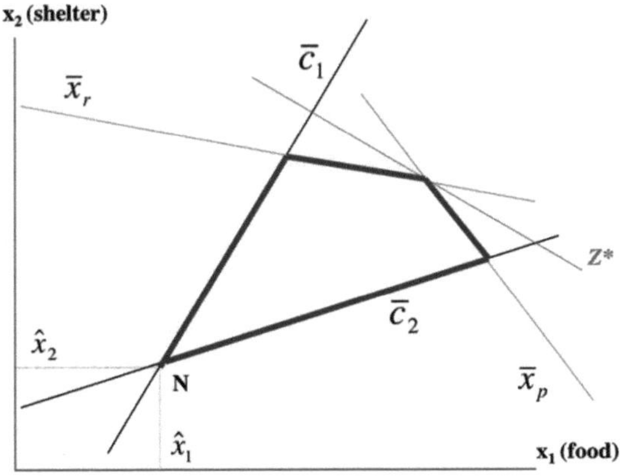

Fig. 12.3 Sustainability constraints in a linear programming model
Source: Based on Bartelmus (1979), fig. 1, p. 260; with permission by the copyright holder, Elsevier.

standards. In practice, interdependent ecological, social and demographic limits are difficult to determine. Consensus on separate limits is only a first step toward rational targets setting as targets might overlap, for instance when determining carrying capacities of human populations at different standards of living.

The practical use of a feasibility space for economic activity is therefore questionable, especially if many more activities, outputs and standards are included. Still, Fig. 12.3 makes the vision of sustainable development visible in terms of minimum inner and maximum outer limits [FR 3.1]. At point N, basic human needs are just met with as the lowest acceptable amounts of total outputs of food and shelter. More importantly, the restrictions for resource availability and emissions turn the original approach of pollution abatement (Equations 12.1) into a precautionary model of producing within preset environmental capacity limits (cf. Section 13.2).

The introduction of an optimizing objective function turns the constrained input-output system into a linear programming model. Figure 12.3 shows the maximum net output (for final consumption) value Z^* for the (linear) objective function

$$Z = v_1 x_1 + v_2 x_2 \stackrel{!}{=} \max \qquad (12.6)$$

For given output weights of unit value added generated by the production of food v_1 and shelter v_2, Z^* represents the highest feasible Z value. This value is indeed another version of a maximum greened GDP (total gross value added), where environmental and social (basic human needs) constraints are taken into account.

Introducing more than one limiting factor of production, notably produced capital, calls for considering substitution in the production functions. It also opens up the possibility of reserving some output and natural resource reserves for future use, i.e. capital formation and maintenance – the next section's topic of dynamic modelling.

12.3 Dynamic Analysis: Optimality and Sustainability of Economic Growth

12.3.1 Dynamic Linear Programming

Section 12.2 introduced limits in the availability of scarce natural capital in a standard linear programming model. Overuse of natural capital, i.e. either running down natural resource stocks or degrading environmental sinks, threatens the sustainability of economic activities. The key questions, asked repeatedly in this book, are how close are these environmental constraints and when are we running out of environmental support functions? The urgency of immediate and radical action, evoked by environmentalists, calls for further scrutiny of the time path towards hitting potential environmental limits. Dynamic linear programming is tailored to answering these questions while adhering to the efficient (optimal) use – now and in the future – of limited produced and natural capital. The challenge is to determine what amount of produced and non-produced goods should be reserved for future use.

The basic approach of dynamic linear programming is to allow for future use of outputs in the static system of equation 12.5. In principle, the use of outputs x_i can then take place either in the current period t or the future period t + 1 as

- Inputs into different industries j during the current period: $x_{ij}(t)$, or
- Net capital formation (including inventories of goods to be used as inputs or final consumption in future periods), increasing the capital stock of industries by $\Delta K_i = K_i(t+1) - K_i(t)$.

Output x_i would now have to be large enough to cover both present and future uses:

$$x_i(t) \geq x_{ij}(t) + \Delta K_i \quad (12.7)$$

Further assuming fixed capital requirements b_{ij} per unit of output of industry j from industry i, and distinguishing final consumption c from capital formation ΔK as components of final demand $\boldsymbol{y}$, one can describe the dynamics of the two-commodity economy as

$$\begin{array}{ll} x_1 \geq a_{11}x_1 + a_{12}x_2 + \Delta K_1 + c_1 & \\ x_2 \geq a_{21}x_1 + a_{22}x_2 + \Delta K_2 + c_2 & \boldsymbol{x} \geq \boldsymbol{A}\boldsymbol{x} + \boldsymbol{\Delta K} + c \\ K_1 \geq b_{11}x_1 + b_{12}x_2 & \equiv \boldsymbol{K} \geq \boldsymbol{B}\boldsymbol{x} \\ K_2 \geq b_{21}x_1 + b_{22}x_2 & \\ \Delta K_1, \Delta K_2, x_1, x_2 \geq 0 & \boldsymbol{\Delta K}, \boldsymbol{x} \geq 0 \end{array} \quad (12.8)$$

Having introduced a new primary factor, capital, the linear programming problem is now maximizing final demand, i.e. final consumption and net capital formation, under the restrictions of (12.8) or as its dual of minimizing capital input costs.[3] Textbooks on linear programming [FR 12.2] provide proof and explanation of the weights attached in the objective functions of our model – either as shadow prices of the goods and services of final demand p_i with the objective function

$$\Sigma\, p_i(c_i + \Delta K_i) \overset{!}{=} \max \quad (12.9)$$

or as the unit shadow cost or rent r_i of the use of the limited primary factor (capital) k_i with the objective function of the dual

$$\Sigma\, r_i k_i \overset{!}{=} \min \quad (12.10)$$

subject to prices not exceeding unit factor costs.

[3] The dual of a linear programming model yields the same optimal value as the primal (in shadow or accounting prices). The dual changes a maximization problem into a corresponding minimization problem and vice versa. Again, we see here the income (factor cost) = net output identity described in Section 12.1.1 for the basic Leontief model.

12.3.2 Optimal Growth and Sustainability

The above discussion of optimization under sustainability constraints helps understand the introduction of environmental concerns in more generic models of maximizing welfare and economic growth. These models are the typical, largely theoretical, response of mainstream economists to the environmentalist critique of ignoring long-term environmental concerns, or dealing with them at best as a matter of short-term cost internalization. Rather than optimizing behaviour of economic agents at the microeconomic level, optimal growth models take the view of an overall social planner, who aims at maximizing national social welfare, now and in the future. Welfare in turn is seen as a function of consumption of goods and services and environmental quality. Optimal growth models thus introduce a social welfare function, whose optimality is determined by maximizing the discounted welfare value in each future period.

Note that in such models all time-bound variables are endogenized rather than estimated econometrically outside the system of interdependent variables (as in CGE models). The models go rarely beyond 'conceptualization' as they abandon linearity and cling to the smooth utility and production functions of neoclassical economics. They do succeed, though, in clearly defining long-term sustainability of an optimizing economy – but within the particular model assumptions [FR 12.3].

To gain insight into the meaning of highly complex multivariate dynamic optimization under environmental restrictions the linear programming model can be reformulated as a general optimization problem under an environmental constraint. Applying the standard Lagrange multiplier method of optimization reveals the multiplier as the shadow price (or cost) in the optimum of the linear programming model (Dorfman et al., 1958). The multiplier thus measures the change in the value of the objective function in a *non-linear* constrained optimization problem, brought about by a marginal change in the constraint. The shadow prices of the linear programming model can therefore be interpreted as weights for marginal changes of final demand and capital use categories in the optimum situation (cf. Equations 12.9, 12.10).

A simplified prototype optimal growth model can elucidate these model features. The model was initially advanced for rejecting conventional net national product as a welfare measure due to environmental constraints (Mäler, 1991). More recently, the model advanced a sustainability criterion, which may differ from the optimization criterion of maximum (discounted) net present welfare. As shown in Box 12.1, the model maximizes a social welfare function, depending on final consumption C, capital use (including natural capital) K, environmental damage Z, and labour input L:

$$W = W(C, K, Z, L) \tag{12.11}$$

generated with given stocks of produced capital $\bar{K}_1$ and natural capital $\bar{K}_2$.

The main rules and conclusions from solving the model are:

Box 12.1 Developing an optimal growth model with natural capital

***STEP 1*:** Introducing natural capital of forests (forest inputs K_2, logging X, afforestation H) and sinks (pollution P and defensive expenditures R) into the production function Y, with

$Y = Y(K_1, L_1)$ flow of final output (aggregate production function)
$X = X(K_2, L_2)$ logging rate
$H = H(L_3)$ net afforestation rate (incl. natural growth)
$P = P(Y)$ pollution from producing Y
R = portion of Y devoted to mitigating pollution damage
$Z = Z(R, P)$ net environmental damage (affecting welfare directly)

***STEP 2*:** Specifying the model dynamics (introducing differential equations for capital formation):

$dK_1 / dt = Y(K_1, L_1) - C - R$ conventional capital formation as the difference of final demand minus consumption C and damage mitigation R

$dK_2 / dt = H(L_3) - X(K_2, L_2)$ net natural capital formation or depletion in forests

***STEP 3*:** Solving the problem of maximizing the discounted flow of social welfare W over the indefinite future. Maximizing the current value Hamiltonian (a multivariate generalization of the Lagrange multiplier method) obtains net social welfare along the optimal time path as

$$W^* = W(C, X, L, Z) + p(dK_1 / dt) + r(dK_2 / dt)$$

where shadow prices p and r reflect the present value of future returns on a marginal change in the availability of the present capital stock. W^* is the sum of current welfare W and discounted future welfare from current changes in produced and natural capital.
Source: Dasgupta and Mäler (1991, simplified).

- *Capital (incl. natural capital) maintenance* rule of sustainability: if the total stock of capital $p\bar{K}_1 + r\bar{K}_2$ is valued in shadow prices along the optimal time-trajectory of welfare generation, non-declining welfare is ensured only if the value of the total capital stock (in constant prices) does not decrease (Mäler, 1991).
- *Intergenerational equity*: the maximum Hamiltonian value, which is the maximum welfare measure (see Box 12.1), represents the maximum feasible consumption value that can be maintained forever. The assumptions for this

fortunate coincidence is that the substitution elasticity between exhaustible natural resources and other inputs is equal or greater than 1, and that the elasticity of the output-over-produced-capital ratio is greater than that of natural capital (Solow, 1974a, 1974b).

- *Hartwick's rule*: for the special case of exhaustible resources, the rule requires the reinvestment of rent (for natural capital depreciation) in reproducible capital to ensure constant (sustainable) consumption under the above assumptions (Hartwick, 1977).

The model outcomes thus depend, apart from the usual perfect market and substitution (in production and consumption functions) assumptions, on what is packed into the welfare function (12.11). In particular, there is a wide variety of different, and differently categorized, primary production factors that can be included or ignored. Moreover, the production factors may interact in many alternative ways in generating widely differing welfare effects. As pointed out by the authors themselves 'no one can seriously claim to pinpoint the optimal level of current consumption for an actual economy' (Arrow et al., 2004). The abstract model serves indeed mainly the conceptualization of sustainability, specifying the need of keeping capital intact for non-declining welfare generation.

In fact, if the welfare package is broad enough, non-decline of welfare can also be viewed as sustainable development (Mäler, 1991). Note however that the search for 'empirical evidence' for the model's sustainability criterion had to resort to the narrowly defined green accounting indicators of 'genuine' investment and wealth (Arrow et al., 2004). These indicators are quite similar to the environmentally adjusted capital formation (ECF) and asset indicators of the SEEA (Section 8.2.2), catering to sustainable economic growth rather than development.

12.3.3 Some General Conclusions

Facing environmentalist adversity to economic growth, economists introduced environmental issues in their growth models since the 1970s. As to be expected, optimal growth analyses come to differing conclusions about the relevance of environmental limits, depending on model assumptions. To illustrate the range of arguments about optimality and sustainability in optimal growth models it may suffice here to summarize the conclusions from models presented in a reader on environmental macroeconomics (Munasinghe, 2002):

- *Technological progress can overcome resource scarcities* through reduction of extraction cost, substitution and discovery, *and environmental degradation* through environmental protection. The 'huge reserve of detailed physical, chemical, geological and physiological relationships' just needs to be unveiled by 'natural scientists and engineers'. There is no 'clear and present case' of a non-substitutable resource 'in limited supply, essential to life and welfare' (Koopmans, 1973).

- *Technological progress, substitution* of natural capital by produced capital *and increasing returns to scale make sustainable growth* of per capita consumption *feasible*, with optimal rates of natural resource use 'of the order of magnitude observed for many natural resources' (Stiglitz, 1974).
- With relative scarcity of natural capital and diminishing returns to technological progress, *a global steady-state economy can be reached during a transitional period* of slowing increase of labour productivity and real per capita income growth (England, 2000).
- Model runs show that *an optimal growth trajectory and a transition to a steady-state economy may not exist.* In the absence of governmental (environmental policy) intervention, the ecosystem collapses, and optimization and forecasting do not produce a feasible solution. 'An ecological economy cannot grow limitlessly' (Islam, 2001).[4]

Technical progress plays a crucial role in arguing the sustainability of economic growth and its welfare effects. Most economists rely on human knowledge and inventiveness as the saviour from environmental and related economic collapse. Environmentalists, on the other hand, point to the physical laws of entropy and complementarity in the use of energy and materials: critical natural capital is bound to run out eventually if current demographic and economic growth patterns continue. Empirical evidence seems to be on the side of the economists, at least as far as natural resource depletion is concerned. Decreasing natural resource prices indicate reduced scarcity for many natural resources. As a result, we could expect an increase in 'effective' natural resource stocks.[5] But all depends, of course, on our ingenuity. Will technology be the saviour? Possibly.

Parts II and III assessed empirically the impacts and repercussions of the environment-economy interaction. In this part we used these assessments, at least in principle, for prediction and policy analysis. However, simplifying model assumptions and selectivity in model variables usually impair practical policy advice. On the other hand, introducing the value-laden vision of sustainable development into economic theory gives us a more rigorous understanding of the paradigm. The result is a pragmatic focus on the sustainability of economic growth in applied and theoretical environmental-economic analyses. The final part of the book makes use of our visionary, empirical and analytical knowledge to offer a few strategic 'conclusions'. Admittedly, these conclusions are far from conclusive, as indicated by a final chapter on remaining 'questions'.

---

[4] One should probably not read too much into the progressive greening of the economists, as time goes by.

[5] Barnett and Morse are among the first to find a long-term decrease in real extraction cost of most minerals. See also the Simon-Ehrlich wager [FR. 11.2]. According to Baumol (1986), 'effective natural resource stock' (even of non-renewable resources) might increase when technological innovation leads to a revision of usable resource stocks at a rate that exceeds resource use.

Further Reading

FR 12.1 Computable General Equilibrium

Munasinghe's (2002) reader on *Macroeconomics and the Environment* gives an overview of environmental-economic analysis and modelling. Computable general equilibrium (CGE) models play a prominent role in this review. Conrad (1999) provides a concise description of the 'principles' of CGE models of environmental-economic policy analyses. Most applied CGE models are based on input-output tables and analysis [FR 10.1] for determining their benchmark situation.

Quite unusual for a statistical office, Statistics Norway seems to have moved from descriptive natural resource accounting to introducing environmental concerns and energy consumption into a multi-sectoral dynamic CGE model (http://www.ssb.no/emner/09/90/rapp_200418/rapp_200418.pdf; Alfsen, 1996). A dynamic CGE model of the USA compares a backcasted scenario without environmental regulation with the actual regulated situation: for 1973–1985 GDP has been reduced by 2.59% owing to environmental protection (Jorgenson & Wilcoxen, 1990).

As part of an EU investigation into green accounting the GREENSTAMP project suggests to replace the green GDP by a modelled 'greened' GDP, i.e. 'a hypothetical national economic product that would be obtainable ... subject to ... a specified set of environmental standards' (O'Connor, 1999). Model results indicate that the combination of technology and 'sustainable consumption' allows standards of living in France to improve while respecting sustainability standards. The model restricts, however, its environmental policy analysis to energy consumption and its pollution effects.

FR 12.2 Linear Programming and Economic Analysis

Dorfman et al. (1958) is probably still the best text on the use of linear programming in economic analysis. Much of Sections 12.2 and 12.3.1 is based on this book. Paris (1991) focuses on duality in economic applications of linear programming such as factor cost minimization for given final demand as the dual of GDP maximization with given primary factors (cf. Section 12.2). Textbooks on economic mathematics (such as Chiang, 1984) may facilitate access to the sometimes-challenging mathematics of linear and non-linear, and dynamic programming. An early call for applying linear programming or activity analysis to the assessment of sustainability limits in 'eco-development' (Bartelmus, 1979) went largely unheeded.

FR 12.3 Sustainability in Optimal Growth Models

Mainstream economists extended optimal growth models of inter-temporal welfare maximization to natural capital endowment. Some of these models, whose main findings are cited in the text, can be found in Munasinghe (2002). Dasgupta and Mäler (1991, 2000) use the model to delineate an environmentally modified net national product indicator as a welfare measure that reflects optimal growth as well as sustainability. Arrow et al. (2004) explain that the maximum welfare value of this model does note have to coincide with sustainability in the sense of perpetual constant per capita consumption. Pointing out this discrepancy may be the reason for co-opting environmentalists like Paul Ehrlich and Gretchen Daily as co-authors of this article. It remains to be seen if some euphoria about the 'friendship' between environmentalists and economists (Christensen, 2005) will stand the test of time, especially when environmentalists obtain a clearer picture of the model assumptions.

Review and Exploration

- Explain the differences and relationships between input-output and CGE models. How do they deal with environmental impacts and policies?
- Is a 'greened' (modelled) GDP preferable to a green (accounted) GDP/NDP for supporting sustainability in policymaking? Compare the different greened GDPs resulting from CGE, linear programming and optimal growth models.
- What is the purpose of dynamic modelling? How does it compare to comparative-static (CGE) analysis? Can it capture the (non)sustainability of economic growth?
- What are your conclusions about the use and usefulness of modelling – vs. direct data use – for policymaking?
- Is technology the saviour from environmental collapse?

Part V
Strategic Outlook

Part IV's analysis of potential limits to economic growth at national and global levels sets the tone for some strategic conclusions about tackling these limits. Chapter 13 presents strategies and policy measures for dealing with impacts of production and consumption that threaten to violate environmental limits. The strategies apply mostly to governmental policy but include also voluntary action by corporations and households, motivated by a new environmental ethics. Global and trans-boundary environmental impacts require international action. Chapter 14 examines, therefore, the need for improving global governance in order to advance sustainable growth and development in a globalizing world.

The concluding chapter raises again the initial questions of Part I and asks what we learned about them. Many conclusions remain tentative and raise further questions. It is thus quite appropriate to end the book as it began with a chapter on 'questions, questions, questions'. This should not be taken as resignation before a host of open issues, but rather as encouragement of further quantitative analyses.

Chapter 13
Tackling the Limits to Growth

None of the above-described indicators and models provides an unequivocal answer to whether economic growth, and what kind of growth, are sustainable. Rather, the dichotomy between pessimistic environmentalists and more optimistic economists persists in measurement and analysis of the environment-economy interaction. So what should and could be done about an undeniable problem, whose significance is judged differently?

To be on the safe side let us set out from the pessimistic view of the Limits-to-Growth (LTG) model. The model explains environmental impacts in terms of the popular IPAT identity as the result of population growth, wasteful affluence, and effects of the energy needs of technology (Meadows et al., 2004).[1] The model's more optimistic, but 'less likely' scenarios reveal 'responses' to non-sustainable resource depletion and pollution, which together would attain sustainable development (op. cit.; see also Section 11.2.1):

- *Population control* by means of birth control, which should limit reproduction to two children per family (scenario 7)
- Plus: *limiting industrial output* by means of moderation in lifestyles and more efficient capital use, in other words greater sufficiency in consumption and greater eco-efficiency in production (scenario 8)
- Plus: *technological progress* in reducing the remaining pollution (scenario 9).

Birth control and sufficiency are the results of changes in individual behaviour. On the other hand, deliberate R&D or spontaneous inventions of creative minds bring about environmental technologies. For generating these behavioural and technological changes the LTG authors leave their mechanistic model and call for 'leadership and ethics, vision and courage', supported by a 'networking' civil society.

As hard-nosed economists we want to go beyond 'heart-felt intuition' (op. cit.) about changes in social values and enlightened leadership. This is not to deny the importance of ethics and 'soft' strategies of moral suasion (Section 13.4). However,

[1] I = PAT defines impacts I as the result of three determinants: (1) size of population P, (2) affluence A as GDP p.c. and (3) technology as 'eco-efficiency' I/GDP. This reveals IPAT as an identity:
$I = P \times GDP/P \times I/GDP \equiv I$.

the objective of this book is to facilitate and evaluate rational policies with the quantitative measures and analyses described in the preceding parts. It would fill another book to detail the effects of different economic, social and environmental policies on economic growth and development. The way to confine the discussion of policy measures, besides leaving much to further reading, is to bundle these measures under four basic strategies of dealing with potential environmental limits:

- *Ignoring* the limits: muddling through
- *Complying* with limits: curbing economic activity
- *Pushing* the limits: improving eco-efficiency
- *Adopting* limits: sufficiency in consumption, corporate social responsibility, environmental ethics.

13.1 Ignoring the Limits: Muddling Through

Tackling environmental symptoms when they occur and relying on past experience for taking action can be seen as a muddling-through policy. One view considers such ad hoc reaction as more realistic than comprehensive (costly and time consuming) analyses of fundamental objectives and policy options (Lindblom, 1959). If past experience includes reliance on market forces for signalling a problem and adjusting to its effects, we have a particular form of muddling through. The stalwart of market liberalism, *The Economist* (of 11 September 1999) argues that experimentation by markets is 'a humbler way of going about things than by following the conceited blueprints of politicians, the hubris of monopolistic businessmen, or the arrogance of scientists'; history shows that governments and pressure groups frequently impose their visions – only to abandon them later as mistaken.

As discussed in Section 11.1, the EKC hypothesis is an attempt to justify non-interference in market activities. The assumption is that unfettered economic performance and growth solve environmental problems automatically, or at least facilitate their solution. However, our review of the hypothesis did not find conclusive evidence for a general correlation between economic growth and environmental improvement in the high-income range of the EKC. The dominant force behind environmental improvement appears indeed to be environmental policy, frequently marginalized, however, even in rich countries. It is thus an open question, whether such policy is driven by affluence or by necessity (cf. Section 11.1.2).

Relying on economic growth alone does not seem to be a valid option. On the other hand, there is some evidence that the price signals of the market do reflect natural resource scarcity as in the case of falling prices of mineral commodities (Section 12.3.3). By the same token, rising prices would indicate increasing scarcity and might stimulate the search for more efficient extraction, harvesting and use of natural resources. Adaptation of car use at the peak of gasoline prices is a case in point. The question is whether such observations can be generalized. Short-sighted non-action looks indeed suspiciously like 'passing the buck to future generations and other regions' (Rothman, 1998).

Again, we see here the environmentalist-economist dichotomy at work when dealing with uncertainty or ignorance about environmental damage. Environmental economists take a wait-and-see attitude. They look first for market signals of new scarcities in environmental source and sink services before internalizing the scarcity costs. They also discount uncertain environmental risks according to their preference for current vs. future benefits and, inversely, cost (cf. Section 2.3.2). Ecological economists, on the other hand, call for urgent precautionary and regulatory action, in the face of imminent disaster.

13.2 Complying with Limits: Curbing Economic Activity

Facing up to environmental disaster most environmentalists show hostility toward economic growth, albeit with some focus on the physical side of economic expansion. Their idea of sustainable development can be characterized as 'development without growth – without growth in throughput beyond environmental regenerative and absorptive capacities' (Daly, 1996). This would indeed leave the door open to *economic* growth (expressed in real, constant-price values) as long as it does not violate environmental carrying capacities. For the limitation of the physical scale of economic activity, the use of popular environmental 'management rules' is the prevailing policy advice (Daly, 1990; Sachs et al., 1998):

- Use renewable resources within their regenerative capacity.
- Use non-renewable resources as far as renewable substitutes can be found.
- Discharge waste and residuals without exceeding the absorptive capacities of natural systems.

For concrete policy measures, these rules require specific targets or (safe minimum) standards of natural resource use and emissions and their ambient concentrations. Setting ecological standards at the national (policy) level is problematic but could delimit economic activities within a normative feasibility space (Sections 3.2.2 and 12.2). From the point of view of an already overloaded full-world economy, regulatory *command and control* (CAC) of economic activity is the preferred policy instrument for forcing economic activity into the feasibility space. CAC rules and regulations aim at directly reducing the scale of throughput and corresponding economic activity as the prime ecological sustainability objective.

However, economic activity can be curbed not only by regulating material flows to and from the economy but also by market instruments. Seeking an optimal level of – monetary – output through environmental costing, output is usually lower than the one generated by unfettered markets (cf. Annex I for the case of a Pigovian eco-tax). It might be higher, though, than the level brought about by CAC, owing to the economic and technological prowess of enterprises in reducing environmental impacts and costs. Both approaches could be combined: CAC measures could set and enforce the feasibility space, and the market could then determine efficient – after environmental cost internalization – production and consumption patterns within this space.

Table 13.1 presents a taxonomy of typically applied environmental policy measures. CAC specify what (which policy target) needs to be achieved *and* how it should be achieved, e.g. by prohibiting the use of specific inputs, prescribing particular technologies, or protecting the use of land from economic development. A popular way of creating protected areas in developing countries, are debt-for-nature swaps. The idea is to grant foreign debt relief in exchange for abstaining from economic land use.[2] The other parts of the table indicate various possibilities of relaxing either the setting of targets or prescribing the way of target implementation, or both, for applying more flexible market instruments (see also Annex I.2).

The reason for using the drastic CAC measures is, besides their simplicity of application, lack of trust in the capability of market forces to reach society's environmental goals. Doubt in market solutions stems from

Table 13.1 Taxonomy of environmental policy instruments

	Policy target specified	Policy target not specified
'How' specified (implementation process prescribed)	CAC: - Prohibitions (of hazardous inputs, discharges and overuse of natural resources) - Environmental standards and technology specified (incl. recycling/reuse) - Land appropriation, purchase or expropriation for environmental protection - Obligatory insurance for specific environmental impacts	- Subsidies for particular equipment - Transfer of technology - Liability (with care standard)
'How' not specified (implementation process not prescribed)	- Tradable pollution and resource use permits (cap and trade) - Design and performance standards - Voluntary agreements (including environmental audits, labelling etc.)	- Emission and product charges - Resource rent capture (royalties) - Deposit-refund system - Technical assistance (open-ended) - Property rights for environmental sinks and sources (bargaining) - Liability (without care standard) - Subsidies (open-ended, grants and removal of subsidies) - Environmental information and education

Source: Russel. Clifford S. (2001), Applying economics to the environment, table 9.3, modified; with permission by the copyright holder, Oxford University Press.

[2] Preventing economic development for the creation of nature reserves meets of course with the resistance of land owners or users facing governmental land appropriation. Typically, international NGOs such as the WWF (http://www.worldwildlife.org/conservationfinance/swaps.cfm) initiate these swaps with some financial contribution; ultimately the swaps require a formal agreement between the creditor and debtor country.

- A tendency of economic agents to underestimate uncertain potential environmental damage
- Possible 'irreversibilities' of environmental damage, for which time-lagged individual responses to market incentives might come too late.

Immediate and fully controlled environmental action makes sense for averting imminent environmental disaster. The *precautionary principle* of the Rio Declaration (United Nations, 1994, Rio Declaration, Principle 15) points in that direction in a less stringent manner: 'lack of full scientific certainty shall not be used as a reason for postponing cost-effective measures to prevent environmental degradation'. A simpler popular formulation is to be better approximately right in time than optimally right too late.

It comes as no surprise that ecological economists adopted this principle as a justification for preferring proactive rules and regulations to reactive market instruments (cf. Section 2.4). The question is, whether governments are indeed in a better position – than individual preferences expressed in markets – to weigh uncertain risks against the cost of reducing the risks.[3] In fact there is no good reason why the above-mentioned management rules could not be relaxed in some cases. Why should we forgo decreasing some of the natural resource stocks, e.g. for current poverty alleviation, when future requirements for the resources are highly uncertain?

13.3 Pushing the Limits: Eco-Efficiency

Typically the political process, rather than rational quantitative analysis, guides CAC action. CAC is thus particularly inefficient when economic agents possess better information than remote and sluggish bureaucracies. The objective of market instruments of environmental policy is to prompt consumers and producers into using this information under competitive pressure. Eco-efficient production and consumption patterns are the expected results.

13.3.1 Eco-Efficiency and Resource Productivity

CAC prescription of existing technologies thwarts human ingenuity in finding innovative and least-cost solutions to environmental problems. This is the reason for letting market forces search for ecologically and economically efficient products and production processes. The World Business Council for Sustainable Development defines such 'eco-efficiency' as 'a management strategy [of corporations] that links financial and environmental performance to create more value with less ecological

[3] A survey by *The Economist* (of January 2004) presented several examples of conspicuous failures of governments to reasonably balance risks and net returns from protection against risk, notably in the areas of hazardous pollution, BSE (mad cow disease) and the US fight against terrorism.

impact' [FR 13.1]. Environmental economists also favour the supply side of market exchange, considering consumers hardly knowledgeable about production and emission processes (Turner et al., 1993). Economic modelling confirms that new environmentally sound technologies can open up the feasibility space for economic activity by pushing outward environmental source and sink limits (cf. Fig. 12.3).

Faith in eco-efficient technology is most pronounced in the concept of metabolic consistency [FR 13.1]. The idea is to imitate nature, which 'does not know the concept of waste'.[4] One of the protagonists of consistency sees the seamless incorporation of industrial metabolism into nature's metabolism as a paradigm shift from quantitative eco-efficiency to new qualitative innovation (Huber, 2004). The purpose is still to maximize production and minimize environmental impact. A more modest view of consistency might see it, therefore, as a particularly efficient type of eco-efficiency. Plate 13.1 is a simplified example of how waste from coffee production can be channelled into a highly profitable side activity – mushroom breeding. In fact, in this case study, revenues from sales of shitake exceeded those of coffee.

Eco-efficiency has also become the basic tenets of industrial ecology, a relatively new field of research on industrial metabolism, i.e. material flow analysis at the enterprise level (Lifset & Graedel, 2002). Sections 2.4.2 and 6.3.1 presented resource productivity (GDP per material input) as the key indicator of ecological sustainability at the macroeconomic level. The connection between micro-level corporate eco-efficiency and macro-level national or regional resource productivity is not straightforward, however. There appears to be some wishful thinking about corporate social responsibility (CSR), which would motivate enterprises to reduce natural resource use and emissions for the sake of the greater social good. In practice, neither corporate environmental accounting nor environmental management are likely to fully embrace any goals beyond cost saving and corporate image improvement (Sections 9.1.1 and 9.2).

Eco-efficiency remains thus most useful as a macroeconomic objective for policy instruments that influence the behaviour of microeconomic agents. To this end, eco-efficiency and its instruments address both sides of the material flow balance with the objectives of

- Increasing resource productivity (GDP/DMI) for the dematerialization of the economy
- Decreasing environmental impact intensity (DPO/GDP) or its inverse, pollution 'efficiency' (GDP/DPO) for the detoxification of the economy.[5]

[4] According to the 'vision' of the Zero Emissions Research Initiative (ZERI) (http://www.zeri.org/index.cfm?id = vision). Considering the 'waste' of large amounts of seeds that do not germinate, Ehrenfeld and Chertow (2002) contest this view and prefer referring to 'nature's bounty … as eco-effectiveness'.

[5] See Section 6.3.1 for the definitions of the material flow indicators. As also discussed in that section, detoxification can either be considered as a supplementary sustainability concept or subsumed under the general notion of dematerialization.

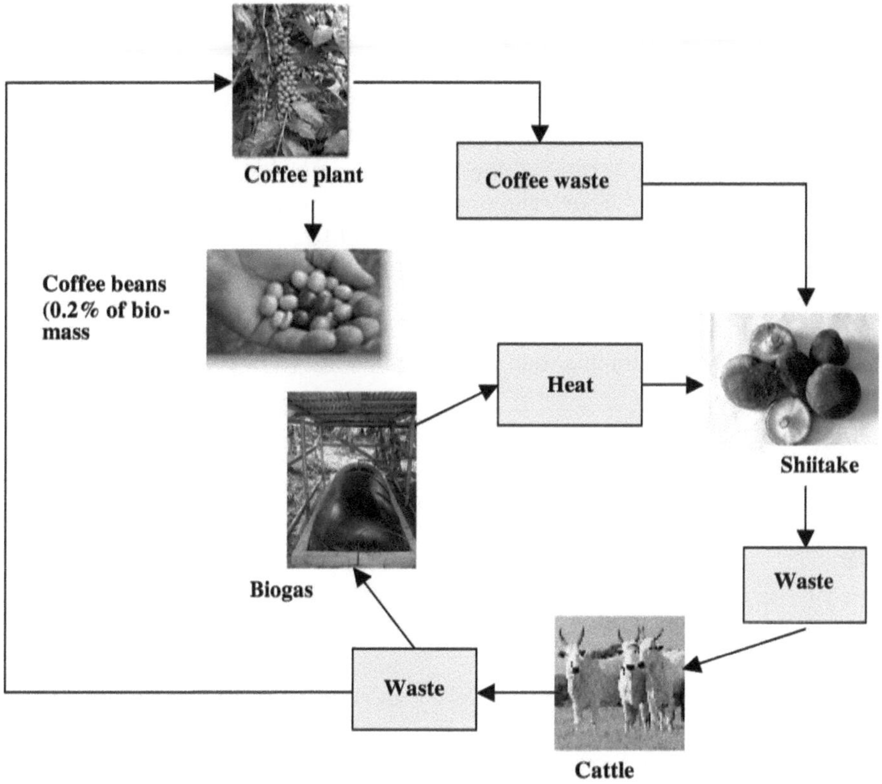

Plate 13.1 Metabolic consistency: coffee and mushroom production
Source: Based on Steinbrink (2001), fig. 2; with permission by the copyright holder, Zero Emission Research Initiative, ZERI (*See Colour Plates*).

The EU strategy on the sustainable use of natural resources (Commission of the European Communities, 2005, annex 3) defines eco-efficiency (EE) as the ratio of resource productivity (value added per material input: VA/MI) and 'resource specific [pollution] impact' (I/MI):

$$EE = VA/MI : I/MI = VA/I \qquad (13.1)$$

Environmental impact, i.e. the generation of residuals over the life cycle of a product, results from direct and indirect ('upstream') material inputs. Eco-efficiency seems thus to be reduced to value added per unit of wastes and residuals, ignoring the potential depletion of natural resources used. This is probably an unintended result of the EU's eco-efficiency definition since the strategy calls for the simultaneous reduction of environmental impacts and the improvement of resource productivity. At any rate, reference to the product life cycle introduces indeterminate time periods into microeconomic impact assessment, complicating annual national-accounts-based macro-analysis of eco-efficiency.

The EU strategy remains thus just this: a strategy that seeks to achieve dematerialization but lacks an operational concept for implementation. The strategy refrains, therefore, from adopting the targets of the EU's Sixth Environment Action Programme due to lack of knowledge and indicators. As discussed in Section 2.4.2, determining the amount of dematerialization needed for sustainability requires the setting of national targets, or at least guardrails, such as Factor 4 or 10. For structural and regional policies, one would also have to specify compatible standards at regional and sectoral levels. A variety of policy instruments, including the above-described CAC measures and market-based 'economic instruments' can be applied for meeting eco-efficiency targets and standards.

13.3.2 Categories and Efficiency of Market Instruments

13.3.2.1 Strategic Principles

Market instruments can improve both ecological and economic sustainability [FR 13.2]. Ecological sustainability would use these instruments for reducing material input and residual output by increasing the cost of material inputs and penalizing wastes and emissions. Stressing, however, the inability of markets to achieve distributive equity or sustainable scale, ecological economists rank the allocative efficiency of market instruments lower than setting scale and equity limits (Daly & Farley, 2004; Costanza et al., 1997a).

Economic sustainability aims at the internalization and eventual reduction of environmental cost according to the *polluter/user-pays principles* (PPP, UPP). Contrary to the precautionary principle, which caters to a preventative CAC approach, the PPP and UPP seek to burden those who caused pollution, congestion and natural resource depletion with the cost of damage mitigation or compensation. To the extent that cost anticipation deters economic agents from polluting or depleting, the two principles may also have precautionary effects. The UPP is less clearly defined. It refers usually to natural resource use by corporations but could also include the responsibility of consumers for their wasteful consumption of environmentally damaging products.

Initial environmental cost internalization and full-cost pricing by enterprises does not mean that producers have to bear all the cost. Depending on price elasticities of supply and demand, enterprises might be able to share the effects of cost-pushed price increase with consumers. At the international level, shared responsibility for outsourcing hazardous production processes and importing natural resources would justify some compensation of sustainability 'exporting' countries by the importers (cf. Section 6.3.2).

A more specific microeconomic formulation of the UPP focuses on the compensation of providers or protectors of ecological services. The International Union for Conservation of Nature and Natural Resources (IUCN) has been promoting *eco-compensation* according to the benefits of ecological services provided, or the – damage – cost of their loss. Considering such benefit or damage as externalities of economic activity, their internalization in the budgets of households and enterprises would be desirable from optimal production and consumption points of view. The drawbacks are measurement and valuation problems of ecosystem services (Sections 2.4.1, 8.1.3). On the other hand, case studies indicate that in particular situations, the carrot of subsidies and the pacifier of compensation (e.g. for giving up land development for eco-system maintenance) may be conducive to 'harmonious' development[6] [FR 13.2].

13.3.2.2 Market (Dis)incentives

Different market instruments show different economic and ecological effectiveness. A brief evaluation of the main categories of these instruments gives a first impression of their use, usefulness and information requirements for setting them at an 'appropriate' level. As discussed in Section 2.3.2 and Annex I, one would ideally seek to set the incentive for environmental cost internalization at the optimal level. At that level, the sum of marginal environmental damage and conventional economic cost equals marginal revenue. In practice, some kind of heuristic standard costing, as applied in green accounting, is probably the only way for an informed setting of market instruments.

Table 13.1 distinguishes market instruments from 'hard' CAC measures by relaxing the prescription of what environmental protection should achieve and/or how environmental measures should be carried out. 'Soft' instruments of education, information, environmental subsidies and voluntary agreements are most 'relaxed' as their application is usually optional. More incisive tools, which set clear standards and disincentives to prevent or reduce the violation of environmental standards, can be categorized as 'semi-soft' (or semi-hard). They are the most promising tools in changing the environmental behaviour of producers and consumers.

Table 13.2 evaluates common environmental policy instruments as to their ecological and economic efficiency and practicality. Market instruments either create new markets, or attempt to influence market behaviour by incentives for environmentally friendly and disincentives for environmentally damaging production and consumption. Actual applications frequently combine different instruments of incentive subsidies and disincentive charges and taxes. Pigovian eco-taxes, deposit-

[6] Harmonious development is the fundamental principle of tackling the social impacts of accelerated economic growth in China (Li et al., 2007).

Table 13.2 Evaluation of environmental policy instruments

	+	–
Command and control *(hard instruments)* - *Prohibition* - *Standards and regulations*	- High and rapid efficiency (in case of uncertainty about impacts) - Effective monitoring and control - Most incisive for high-risk impacts (precautionary principle) - Recycling/reuse applications - Transparency facilitates acceptance	- Economic inefficiency (in finding least-cost solutions) - 'Freezing' existing (best available) technology - Delays in application from legislative process
Market disincentives - *Charges/taxes on emissions, products and natural resource use* - *Removal of environmentally damaging subsidies* - *Deposit-refund systems (recycling)*	- Economic and ecological efficiency - Prompting innovation - Generation of revenue - Fiscal neutrality of eco-tax reform	- Politically set levels of disincentive - Sectoral rather than microeconomic application - Implementation delays - Limited acceptance - Limited coverage of pollutants - High cost incidence with inelastic demand - Difficult ('optimal') damage estimation and valuation - Regressive taxation
Market creation		
- *Property rights and bargaining*	- Greater care for owned assets	- Transaction cost of bargaining (identification of agents and impacts) - Imperfect markets
- *Tradable permits*	- Property rights for sink functions - Cap-and-trade paradigm (combining standards and market forces/preferences) - Application at national and international levels	- Political standard setting - Limited coverage of pollutants - Ignoring local effects - Imperfect markets
Soft instruments - *Subsidies* - *Education* - *Information* - *Voluntary agreements*	- Changing consumption patterns - Non-economic sustainability concerns (equity, ethics) - Acceptance of policy measures - Supporting innovation - Low implementation cost	- Limited adherence to voluntary agreements - Negative economic and ecological effects of subsidies - Advocacy by interest groups (moral suasion)

refund systems and tradable pollution permits are the most commonly used instruments. After its success in London, congestion pricing is being considered in many other cities for reducing rush hour traffic and pollution. Daly (1996) even considers cap-and-trade policies, extended beyond emission trading to natural resource use, as a 'paradigm' for ecological economics: the initial capping caters to the primary goal of ecological sustainability as a 'scale limit' while trading of environmental credits allows for allocative efficiency of conventional economics.

13.3.2.3 Market Creation vs. Public Goods Management

Establishing property rights for public environmental goods is the prerequisite for creating a market for these goods (see Annex I, Section I.2). The purpose is to generate better maintenance of nature's assets, as well as more efficient supply and use of their services through market negotiation and pricing. However, as also indicated in the Annex, Coase-type negotiations rarely work in practice.

The typical features of non-exclusion and non-rivalry of an environmental public good, together with short-sighted overexploitation by economic agents (the tragedy of the commons: Section 2.3.2), would make it necessary to maintain governmental ownership. In this case, cost-benefit analysis is the only tool to introduce some rationality in determining the social value and cost of providing the public good. However, controversial valuation techniques and limited validity from an overall Pareto optimality point of view impair the efficient governmental supply of public goods and services (Section 2.3.2 and Annex I).

An effective way of managing environmental assets is maintaining public ownership, but leasing out the use of the assets to private corporations. In this case, the capture of resource rent through royalty payments (basically a taxation of profit from resource extraction and sale) can be justified from a long-term capital maintenance point of view. The absence of individual property rights indicates governmental responsibility for reinvesting the royalties for the maintenance of the 'common wealth'. Of course, reinvestment into any type of capital caters to the weak sustainability concept only (Section 2.3.1). While applying *prima facie* to scarce natural resource stocks, the case can also be made for the use of absorptive capacities of the environment as argued in the greened national accounts (Section 8.1.2).

The inability of many natural-resource-rich developing countries to transform their natural wealth into economic growth has become known as the 'resource curse' (Auty & Mikesell, 1998). The symptoms include

- Increase in currency value because of resource exports and corresponding slowdown of export-led development (Dutch disease[7]).
- Use of natural resource revenues in public and private consumption rather than for reinvestment.
- Social problems of conflict and corruption dogging the use of resource revenues.

Figure 13.1 contrasts two countries' differing success in turning natural wealth into economic growth. Botswana's government managed to recover a much higher percentage of natural resource rent; it also reinvested these rents following a formal investment rule based on a Sustainable Budget Index. The result

[7] The name stems from oil discovery in the North Sea, which created an increase in the Dutch currency value, reducing the competitiveness of the Netherlands in international markets.

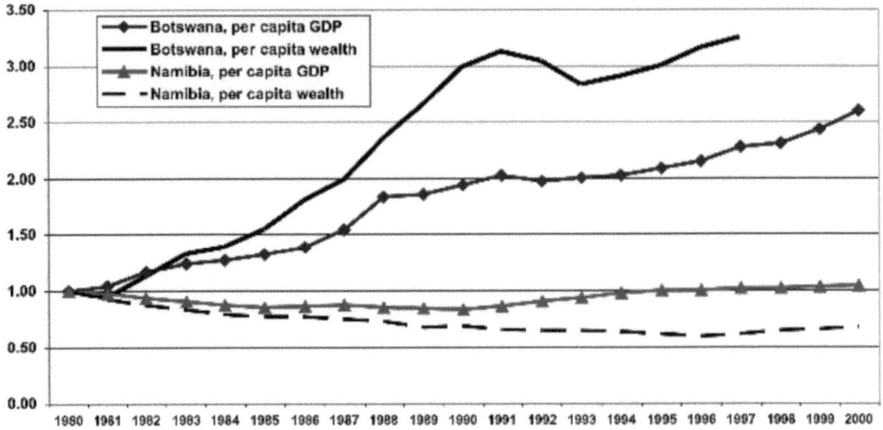

Fig. 13.1 Natural wealth and economic growth in Botswana and Namibia
Source: Lange (2004), fig. 3; with permission by the copyright holder, Springer.

Box 13.1 Averting the resource curse: The Chad-Cameroon pipeline project

In 1999 the World Bank financed a pipeline from landlocked Chad through Cameroon to deliver its oil to an export terminal on the Atlantic Ocean. The deal was to invest most of the country's oil revenues in a development fund under the control of the World Bank. In January 2006 the Bank froze the oil revenue accounts because of a new national law, giving the government unencumbered access to the oil revenues. In response, the government ordered two oil companies out of the country, dismissing at the same time several ministers. In July 2006, the compromise found was a memorandum of understanding, committing the government to spend 70% of its budget on 'poverty reduction programs'.
Sources: http://web.worldbank.org/WBSITE/EXTERNAL/COUNTRIES/AFRICAEXT/EXTREGINI/EXTCHADCAMPIPELINE/0,,contentMDK:20531903~menuPK:1104029~pagePK:64168445~piPK:64168309~theSitePK:843238,00.html; public news.

was a build-up of productive and financial wealth, and average annual economic growth (GDP p.c.) of 5%. Namibia, which does not have an explicit reinvestment policy, showed stagnant wealth and no discernible economic growth. The comparison is based on case studies of green accounting in the two countries (Lange et al., 2003). The authors conclude that such accounts 'provide a framework for a new way of thinking about environmental and natural resource management'. Such thinking may not come easy as illustrated by a World Bank project attempting to avert the resource curse in a newly oil-rich country (Box 13.1).

13.3.3 From Theory to Practice: Ecological Tax Reform in Germany

Germany's ecological tax reform illustrates how governmental policy may water down an ambitious programme of cost internalization and a much-touted 'win-win' or 'double-dividend' strategy for sustainable development (Box 13.2). The idea of the tax reform is to use the revenue from a Pigovian eco-tax to reduce the labour tax burden (social security contributions) of both employers and employees. In other words, the purpose is a tax shift off the 'goods' of labour (in a situation of unemployment) and onto the 'bads' of environmental impacts.

Table 13.3 sketches the transition from a theoretically desirable tax base of environmental damage to more practical emission and energy-use bases. The table also indicates the increasing de-ecologization of an eco-tax down to its current exemption-riddled implementation. In the end, the ecological tax reform looks more like a revenue-spinning tax increase for vague social and economic objectives than an effective instrument of environmental policy. Reading the table bottom-up points to possibilities of re-ecologizing the tax by relating its base to emission or preferably emission cost, instead of energy use.

Tax rate differentiation and exemptions, notably for lignite and high-energy consumers, serve social and economic objectives rather than environmental ones.

Box 13.2 Ecological tax reform in Germany

Discussion of the double-dividend of an environmental tax dates back to the late 1970s in Germany. It was up to a red-green coalition of the social(ist) and green parties to implement the first stage of the ecological tax reform (ETR) in 1999. As of now, it is not clear whether the new government will continue implementing the further stages of the ETR [FR 13.2].

The ETR introduced a new electricity tax and increased the tax rates for gas, heating oil and gasoline (the latter by 3.07 euro-cent per litre). Most of the tax rates were intended to increase even beyond the legislative period of the (now defunct) coalition. Electricity from renewable resources and other environmentally friendly energy production is exempted as are coal and nuclear energy. A reduced tax rate applies to industrial and agricultural energy use beyond 50 megawatt/hr.

Tax revenues are used to lower equally social security payments of employers and employees, in order to realize the double dividend of the tax reform. High-eco-tax payers and generally all households receive further tax relief. The government claims that lower fuel consumption and corresponding CO_2emissions, and greater use of public transport (during 2000–2003) are the result of the eco-tax reform. Rising oil price in world markets might have played a more important role.

Table 13.3 Eco-tax in theory and practice

Tax base	Rationale	Information requirement and critique
1. Marginal damage values ⇓	Pareto optimality, allocation of externalities according to the polluter-pays principle (PPP)	- Measurement of (optimal) environmental damage - Allocation of damage cost to causing agents - Inconsistency with the cost concept of the national accounts
2. Average avoidance and mitigation cost ⇓	- Practical (standard) costing of externalities - Iterative cost adaptation for compliance with standards	- SEEA application of environmental maintenance costing - Sectoral rather than individual cost allocation - Normative standards for environmental cost calculation
3. Physical impact indicators ⇓	Assumed correlation of physical indicators with damage and/or avoidance cost	- Physical input-output table (PIOT) - Uncertain emission-damage correlation due to politically set tax rates
4. Energy consumption (basic rationale of eco-tax reform in Germany) ⇓	Assumed correlation of energy consumption with emission potential of primary energy carriers	- Assessment of emission potentials of different energy carriers - Energy consumption and its emission potential as placeholder for environmental impacts
5. Energy consumption of selected sectors (implementation of eco-tax reform)	Exemptions for maintenance of competitiveness and regional employment	Loss of eco-efficiency: dominance of economic and social objectives

Source: Bartelmus et al. (2003), table 4.

Projected annual tax revenues of about 30 billion DM for 2002 and 2003[8] amount to about half the environmental cost estimates, assuming a continuing downward trend of environmental costs (see Section 8.3 and Annex II). Even this cautious estimate of environmental cost development indicates a significant under-coverage of environmental costs by a loopholed and narrowly defined (energy-consumption-based) eco-tax.

The ecological tax reform seems to have missed by and large the target of hitting the environmental bad. But what about improving the good of employment? Market instruments should bring about Pareto-efficient, welfare maximizing resource allocation through environmental cost internalization. Market forces then

[8] The Euro (€) replaced the Deutsche mark (DM) in 1999 at a rate of 1 € = 1.95 DM.

determine any cost-incidence and price formation. There is thus a priori no compelling argument for a particular use of the tax revenues, such as reduction of labour cost, compensation for environmental damage, investment in natural capital (environmental protection) or reducing governmental debt – to name the typically advanced suggestions for realizing double or triple dividends. One could however reason that high-risk delays in implementing the eco-tax require the fast use of revenues for environmental protection. Economic efficiency would have to be sacrificed in this case for ecological efficiency by speedy environmental action.

The two case studies of resource rent absorption and ecological tax reform suggest alternative uses of the revenues from environmental fiscal instruments.

Governmental owners of environmental assets should take the long-term view of sustaining economic activity by means of capital maintenance. Their objective should be to reinvest the natural capital rent in productive capital in order to maintain – as a minimum – the total capital value (in constant prices) for achieving weak sustainability of economic performance. Any excess revenue could then be used to increase capital for future economic growth and development or be spent for other purposes such as (public) consumption or social transfers.

Private owners, on the other hand, should be either compensated for ecological services made available to others, i.e. for positive externalities, or penalized for the destruction or degradation of natural assets and corresponding negative externalities. This policy of internalizing positive and negative externalities takes the short-term view of changing the behaviour of economic agents towards greater economic and ecological efficiency in handling environmental assets. Revenues from eco-taxes might in this case be used for any governmental purpose ('for the public good'). There is no reason (except offsetting the delayed efficacy of fiscal disincentives) why these revenues would have to reduce labour cost and income taxes rather than ending up in the general tax-pot. The main purpose of the widely publicized win-win strategy looks indeed more like selling the eco-tax to a tax-averse general public.

13.4 Adopting Limits: Sufficiency, Corporate Social Responsibility, Environmental Ethics

13.4.1 Voluntary Action: Sufficiency and Corporate Social Responsibility

Section 3.2.1 addressed the concept of sufficiency as a necessary contribution of consumers to sustainability. Environmentalists consider sufficiency as a means of counteracting the greed of corporations through restraint from the demand side of markets [FR 13.1]. In their opinion, eco-efficiency in production is necessary to tackle environmental problems but cannot bring about, on its own, lasting environmental improvement. The reasons are

- *Rebound effects* from natural resource (cost) savings, which tempt consumers to increase their consumption, e.g. by driving more and at higher speed in low-gas vehicles
- *Failure of cost internalization* policies for both enterprises and households when fiscal charges or pollution caps are set too low or exemption-riddled
- *Lack* or limited effectiveness *of eco-efficient technologies* for imminent or increasing environmental risks
- The lure of advertisements and social pressures, stimulating a mutually reinforcing cycle of *overconsumption*, income generation and economic growth.

The first three arguments refer to the need for *ecological sufficiency*. Ecological sufficiency deals with environmental impacts that are out of the reach of eco-efficiency. The last argument rests on broader psychological, physiological and ethical grounds. It calls for balanced consumption ('doing the right things') rather than efficiency in production ('doing the things right'). This broader concept of an *ethical sufficiency* is expected to bring further rewards of a 'good life' [FR 3.2] by

- Improved health and well-being from a simplified lifestyle
- Spiritual awards from solidarity with poor countries and future generations, whose needs could be satisfied by foregoing consumption and redistributing income and wealth [FR 3.2, 13.1].

Figure 13.2 shows the introduction of a maximum consumption limit (for food) $\bar{c}_1^{max}$ into the linear programming model of Fig. 12.3. The figure illustrates how this additional limit could indeed make the original pollution limit $\bar{x}_p$ redundant by reducing the feasibility space. It also shifts the maximum net product value Z^* to a lower level Z^{**} as the tangent to the reduced feasibility space (in highlighted borders).

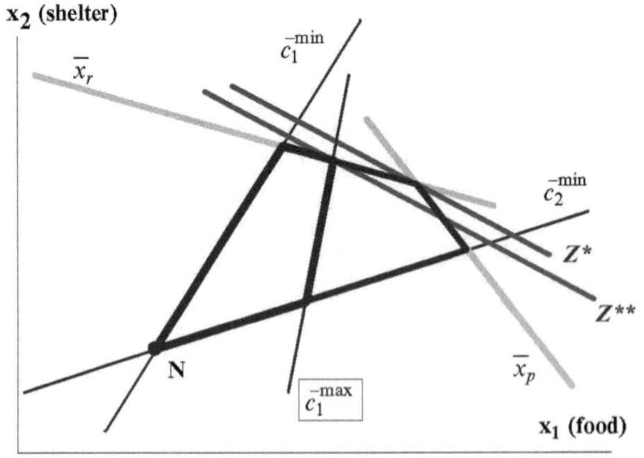

Fig. 13.2 Maximum consumption limits in the feasibility space

Section 9.1.1 described corporate social responsibility (CSR) as a voluntary restraint in the pursuit of economic gain for the benefit of the common good. CSR could thus be a sign of a corporate version of sufficiency. However, the section also cautioned about taking CSR as a guiding principle for social and environmental policies by profit- and growth-oriented enterprises.

Soft instruments of moral suasion, information and education, and perhaps more effective financial incentives (subsidies, tax relief) can encourage the voluntary adoption of sufficiency by producers and consumers. New growth theory claims that R&D and education policies can indeed influence the pace and direction of technological development and economic growth (e.g. Solow, 2000). If, on the other hand, environmental technologies fall out of the blue, as argued by conventional (neo-classical) growth theory, then policy would not be able to induce and shape environmentally sound innovation.

Policies of changing consumption patterns focus therefore on educating consumers about their environmental and social impacts. The softness of these instruments lies in the relatively mild appeal to lifestyle changes without prescription of targets and measures, as indicated in the south-east corner of Table 13.1. Of course, if stronger measures of coercion would be used, sufficiency would lose its voluntary features, with education being turned into brainwashing.

13.4.2 The Driving Force: Environmental Ethics

Calls for sufficiency, moderation and responsibility lead back to advocacy of non-material values and norms of solidarity and equity. Part I summoned these values under the headings of deep ecology and intergenerational equity of sustainable development. One can question discussing philosophical disciplines of environmental and sustainability ethics in a book on *quantitative eco*–nomics. Still, the relevance of ethical and even religious advocacy in ecological economics [FR 1.1] justifies a brief reference to this topic.

As discussed in Section 3.3 ecological economists seek moral or ethical justification for a change in dominant social values, both of individuals and policymakers. To this end they employ the help of institutional economics, extended to the environment in co-evolutionary economics. In their opinion this will also bring about the demise of puzzle-solving mainstream economics. Environmental economists, on the other hand, would argue that environmental ethics and normative economics lack rigour in refuting established economic analysis, run contrary to the dominating utilitarian behaviour of economic agents and governments, and, like religions, fail to reach general consensus on different philosophies. Note, in this context, that the Rio Earth Summit thwarted an attempt to develop an Earth Charter and presented instead a watered-down Rio Declaration.

This book contends that measurement supports efficient management. It does not elevate measurement to paradigmatic level in the sense of 'we measure what we value and value what we measure' (United Nations, 2001b; World Bank, 2003). Clearly, one cannot deny the role of non-measurable ethical or religious convictions in the all-encompassing notion of sustainable development. Individual beliefs drive the acceptance or rejection of this concept. In a democracy, the implementation of integrated environmental-economic policies has to take social values and their change into account, even when focusing more narrowly on the sustainability of economic growth.

However, judgemental analysis of what *ought* to be in our value system and what *should* be our response to moral imperatives resists empirical assessment and analysis. Mixing the two blurs the rational assessment of any threat to the sustainability of human activity and of the adequate response to this threat. This book focuses, therefore, on facts and figures rather than norms, faith, and faith-based advocacy. Ethical questions are out of focus here; elsewhere, other eyes or lenses could and should redirect their focus on social values of equity and care in human relationships with nature [FR 13.3].

Further Reading

FR 13.1 Strategic Principles: Eco-Efficiency, Consistency, Sufficiency

The World Business Council for Sustainable Development, a coalition of 175 international companies 'committed to sustainable development', has been the international protagonist in defining and advocating *eco-efficiency* (http://www.wbcsd.org/templates/TemplateWBCSD5/layout.asp?type = p&MenuId = NzA&doOpen = 1&ClickMenu = LeftMenu). The more ambitious idea of imitating nature's production processes in economic activity is sometimes called biomimicry or bionics (http://en.wikipedia.org/wiki/Biomimetics). It has become more popular under the name of *consistency* (Huber, 2004). The web site of the Zero Emissions Research Initiative (ZERI) presents practical (but limited to mostly agricultural production processes) case studies (http://www.zeri.org/index.cfm?id = vision). They are similar to the eco-techniques of local eco-development (Section 3.3.2) in harnessing local knowledge and resources. McDonough and Baumgart aim at the 'next industrial revolution' with their design firm for wasteless products and production processes (http://www.mbdc.com/overview.htm).

Section 3.2.1 introduced *sufficiency* as a response to overconsumption. Opinions differ on whether sufficiency is a necessary complement to eco-efficiency in sustainability strategies, or just a supportive supplement. Ecological economists tend to argue the strategic complementarity of sufficiency (Sachs, 1995; Daly, 1996; Sachs et al., 1998). Huber (2004) ranks consistency first, with eco-efficiency next (for short-term solutions), and sufficiency last (as a stopgap measure).

FR 13.2 Use and Usefulness of Market Instruments

The polluter-pays-principle (PPP) and the user-pays-principle (UPP) (OECD, 1989) are widely accepted rationales for applying instruments of environmental cost internalization. Principle 16 of the Rio Declaration (United Nations, 1994) caters to the application of the PPP by means of 'economic instruments' (a synonym for market instruments). The Declaration precedes, however, such advice by warning of 'threats of serious or irreversible damage' that would call for a 'precautionary approach' (Principle 15). Most environmental economics textbooks [FR 2.2] describe and evaluate these instruments. Barde (1994) gives an overview of OECD countries' experience with the use of economic instruments.

The IUCN propagates the user/beneficiary-oriented 'eco-compensation', notably in China (http://www.iucn.org/en/news/archive/2006/09/04_china.htm). The EU launched the so far largest multi-country, multi-sector emission trading scheme in January 2005; however, emission caps have so far been too low for effective trading and emission reduction (http://ec.europa.eu/environment/climat/emission.htm). Goulder (1995) coined the popular term of a 'double dividend', the guiding principle of the German Ecological Tax Reform (http://www.ecologic-events.de/oekosteuer/en/) (mostly in German). Case studies in Southern Africa demonstrate the merits of scarcity pricing (notably of water use) and rent capture for reinvestment in productive assets (Lange et al., 2003).

FR 13.3 Environmental Ethics

Environmental ethics stresses human rights of access to and equity in the distribution of nature's resources and amenities: see, e.g. the contributions to Tolba (2001, chs. 2.19–25). Contrary to a spiritual or religious view, ethics presents a philosophical analysis, notably of the 'intrinsic' value of nature (Elliot, 2001). Sylvan and Bennet (1994) distinguish three levels of environmental ethics: shallow (anthropocentric), intermediate (allowing for some value of non-humans) and deep (rejecting priority of human over non-human values). Readers on environmental ethics present the wide range of arguments about the role of equity and equality in social values and sustainable development (O'Neil et al., 2001; Light and Rolston III, 2003). FR 2.1 describes the environmentalist call for replacing the economic rationale of the *homo oeconomicus* by a more altruistic *homo politicus* (and other *homines*).

The flagships of internationally agreed ethics, the United Nations Charter (http://www.un.org/aboutun/charter/) and the Universal Declaration of Human Rights (www.un.org/Overview/rights.html) do not refer to environmental concerns, since environmental protection came to the United Nations' agenda only through the 1972 Stockholm Conference (United Nations, 1973). Ten years later the Rio Summit rejected proposals for an *Earth Charter* in favour of a weaker, anthropocentric Rio Declaration that focuses on *human* needs (United Nations, 1994).

Non-governmental organizations, upon the initiatives of the Earth Council and Green Cross International, have since set out to develop the Earth Charter from the grassroots by means of a popular movement (Lubbers & Morales, 2001; www.earthcharter.org).

Review and Exploration

- Is muddling through an option?
- Is eco-efficiency a useful (operational) concept for attaining sustainable economic performance and growth?
- How realistic is consistency for avoiding environmental impacts?
- Should we change our life(style) to attain sustainable development? Or can we rely on environmental policy?
- Compare the economic and ecological efficiency of policy instruments. Why should we use command-and-control measures?
- How realistic are win-win strategies such as an ecological tax reform?
- Compare the anthropocentric and eco-centric approaches to environmental ethics. How do they foster environmental protection? Should they be part of *eco*–nomics?
- Do we need an Earth Charter?

Chapter 14
Globalization and Global Governance

Globalization is not a new phenomenon. Crusades and other forms of proselytizing spread faiths and cultures over the planet. Colonization was a less disguised approach of imperialism and exploitation of the natural wealth of 'undeveloped' countries [FR 1.2]. What changed towards the end of the last century is the reach, speed and intensity of global interaction and interdependence among countries and their citizens.

Depending on the scope of this interaction, a broad multidimensional, and narrow economic definition of globalization can be ventured as

- An accelerated process of economic, social, cultural and epistemic exchange and interdependence among countries and communities, owing to advances in transport and information/communication technologies, and the lowering of trade barriers; and more narrowly
- Trade liberalization and transnationalization of the economy,[1] i.e. the internationalization of markets for goods, services, labour and capital, and the integration of production and marketing by transnational corporations.

Does this mean that we have entered a 'global age', in which political movements, corporate management and a 'world state' will replace the nation-state (Albrow, 1997)? Are downsizing governments losing control over social movements and losing sight of their social and environmental mandates? Are national and international civil societies entering the power vacuum, or is it the transnational corporation? Most of the answers given to these generic questions remain theoretical or speculative [FR 14.1]. At the same time, global environmental problems and policy failures in tackling them triggered much of the new 'globalism' (op. cit.) so visible in anti-globalization protests.

Globalization has been blamed for the failure of achieving sustainable development; others argue that it may facilitate such development. After examining the

[1] As pointed out by Tilman Santarius of the Wuppertal Institute for Climate, Environment and Energy, 'transnationalization' of production chains is a defining feature of recent globalization, in contrast to long-standing 'internationalization' of markets.

claims and counter-claims about the effects of globalization, this chapter discusses what should be done about it. The World Trade Organization (WTO) has become the rallying point of the anti-globalization movement. Greening the organization or creating countervailing power represent the range of suggestions for curbing WTO's trade liberalism.

14.1 Sustainability Effects of Globalization

Extending market forces beyond national borders in order to reap the fruits of Ricardian (Ricardo, 1817) 'comparative advantage' is the basic rationale for foreign trade. The purpose is efficient use of different know-how and resource endowments of countries through specialization in production and trade. The relatively peaceful time after the Second World War, together with technological innovations in transport, information exchange and communication, and international support for trade liberalization, increased the scope, magnitude and speed of foreign trade. The result has been unprecedented economic growth and interdependence of countries, which many take as a sign of inevitable globalization.

Opinions differ strongly, though, as to the positive and negative effects of globalization. On one side, the above-discussed EKC hypothesis (Section 11.1) re-emerges in defence of trade liberalization: trade-driven economic growth brings not only prosperity but also environmental protection. This supposedly explains 'why greens should love trade'.[2] Others caution about opening the borders to the unfettered pursuit of economic profit by powerful uncontrollable transnational corporations. In their view, calls for corporate social responsibility and public-private partnership (Section 9.1.1) can do little to prevent these corporations from 'disembedding' economic activity from its – national – institutional framework. National social achievements, including social security, environmental quality and the preservation of cultural heritage risk being lost in a competitive 'race to the bottom'. Once 'stuck to the bottom' of lost social and environmental institutions (Porter, 1999), any revival of social and environmental values becomes unlikely under continuing competitive pressure.

Table 14.1 lists the main arguments for and against globalization in terms of its effects on the three basic dimensions of sustainable development. The table includes the social dimension because of the importance attached to entrenched poverty in the globalization debate. On the positive side, mainstream economists argue that trade-triggered growth facilitates environmental protection and alleviates poverty in the trade-partners. In their view, global trade liberalization will also lead to greater efficiency in the use of natural resources, and will remove mismanagement and corruption by introducing environmental and social standards to developing countries. Also, there appears to be little evidence for any country lowering its

---

[2] Headline of *The Economist* of October 9–15, 1999.

Table 14.1 Sustainability effects of globalization: Pros and cons

+	–
Economy	
Welfare gains: - Trade is good for economic growth (efficiency gains from international specialization, foreign direct investment spurs economic growth) - Increase in product variety	Welfare/wealth losses from: - Outsourcing (job losses) - Externalization of social costs (see 'environment', below)
Efficiency gains: - Comparative advantage of trade (benefits all countries) - Removal of mismanagement and protective subsidies (through competition) - Enforcement of TRIPS (trade-related aspects of intellectual property rights) (incentive for innovation) - Access to new technologies and larger markets	Efficiency losses: - Market dominance by TNCs (trans-national corporations) (affects national and international competitiveness) - Currency and financial market speculation (volatile capital flows) - TRIPS (creating and maintaining technological monopolies)
Social values	
- Growth is good for the poor (trickle-down effect, compensation of losers by winners) - Competition removes parasitic groups - Transfer of social (labour) standards - Increase in minimum wages - Creation of inter-governmental organizations (United Nations and its specialized agencies, Red Cross etc.) and global civil society - Information exchange (epistemic society)	- Marginalization of poor countries (technological divide by TRIPS, concentration of wealth in TNCs, one-sided removal of trade barriers) - Disembedding of markets from national social achievements and institutions (abandoning labour, health, social security and human rights standards under competitive pressures) - McDonaldization through fordism (cultural homogenization) - Asymmetric mobility of production factors (generating unemployment from 'unfair' competition by low-wage countries) - Uncivil society exploits power vacuum (terrorism, corruption)
Environment	
- Growth through trade is good for the environment (EKC-hypothesis) - Natural resource saving (from least-cost sources) - Use of free local carrying capacities - Transfer of sustainability model (environmental preferences, spread of environmental awareness, debt-for-nature swaps, environmentally sound technologies) - Removal of environmentally damaging subsidies (in particular, fishing, energy, agriculture)	- Overexploitation of natural resources by TNCs (tragedy of the commons) - Spread of wasteful consumption patterns - Rebound effects from resource (cost) saving - Import of sustainability from poor (weak) countries (externalization of environmental costs) - Environmental dumping (EKC rejected: competitive race to the bottom)

environmental standards or for relocating its business to pollution havens, as a race to the bottom would have it (Wheeler, 2002).

Environmentalists and other anti-globalization groups dispute these effects. They point to the lack of evidence for the EKC hypothesis, the propagation of Western wasteful lifestyle, sustained by environmental exploitation of the South, and failure of trickle-down growth policies in the development decades (Box 3.2). An alliance of weak governments and profit seeking business, the much-touted public-private partnership, will not hinder the dumping of environmental and social standards and policies. Weaker strata of society, developing countries and future generations will have to carry the burden of this 'externalization' of previously internalized social costs of environmental impacts and inequity.

The details of the violently debated – just think of the events surrounding the WTO conferences in Seattle, Cancún and Hong Kong – arguments are left to further reading [FR 14.1]. The purpose of this chapter is to draw attention to the need for

- Replacing clamorous protests and advocacy by a rational and comprehensive quantitative assessment of the different assertions
- Examining the dominance of neoclassical market economics and its growth paradigm in the established global governance regime.

To this end, the measurement and accounting tools described in this book should be further harmonized in international recommendations and included in the regular programmes of the national statistical services. For instance, the worldwide use of greened national accounts could provide the information for confirming or rejecting the claim that market imperfections (together with capital mobility) render comparative advantage in foreign trade irrelevant (Daly & Farley, 2004). Full cost accounting, including environmental and other trade externalities, might reveal that the total cost of natural resource exploitation exceeds commodity prices. In this case, resource use and trade would indeed have to be curbed – either directly or by international tools of cost internalization (Bartelmus, 1994a). Unfortunately, the paucity of data does not permit so far an authoritative conclusion about the net benefits or damage of globalization.

The next question is whether and how globalization's sway over social and environmental values can be checked. Should we change the nature of globalization? Or should we let globalization take its inevitable course while separate social and environmental policies could tackle negative effects and reinforce positive ones? The following section addresses these questions. It examines the international institutions and actors that foster or could rein in globalization.

14.2 Global Governance for Sustainable Development

As discussed, environmentalists and economists offer partisan opinions about globalization, facilitating or impeding sustainable growth or development. As before (when setting out from the pessimistic outlook of the LTG model in Chapter 13), let us take the positive effects for granted, focus on the negative side, and ask what could be done to 'civilize' (Kates, 2003) globalization.

Chapter 13 reviewed strategies and policies at the national level. All of them could also apply at the international level. Most multilateral environmental agreements (MEA) are regulative in nature. Some like the Kyoto Protocol also propose market instruments for emissions trading. The Johannesburg Summit's implementation plan calls for voluntary partnerships between governments, civil society and corporations. However, at the global level, there is no similarly powerful institution as the national government, which could enforce with equal strength and efficiency the use of these instruments. Rather we have a 'world of states' ('*Staatenwelt*': Messner & Nuscheler, 2000), which is a far cry from a world state. States do collaborate though in agreeing on a multitude of treaties, conventions, regimes and institutions. Together, these agreements make up global governance rather than government [FR 14.2].

There are over 200 MEA; they are fragmented and overlapping, and lack implementation, control and funding. International consensus is much stronger in the powerful Bretton Woods institutions (IMF, World Bank) and the World Trade Organization (WTO). The recent failure of the Doha round of negotiations might indicate some fraying of the WTO, though. Still, the WTO has been the envy of environmentalists, who would like to obtain its powers of arbitration and sanctions. At the same time the WTO, more than any other organization, has become the symbol of globalization and capitalist dominance. This section discusses the need for and possibilities of rectifying this imbalance in global environmental and economic governance.

14.2.1 Greening the WTO

The WTO is the successor to the General Agreement on Tariffs and Trade (GATT). Its main function is to ensure free trade of goods and services among member states by removing or reducing trade barriers and settling trade disputes. Box 14.1 lists the main WTO rules and stipulations, which refer to environmental protection. Almost all WTO rules address the effects of environmental protection *on* trade. They ignore largely the effects *of* trade liberalization on the environment. One exception is the statement of the WTO Trade and Environment Committee (TEC) that trade liberalization 'can yield benefits both for the multilateral trading system and the environment'.[3]

The TEC was established in the wake of the 1992 Earth Summit to examine the relationships between environmental policy and trade. In line with WTO policy, the Committee left dealing with the environmental impacts of trade to environmental agencies and policies. The TEC concluded that

[3] http://www.wto.org/english/thewto_e/whatis_e/tif_e/bey2_e.htm.

Box 14.1 Main environmental provisions of the WTO

Marrakesh Agreement Establishing the WTO (1995, preamble): '...allowing for the optimal use of the world's resources in accordance with the objectives of sustainable development, seeking both to protect and preserve the environment and to enhance the means for doing so...'.

Subject to the non-violation of WTO principles of non-discrimination, the following exemptions to WTO rules apply:

- *GATT 1994, Article 20* (trade in goods), *TRIPS, Article 27* (property rights), and *GATS, Article 14* (trade in services) for policies 'protecting human, animal and plant life or health'
- *Technical Barriers to Trade Agreement* and *Sanitary and Phytosanitary Measures* (preamble): 'no country should be prevented from taking measures necessary to ensure the quality of its exports, or for the protection of human, animal and plant life or health, of the environment...'
- *Agreement on Subsidies and Countervailing Measures*: allows subsidies, up to 20% of firms' costs, for adapting to new environmental laws.

Sources: [FR 14.2].

- Countries, which are signatories of an environmental agreement, should settle any disputes (including trade-related ones) under that agreement.
- When only one country is a signatory, an unsettled issue is 'open for debate'.
- WTO rules, with the exemptions described in Box 14.1, apply when there is no environmental agreement.

Box 14.2 presents an example of the case where an Article 20 exemption applies but WTO trade principles are violated. Note that in this case the WTO ruling was in favour of the exemption of Article 20, but had to decide against the environmentally motivated import ban of shrimps because of a violation of the basic principle of non-discrimination.

The preamble to the Agreement on Establishing the WTO does refer to sustainable development (see Box 14.1). Focusing on the 'optimal use' of natural resources this reference looks more like a call for efficient natural resource exploitation than a desire to implement sustainable development. The Ministerial Doha Declaration (2001) did confirm WTO's commitment to sustainable development. However, suspension of the Doha negotiations at the July 2006 meeting of the General Council (albeit recently resumed) indicates that national economic interest may still overpower both, free trade and concerns for related environmental impacts.

All in all it appears that WTO rules do respect national environmental policies and standards, even if proclamations on sustainable development appear to be rhetoric. Still, environmentalists call for the reform of the WTO 'to make it more open and broadly accountable' (Speth, 2003). Eager to employ the powerful tools of

Box 14.2 WTO ruling: The shrimp-turtle dispute

In 1997 India, Malaysia, Pakistan and Thailand brought a complaint against a US ban on shrimp imports from their countries. The US enacted the ban in compliance with national law, which prohibits the importation of shrimp harvested with (netting) techniques that endanger sea turtles.

The WTO Appellate Body confirmed that the import ban is justified according to GATT Article 20. It ruled against the US, however, because the ban violated the basic WTO principle of non-discrimination. The US discriminated against the four Asian countries by not giving them the same advantages as those granted to Caribbean nations. The advantages included technical and financial assistance and longer transition periods for installing 'turtle excluder devices'.

Source: http://www.wto.org/english/thewto_e/whatis_e/tif_e/bey2_e.htm#turtle.

trade sanctions for enforcing environmental agreements, they seek to instil environmental concerns right into WTO's existing rules and policies. This would ensure that the multilateral trading system addresses the hitherto ignored environmental impacts of trade. The Wuppertal Institute for Climate, Environment and Energy advanced the idea of a 'Strategic (environmental) Impact Assessment (SIA) Body', at a par with the Trade Policy Review and Dispute Settlement Bodies of the WTO. The SIA Body could 'rationalize' WTO policy by factoring in scientific – environmental – knowledge (Santarius et al., 2004).

It is far from certain that such infiltration would succeed in turning WTO into an agency that fosters sustainable development. Changing directly some of the WTO rules might be more effective as a tool of restricting trade for actual and probable environmental impacts. Taking probability into account would also cater to the precautionary principle recommended by the 1992 Earth Summit (Section 13.2). Adjustments to WTO rules could include (UNDP, 2003b; Santarius et al., 2004)

- The use of 'waivers' of WTO rules for specific environmental policies
- Shifting the burden of proof from environmental complainants to the presumed violator of environmental law
- Extending the exemptions of Article 20 to environmentally damaging production processes
- Observer status for environmental organizations at WTO conferences to render WTO's decisions more transparent, and most of all
- Restricting the power of WTO's Dispute Settlement Board when dealing with environment-trade effects, possibly by shifting arbitration and ruling to an international court.

Such softening of WTO rules risks protectionist trade rulings for economic gain under pretence of environmental protection. As a first step towards a possible compromise, the Doha Declaration agreed in 2001 to launch negotiations on trade and environment. The main issues on the Doha agenda are clarification of the relationships between WTO rules and MEA,[4] information exchange with, and observer status of, environmental organizations, trade barriers to environmental goods and services, and fishing subsidies. Missed deadlines and disagreement, notably on export subsidies and market access, make the full implementation of the Doha Action Plan and the adoption of sustainability criteria in WTO rules and policies unlikely. Other measures aim therefore at curtailing the scope and domination of the WTO by the creation of countervailing power in environmental agencies or citizens' movements.

14.2.2 Creating Countervailing Power

Global environmental governance is weak in comparison to international trade agreements under the WTO. Mandates are spread over numerous global and regional organizations and 'specialized' UN agencies. The agendas of the global environmental conferences lack implementation and control, as do most of the MEA. UNEP, the main coordinating body, is one of the smallest organizations of the United Nations. Various proposals suggest, therefore, changing the imbalance of power between the international trade regime and environmental governance by

- Curbing the scope and enforcement powers of the WTO, possibly even up to the point of dismantling the organization
- Strengthening existing environmental organizations, or creating a World Environment Organization, on a par with the WTO
- Opposing the top-down global regime of the WTO with grassroots movements of concerned citizens and non-governmental organizations (NGOs).

Environmentalists argue that WTO's objective of creating a global economy opens national borders to the dominance of private over community benefits (Daly & Farley, 2004). The result of such 'borderlessness' is a 'power shift … from national, state, and local governments and communities, toward …global corporations, banks, and the global bureaucracies they helped to create…' (Mander, 2003).

Environmentalists differ on how to deal with corporate dominance in a borderless world. Daly and Farley (2004) favour 'internationalization' over globalization. Internationalization is to improve international relations between sovereign nation-states by means of trade, treaties and alliances. Obviously, but without saying so, this would make the WTO redundant. Some deep ecologists and activists seek to reverse globalization by localization. This approach relies on a grassroots

[4] Conflicts could arise from trade restrictions in MEA, notably the Convention on International Trade in Endangered Species of Wild Flora and Fauna, the Montreal Protocol and the Basel Convention on Hazardous Waste. To date, however, there have been no disputes in this regard.

ground swell to curb globalization, led by 'hundreds of activist organizations' (Mander, 2001):

> There is nothing inevitable about the present system [of globalization]. It is just a set of rules and institutions that we can change if we want to. If we have a democracy (do we?) then we can change it. A lot of people understand this, and are mobilizing to change it, as we have seen in Seattle and other places. (Mander, 2003)

The question is, should grassroots movements of civil society and its institutions become major players in national and international policymaking? Non-governmental organizations are likely to answer in the affirmative. Representatives of governments and intergovernmental organizations point to the 'democratic deficit' of organizations that are not elected and can, therefore, not be held accountable for their actions. The discussion of communitarian thought and local eco-development (Section 3.2.3 and FR 3.3) also indicated a lack of local government commitment and diversity of local conditions and programmes as obstacles to the general acceptance of local-level development strategies. Add to this the need for transferring central power to local organizations for attaining a certain degree of self-reliance, and the prospects for localization replacing globalization look unlikely indeed.

If globalization and its captain cannot be sufficiently tamed from the grassroots, the only solution for rectifying the institutional imbalance would be to strengthen existing environmental agencies or to create a new one that holds its own against the WTO. Strengthening UNEP has been mantra in sessions of its Governing Council. Proposals for a World Environment Organization have run into opposition by almost all countries [FR 14.2]. After exhorting the 'entitlement' of human beings 'to a healthy and productive life in harmony with nature' (Principle 1), the Rio Declaration (United Nations, 1994) hastens to stress in the second Principle the 'sovereign right [of nations] to exploit their own resources pursuant to their own environmental and developmental policies'. Clearly, this reflects antagonism against a new powerful environmental or developmental organization. In the end, governments settled at the Rio Summit for cooperation 'in the spirit of global partnership' (Principle 7).

14.2.3 Towards a Global Compact?

Calling for global partnership while stressing national sovereignty does not augur well for governmental collaboration. On the other hand, improved collaboration of international organizations might be, at least in the short term, the only realistic way to tackle the worst symptoms of globalization and to calm anti-globalization protests. This is indeed the above-mentioned internationalization strategy without the implied removal of the WTO.

As discussed in Section 9.1.1, governments also tried to share their responsibility for environmental and social concerns with the private sector; they appeal to corporate social responsibility and bet on the knowledge and financial support of corporations. Such public-private partnership could be seen as a power shift to already dominating multinationals. Possibly in anticipation of this critique, the Secretary General of the United Nations launched in 2000 a 'Global Compact – that would bring companies together with UN agencies, labour and civil society to support universal environmental and social principles'. The ten principles apply to human rights, environment and the fight against corruption. 'Unit[ing] the power of the market with the authority of universal ideals' is expected to help the spread and implementation of these ideals. Parallel to this effort, governments reiterated in their Millennium Development Goals the need for sustainable development, but with a new focus on global partnership and globalization (see Section 3.1.1).

One can hardly criticize the good intentions of partnership, cooperation and sharing responsibilities to tackle global problems. All these initiatives are voluntary, though. As so often before, these proclamations might remain rhetoric, serving the appeasement of demonstrators and gaining cheap credits with the respective constituencies.

What can be done and by whom? The preceding sections offered the alternatives of changing the globalization process or managing its outcomes, greening or removing the WTO, and increasing environmental power or improving public-private collaboration. Much of the arguments, especially on the more radical side, are based on anecdotal or biased evidence, or no evidence at all. One cannot but suspect that in these cases ideologies and preconceived views about what ought to shape society and the planet are behind such argumentation.

Perhaps one conclusion from all these arguments and counter-arguments is that mixing social, environmental and trade policies *at the global level* is not a good idea. Should we really use trade policies to solve environmental problems, e.g. by trade sanctions? Or should environmental policy determine the rules of international trade? Given the fiercely held convictions of environmentalists and (trade) economists, overloading the WTO and environmental organizations might hinder rather than foster policy and regulation. Tackling the worst symptoms of globalization separately while 'unleash[ing] the creative energies of private entrepreneurship' (Rodrik, 1997) seems to be a reasonable approach. As a minimum, such a strategy should be constantly on the outlook for negative impacts of globalization by means of effective early warning indicators. Preferably, these indicators should also show the significance of environmental and social impacts – as compared to the significance of economic benefits – the task of green accounting (Ch. 8).

We have come full circle in the analysis of the interaction of environment and economy and, to a lesser degree, of related social issues. Initially we found near-religious warning about environmental doom from demographic and economic growth policies of governments. New villains, accused of generating social and environmental disaster, showed up later: the multinational corporations and their supportive international organizations. Unfortunately, the assessments of Parts II to IV do not present irrefutable facts and figures about the sustainability or non-sustainability of economic activities, especially at the international level. As a consequence, we are left with a number of questions, for which there may be only partial or no answers at all. The next and final chapter raises some of these questions.

Further Reading

FR 14.1 Globalization

The Brundtland report (WCED, 1987) was among the first to draw attention to 'economic and ecological interdependences among nations' and corresponding 'interlocking crises' in the fields of environment, development and energy consumption. Stiglitz (2003) focuses on the economic and distributional effects of globalization, which increase the divide between rich and poor countries; he argues that 'free market ideology, propagated by the "Washington consensus"' (IMF, World Bank and the US treasury), is the main cause. A UNDP (2003b) report comes to similar conclusions with regard to global trade and human development.

Rodrik (1997) gives a concise and cool assessment, among the vast and polarized literature on the economic, social and environmental effects of globalization. For instance, Daly (1999) warns of '... "globalization" which ... turns out to be unfettered individualism for corporations'. In a similar vein, Mander (2003) holds the power shift to global corporations, banks and global bureaucracies responsible for 'grave consequences for national sovereignty, community control, indigenous cultures and ... the natural world'. In contrast, Bhagwati (2002) calls for the defence of corporations 'against ignorant, ideological, or strategic assaults' by the anti-globalization movement. He also argues that globalization actually '*has* a human face', which can be enhanced by institutional change (Bhagwati, 2004, 2007). Speth (2003) presents a broad range of views on globalization and the environment in a collection of lectures and articles. The International Institute for Sustainable Development (IISD) seeks ways of making trade and sustainable development compatible and offers a primer on trade and environment (http://www.iisd.org/trade/).

FR 14.2 Global Governance

Global environmental policy is enshrined in international treaties and agreements. UNEP keeps a listing of MEA on its environmental law web site: http://www.unep.org/DPDL/Law/Law_instruments/multilateral_instruments.asp.

The well-organized site of the WTO (http://www.wto.org/english/thewto_e/whatis_e/whatis_e.htm) provides for a discussion of its trade and environment policy, of course from a WTO point of view: http://www.wto.org/english/thewto_e/whatis_e/tif_e/bey2_e.htm. The site also describes the current state of the Doha negotiations: http://www.wto.org/english/thewto_e/whatis_e/tif_e/doha1_e.htm. UNDP (2003b) presents a detailed but diplomatically tamed overview of WTO agreements and policy.

Simonis (2005) calls for establishing a World Environment Organization (WEO) for global environmental governance. Biermann and Bauer (2005) present arguments for and against a WEO. Jerry Mander, the president of the International Forum on Globalization, who is also programme director for the Foundation for Deep Ecology, calls for localization as the principal force opposing globalization (Goldsmith & Mander, 2001, introduction). Hines's (2000) even suggests replacing the GATT rules by a GAST (General Agreement on Sustainable Trade) and the WTO by a WLO (World Localization Organization).

A web site of the United Nations presents and explains the Global Compact: http://www.unglobalcompact.org/AboutTheGC/index.html. See for a critique of the Compact: http://www.laetusinpraesens.org/docs/globcomp/globcom2.php.

Review and Exploration

- What is so new about globalization?
- Does globalization help or hinder sustainable development?
- Is comparative advantage a valid argument for trade liberalization?
- WTO – an enemy of sustainable development? Do we need a World Environment Organization?
- Can we, and should we, green the WTO?
- Will the Global Compact mitigate negative and generate positive globalization effects?

Chapter 15
Questions, Questions, Questions – and Some Answers

We are now about to end a long journey that took us from vague anecdotal and even spiritual descriptions of environmental problems to more systematic measurement and assessment. We then extended these assessments into prediction and policy analysis. Figure 15.1 illustrates how we set out from viewing environmental problems as symptoms of overcharged ecological carrying capacities (what's the problem?). Economic activities are responsible for this overuse of environmental services, but they also offer solutions to mitigating environmental impacts (what's economics got to do with it?). Integrative environmental-economic policies face trade-offs and need to set priorities. Before taking action (what can be done?), one has to compare, therefore, the significance of environmental impacts with the benefits of economic activity – now and in the future (how bad is it?).

The interaction of environment and economy spans nearly everything under the sun, and in fact the sun as well. It is no surprise that there are no definite answers to the four questions of Figure 15.1. Lack of knowledge and empirical evidence are

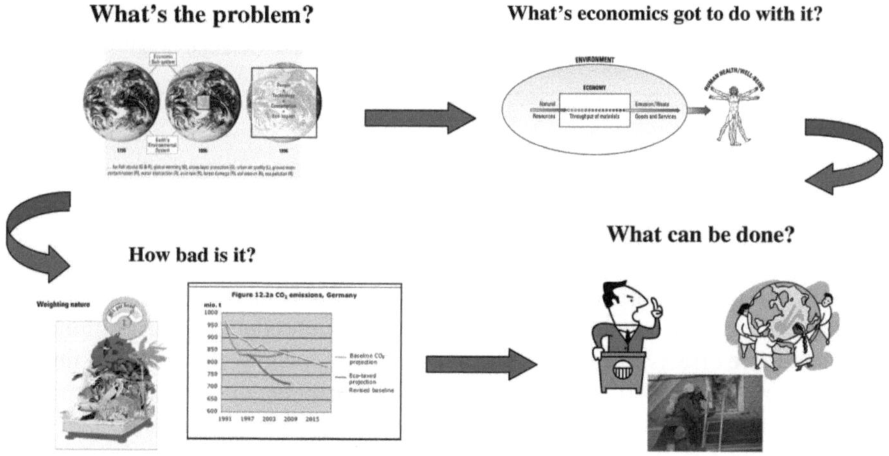

Fig. 15.1 Towards sustainability – conclusive questions

P. Bartelmus, *Quantitative Eco-nomics,*

the reasons. All one can do now, at the end of the book, is to raise these questions again and summarize the – sometimes contradictory and partial – answers in a brief synopsis of 15 questions. The solution of remaining problems will have to be left to future research.

15.1 What Is the Problem?

History tells about the downfall of societies because of overuse of natural resources and local climate changes, but also because of overpopulation, excessive taxation and war (Section 1.1). Current data reveal environmental problems of natural resource depletion, pollution and deteriorating health of humans and ecosystems (Table 1.1).

Q1: Do available data describe the 'problem'?	Environmental indicators assess different symptoms of environmental deterioration. They *alert* us to actual or potential violations of environmental standards. They *do not provide a comprehensive picture of the overall sustainability* of economic growth or development (Sections 1.3, 4.2.3).
Q2: Is it a matter of limits?	The ecological point of view sees non-sustainability as the *transgression* of planetary and local *carrying capacities* (Sections 1.3, 2.4.1). Available data do not assess unequivocally the closeness to, or ultimate violation of, global or regional limits (Sections 1.3, 4.2.3).

15.2 What Has Economics Got To Do with It?

The initial assessment of selected environmental problems (Section 1.3) indicates interdependence of economic activity and environmental deterioration, and human welfare effects from both. The question is if and how the powerful integrative concepts and tools of economics apply to environmental and social concerns. Environmentalists and ecological economists reject the commodification of environmental and social services through market valuation. Environmental economists, on the other hand, describe environmental impacts as the result of market and policy failures, which can be remedied by market instruments and policy reform (Sections 2.1, 13.3). The following questions reflect this *dichotomy*.

Q3: Is the assumption of a rational, utility maximizing *homo oeconomicus* of practical use in an imperfect world?	*Yes*, because - Economics in a vacuum throws light on complex problems (Section 2.1) - Markets reveal individual preferences, which experts or governmental fiat should not override (Sections 2.3.2, 13.3.2). *No*, because - Economic rationality ignores the altruistic side of a *homo politicus* [FR 2.1] - Of the history of wrong diagnosis and projections, and of misleading policy advice by economists (Section 2.1) - Environmental experts are better equipped to see the looming violation of planetary limits than short-sighted economists and policymakers (Sections 1.3, 2.2.3).
Q4: Is economic growth the solution of the environmental problem?	*Yes*, because - Post-industrial service economies are dematerialized and can afford environmental protection (Section 2.2.2) - Policy tools of environmental cost internalization foster innovation and seek optimality and sustainability in economic performance and growth (Section 2.3.2, 12.3, 13.3.2). *No*, because - The environmental Kuznets curve hypothesis is generally rejected (Section 11.1) - Vital environmental thresholds have been transgressed in a full-world economy (Sections 1.3, 2.2.2) - Complementarities of critical natural capital prevent substitution by other production factors (Section 2.4.2).
Q5: Is sustainable development the solution?	*Yes*, because - The paradigm alerts us to interactions with other, notably social, development goals, beyond economic and environmental ones (Section 3.2.2) - Of the need to introduce ethics and social values into policymaking so as to counteract irrelevant or misleading advice by puzzle-solving economists (Sections 2.1, 13.4.2) - Economic wealth does not make us happy (Section 3.2.1). *No*, because of: - Failure of development strategies (Section 3.1.1) - Lip service to, and hidden agendas behind, the opaque cornucopian development paradigm (Section 3.2.1) - Normative (expertocratic or governmental) targets and standards that blur scientific analysis (Section 3.3, 13.4.2) - Lack of comparable measures of social (and other) sustainability effects (Sections 3.1, 3.3.2).

15.3 How Bad Is It?

Assessing the comparative significance of environmental and economic costs and benefits is the core issue of this book. To this end, one can either look back and ask how bad it has been, or look forward and see how bad (or good) it will be.

15.3.1 How Bad Has It Been?

The environmental-economic dichotomy trickles down to measurement. On the one hand, we have physical statistics, indicators, and material and energy balances (Chs. 4 to 6). On the other hand, welfare indices (Section 7.1) and greened national accounts (Ch. 8) attempt to synthesize the physical data in money terms. Physical and monetary accounts confirm (rather than overcome) the dichotomy. The question is, what does this mean for assessing sustainable growth and development?

Q6: Can non-economic indicators assess sustainable economic growth and development?	*Indicators* warn us about environmental and social risks and measure progress in reaching particular targets (Section 4.2.3). Aggregation of indicators into *indices* fails to assess the question of how bad 'it' (the overall environmental-economic situation) is. The reasons are judgemental indicator selection, aggregation problems and lack of operational sustainability concepts (Sections 5.3, 7.1).
Q7: Accounting for sustainability: weighting by weight or pricing the priceless?	- Environmentalists opt for physical measures of the violation of collective sustainability targets (Sections 2.4, 6.2.3, 6.3.1). Economists object to overriding individual preferences and rely on market prices to evaluate environmental scarcities and sustainability (Section 2.3, 8.1). - *Physical balances indicate pressure* on natural systems (Section 6.3.1) but fail to integrate environmental and economic effects (costs and benefits) (Section 6.3.3). - *Integrated (monetary) accounts assess* the integrative notion of *sustainable economic growth* (Section 8.2.1).
Q8: Has economic growth been sustained?	- The short answer *is yes* – weakly, and *no* – strongly. - For *strong sustainability*, most countries show only relative dematerialization, i.e. delinkage of material input from economic growth at levels below sustainability targets (factors 4 or 10) (Sections 6.3.2, 10.1.2).

(continued)

(continued)

	- For *weak sustainability*, case studies of green accounting generally indicate capital maintenance, i.e. positive environmentally adjusted net capital formation, except for some low-income African countries (Sections 8.3, 10.1.3).
Q9: Are we better off?	- *No*: some welfare calculations (ISEW/GPI) confirm the threshold hypothesis of declining welfare at high levels of economic growth (Section 10.1.1). - *Probably*: several GPI calculations fail to confirm the hypothesis (Section 10.1.1). - *Yes*: in terms of wealth and consumption, which might not make us happy, but reflect the continuing human quest for more prosperity (Section 3.2.1).

15.3.2 How Bad Will It Be?

Descriptive models of structural analysis (Section 10.2) look backward: they seek to quantify the causes (driving forces) of past developments, whose influence can be expected to reach into the future. *Predictive models* such as the Environmental Kuznets Curve (EKC) and the Limits-to-Growth (LTG) model (Ch. 11) make such implicit trend analysis explicit. *Prescriptive models* try to give a direct answer to the final question of 'what can be done?'

Q10: What are the causes of environmental deterioration?	- Structural profiles (Section 10.2.1), input-output analyses (Section 10.2.2) and decomposition analyses (Section 10.2.3) show *economic growth* and growth-based energy use as the principal cause of CO_2 emission, counteracted to some extent by eco-efficient technology. - *Critique*: model assumptions and use of placeholders (CO_2) impair the assessment of total environmental impact (Sections 4.3, 10.2.3, 12.1.2).
Q11: Will economic growth be sustainable?	- *Yes*: some economists support the EKC hypothesis of environmental improvement (after initial decline) with high standards of living (Sections 2.2.2, 11.1); they reject the LTG model's predictions of environmental and social collapse, citing model flaws (Section 11.2.2). - *No*: ecological economists and environmentalists reject the EKC hypothesis on empirical (valid only for selected pollutants) and moral (inaction in a full world invites disaster) grounds (Sections 11.1.2,3; 13.1); they adopt the LTG model as the theoretical and empirical underpinning of likely environmental disaster (Section 11.2.1).

(continued)

(continued)

Q12: Does globalization help or hinder sustainable development?	- *Helps*: trade liberalization accelerates economic growth, which facilitates environmental protection and equitable distribution of income and wealth (EKC and trickle-down hypotheses) (Section 14.1). - *Hinders*: competitive pressure (race to the bottom) and global bureaucracies (WTO, Bretton Woods organizations) force governments to sacrifice social and environmental goals for economic ones (Section 14.1). - The evidence is *inconclusive*, in the absence of a comprehensive database and model for assessing the effects of globalization.

15.4 What Can Be Done?

Part I raised the possibility of environmental disaster. The international reaction was to advance sustainable development as a balanced, integrative approach to economic, social and environmental policies. However, the opaque concept (Section 3.2.1) leads to different conclusions about both the severity of the situation and what should be done about it. The pictogram of the last question in Fig. 15.1 indicates policy options of sermonizing rhetoric, fighting the worst symptoms, and national and global partnerships. Chapter 13 advanced strategic principles of dealing with environmental limits. They include *laissez-faire*, command and control, eco-efficiency in production, and sufficiency in consumption. The question is, what works best for sustainability at local, national and global levels? This book focuses on assessment rather than policy advice; it can only raise a few generic questions and conditional answers for stimulating further policy analysis and its feedback to the assessment process.

Q13: What is the policy advice of models?	- *EKC*: correlation between economic growth and environmental improvement is mostly rejected: active policy intervention is needed (Section 11.1). - *LTG*: business as usual leads to collapse: we need a *radical* change in social values; questionable model assumptions and denial of adaptive behaviour (Section 11.2; Ch. 13, Introduction) raise doubts about this policy advice. - *CEG*: market instruments obtain optimality and market equilibrium, taking environmental scarcities into account; the unrealistic assumption of perfect markets (Sections 2.3.2, 12.1.2, Annex I) impairs the claim of optimality in reality. - *Linear programming and optimal growth models*: depending on assumptions, the models indicate compatibility of optimality and sustainability (mainly through technological progress) or incompatibility (due to complementarity of critical capital) (Section 12.3).

(continued)

(continued)

	- *Conclusion*: models help to conceptualize sustainability and optimality, show theoretical connections and options but fail to provide unequivocal policy advice (cf. Section 12.3.3, Ch. 13, introduction).
Q14: Tackling market or policy failure?	- *Market instruments* of environmental policy include soft and hard measures (Table 13.3). High and imminent environmental risks require the use of harder and faster instruments at the cost of decreasing economic and ecological efficiency. Sufficiency in consumer behaviour can supplement eco-efficiency in production (Section 3.2.1, 13.4.1). - Correcting *policy failure* requires critical monitoring by civil society. Sharing policymaking with NGOs and/or corporations risks abdication of legitimate governmental power and accountability (Sections 9.1.1, 14.2.3). - *Conclusion*: we do not have a blueprint for attaining sustainable economic growth and development.
Q15: From vision to mission?	- *No*: jumping from vision to advocacy foregoes quantitative assessment and analysis and opens the door to unreflected advocacy and agitation (Sections 1.2,3, 3.3). - *Yes*: non-countables count: development goals of equity, security, or environmental and cultural heritage require collective agreement and policy (Sections 3.2.2, 13.4.2). - *Yes but*: alarmist doomsday proclamations (Sections 1.2, 11.2.1) may cause costly overreactions (Section 4.3). The rhetoric of cornucopian sustainable development (Section 3.2.1) carries the risk of inaction but may alert to non-economic trade-offs. Institutionalized social and environmental goals may support particular social and environmental policies (Sections 1.2, 3.3).

15.5 Some Non-conclusive Answers

Two main issues emerge from the scrutiny of the environment-economy interaction:

- The all-pervading economic-ecological dichotomy in measuring and analysing sustainable economic performance and growth
- The question of addressing the sustainability of economic growth or of development.

Even if there are no definitive answers, one can take some plausible positions based on the lessons from this book.

Chapter 2 describes the opposing views of environmentalists and economists about environmental concerns. Environmentalists and ecological economists distrust materialistic market preferences for assessing environmental impacts. Lacking a comprehensive theory they rely on their insight and selective evidence for setting limits to economic activity. Environmental economists, on the other hand, reject the mixing of normative environmental goals with positivist economic analysis. One way to overcome the deep-rooted dissent is to make the normative vision more visible in terms of explicit standards and targets, and see if economic activities can play out within this normative framework.

A more realistic step towards *bridging* – rather than overcoming – the dichotomy is compiling hybrid accounts. At least, these accounts connect the two, physical and monetary, sides of the sustainability coin. The accounts cater to the *ecological sustainability* concept of decoupling physical material throughput from monetary GDP growth. The problem with this concept is the question of how much dematerialization do we need? As long as we leave this question open our bridge might lead us either into denial of environmental problems or into the visionary fog of looming disaster. Setting targets, e.g. of reducing overall resource productivity by certain factors, lifts the fog but remains judgemental. Moreover, the weight of material flows cannot capture the significance of environmental effects – a prerequisite for the rational setting of policy priorities.

Referring to market preferences and prices in the integrated environmental-economic accounts appears to be the only way of assessing the – integrative – concepts of *economic sustainability*, i.e. produced and natural capital maintenance. However, the pricing of priceless, i.e. non-marketed, environmental services faces its own problems. Most environmentalists consider human preferences and markets – whether influenced by environmental policy instruments or not – as incapable of grasping the importance of deteriorating life support systems. On the other hand, meeting human needs and wants with scarce economic and environmental resources requires choices among production, consumption and saving/investment options. Efficient choice requires, therefore, the comparative quantitative assessment of environmental and economic costs and benefits.

Mixing normative and factual information in an ethically committed 'soft science' (Funtowicz & Ravetz, 1991) opens the door to advocacy and proselytizing. It also prevents transparent scrutiny and discussion of environmental concerns by individuals, experts, governments and civil society. This is not to deny the significance of visionary and ethical beliefs in shaping human motives and convictions. In fact, the power of such beliefs makes it essential to separate them from 'hard' scientific assessments of environmental conditions and trends.

Both concepts of ecological and economic sustainability refer to narrowly defined environmental sustainability of economic activity and growth. They ignore the achievements in other non-economic areas of multidimensional development. For *sustainable development*, one could, in principle, extend the linear programming framework to introduce further social (and other) limits to economic activity. However, any standard setting and suppositious modelling are bound to be judgemental. Expressing, alternatively, overall progress or regress with regard to these standards as welfare gains or losses is hardly possible given the problems of utility measurement and aggregation.

All these drawbacks in defining and measuring a comprehensive concept of sustainable development suggest repeating our initial question (Section 3.3.2): *has the paradigm run its course?* The answer is again a guarded yes – guarded because of the goodwill attached to the concept in national and international constitutions and conferences, and in participative implementation at local levels. Considering the rhetoric surrounding the cornucopian paradigm we might lower our guard, though. Packing everything in hardly comparable indicators or opaque indices may indeed yield nothing. Worse, the hazy paradigm might conceal 'hidden agendas' of public and private agents and institutions. This book concentrated, therefore, on what can be measured, compared and combined, i.e. the *environmental sustainability of economic performance and growth*, in other words: realistic *eco*–nomics.

Annexes

Annex I
Market Failure and Environmental Cost Internalization – A Primer

I.1 Market and Policy Failure

Market failure in dealing with environmental problems is the *raison d'être* for *eco*–nomics. The failure lies in diverting economic activity from Pareto optimality, a situation where nobody's welfare (utility) can be improved without lowering the welfare of anybody else. It is textbook knowledge that such optimality is achieved under perfect market conditions in a state of general equilibrium. Taking this ideal as a starting point, market failure calls for market intervention by policymakers. The problem is that governments were not very successful in solving the environmental problem and may have aggravated it, for instance by subsidizing environmentally damaging activities. This is the case of policy failure. So back to the invisible hand of the market?

It is no surprise that policy recommendations range from highly interventionist regulations to less intrusive market adjustments. Ecological economists favour direct policy intervention in the economy by setting constraints and regulations for economic activity. Environmental economists, on the other hand, seek to adjust markets for letting them decide about the importance of environmental costs (see Ch. 13).

Let us first clarify the main causes of market and policy failures due to environmental and related externalities, open-access natural resources and the need to provide public goods. Table I.1 is a schematic categorization of the main areas of potential market and policy failure with regard to environmental and, to a limited extent, social concerns of non-sustainability.

Most environmental impacts are so-called *externalities*, i.e. unintended side effects of consumption and production. They are mostly negative, i.e. welfare-impairing effects, marked by a minus (–) sign in segment I. Some externalities are positive such as benefits of agriculture for land and landscape conservation. Most positive effects are however intentional, marked by a plus (+) sign in segment II of the table.

By definition, external effects have been insufficiently considered, if at all, in the market system. They have the following definitory properties (Das Gupta and Pearce, 1972; Mishan, 1973):

Table I.1 Non-market effects conducive to market and policy failure

To	Interdependent, unintentional				Interdependent, intentional		Independent, unintentional	
From	P_E	P_G	P_H	C	P	C	P	C
P_E	–		(–)	–		+		
P_G	(–)		(–)	–	+(–)	II +(–)		
P_H	–	I	–	–		+		
C	–		–	–				
NEA							–	III +(–)

Explanations: P_E = Production of private enterprises
P_G = Production of government
P_H = Production of private households (subsistence)
C = Private consumption
NEA = Non-economic activities (civil and uncivil society)

- Interdependence of economic (productive and consumptive) activities, affecting the production, cost and utility levels of other producers and consumers
- Non-price and non-compensation condition
- Non-purpose (or control) condition.

Segment I of Table I.1 represents the externalities that meet all three conditions. They include 'diseconomies' among enterprises, resulting from pollution ($P_E \rightarrow - P_{E,H}$), impairment of consumption by production activities of enterprises, households and government ($P_{E,H,G} \rightarrow -$ C), and external effects of individual private consumption on other consumers and the private productive sector (C $\rightarrow$ – C, $P_{E,H}$).

Dropping the non-purpose condition obtains intentional, but *related non-market activities*. Interdependent-intentional activities of economic agents produce public and private non-marketed goods and services (segment II). *Private non-market goods* are especially relevant in the case of developing countries where a large amount of production is undertaken as subsistence ($P_H \rightarrow$ C). Other private non-market activities include corporate services to neighbourhood communities, reflecting corporate social responsibility (Section 9.1.1) ($P_E \rightarrow$ C). *Public goods* and services ($P_G \rightarrow$ P, C) such as environmental protection, defence or traffic regulation possess characteristics of non-exclusion of users and joint consumption that does not diminish the availability of the public good (non-rivalry condition).[1] The national accounts include environmental protection and other non-marketed

[1] Non-excludable and non-rival environmental sinks and (re)sources are sometimes considered to be public goods (in the public domain): in general, however, only produced (usually by the government) such goods are deemed to be public. Non-produced environmental assets are more in the nature of open-access resources, especially when their use reduces availability.

output such as own-account production of enterprises for the comprehensive coverage of productive performance. They do not attempt to measure their welfare effects (see Ch. 8).

Segment III refers to similar activities of 'independent', non-economic agents (NEA → ± P, C). They include, in particular, intentional altruistic services of civil society, and corruptive and criminal activities of 'uncivil' society. The measurement of these activities poses considerable conceptual and practical problems, but has been taken up in questionable welfare indices (Section 7.1.1). The segment also includes the impacts of military actions, which may affect current and future economic growth and welfare through the destruction of natural, produced, human and social capital.

A murky issue is the treatment of *government (policy) failure*, owing to short-sightedness (limited legislative mandates), lack of knowledge and delayed action because of the distance of central government from local conditions. If these failures are deemed to be unintentional they would qualify as a segment-I externality of policymaking ($P_G \rightarrow -C, P_{E,H}$). Once you consider corruptive or discriminatory yield to lobbying, the activities become more purposeful, turning them into a segment-II public bad ($P_G \rightarrow -P, C$).

Depletion of natural resources is another difficult-to-categorize activity. On the one hand, the exploitation of natural resources is a deliberate act of production and consumption. On the other hand, the actual loss of an (open-access) natural resource due to overuse – the tragedy of the commons described in Section 2.3.2 – might not be the intention of any of the users. Acting out a prisoner's dilemma, resource depletion for current use could then be seen as $P_E \rightarrow -P_E$ diseconomy. The loss of a cherished natural asset to future generations, without actual intent of harming these generations, would be a $P_E \rightarrow -C$ externality.

I.2 Internalizing Externalities

Environmental externalities can generate actually costed effects. Additional cleaning cost in the smokestack-laundry case is a classic example. Probably more importantly, externalities include non-priced effects on human health, recreation and other (ethical and aesthetic) values from environmental deterioration. Whether priced or not, all these effects distort an otherwise Pareto-optimal situation, in which relative prices allocate scarce resources in the most efficient manner.

To correct the misallocation of resources, governments may prompt economic agents to internalize the environmental (damage) costs they generate by market (economic) instruments. Typical means of cost internalization by market instruments are

- The establishment of individual property rights over open-access resources
- Fiscal incentives for developing and applying environmentally friendly technologies
- Fiscal disincentives replacing or curbing harmful production and consumption processes.

Fiscal disincentives include effluent charges and taxes or royalties on natural resource use. Alternatively, especially in cases of imminent environmental hazards or irreversible environmental impacts, governments should choose the faster command-and-control solution of top-down environmental regulation (cf. Ch. 13).

Pigou (1920) advanced (formally) the internalization of externalities. His suggestion was turning unpriced social costs into private ones by appropriate taxation. Plate I.1 shows the – stylized – textbook example of determining the level of an eco-tax on an enterprise by costing the environmental damage generated by the emission of a pollutant E in the production of Q. In part A, the enterprise is a price taker who cannot influence the market price through production decision or in any other way. The enterprise faces a horizontal demand curve D at the price p for its product Q. Marginal costs represent the supply curve S_1, intersecting with D at point *A* to bring about an equilibrium output of q_1 and corresponding emissions of e_1.

Part B introduces the marginal damage cost curve (MDC) to convert physical emissions (measured in tons) into monetary ($) values. Note that emissions start at zero, but are safely absorbed up to level e_0. When producing q_1, emissions e_1 generate a marginal damage value or marginal social cost MDC_1, which is more than double the marginal (private) cost and price of the product.

The polluter-pays and efficiency principles suggest that the entrepreneur take these social costs into account. Market forces would then reflect not only his/her own production possibilities but also society's (and consumers') dispreferences for environmental and health effects of production. In other words, marginal social costs should be added to the private marginal costs of production, which – in part B – obtains a new supply curve of S^*, intersecting the demand curve at point *B*. A reduction of production and emission from q_1 and e_1 to socially optimal levels of q^* and e^* is the result. At this – optimal – level of production, environmental damage still occurs, albeit at a reduced level of 'optimal' emission e^*. Consumer preferences have now found a balance in the trade-off between the benefits of the product (inherent in the demand curve) and environmental damage (part of the total, social and private, cost curve S^*), under given market conditions and with available production technologies.

The question is, how can we make this hypothetical situation a reality? In other words, how can we prod the enterprise into actually internalizing environmental costs so as to attain the optimal levels of production and emission shown in part B? Part C illustrates the answer. The *Pigovian tax* rate t should be at the level determined by the intersection of the total marginal cost curve S^* with demand D=p as the difference between private and total (private and social) marginal cost at this point (*B*). Imposing this tax rate on the private costs at all production levels obtains the supply curve S_2, intersecting obviously at optimality point *B*. The difficulties of determining this point and the corresponding marginal social cost led to the search for practical solutions. They include

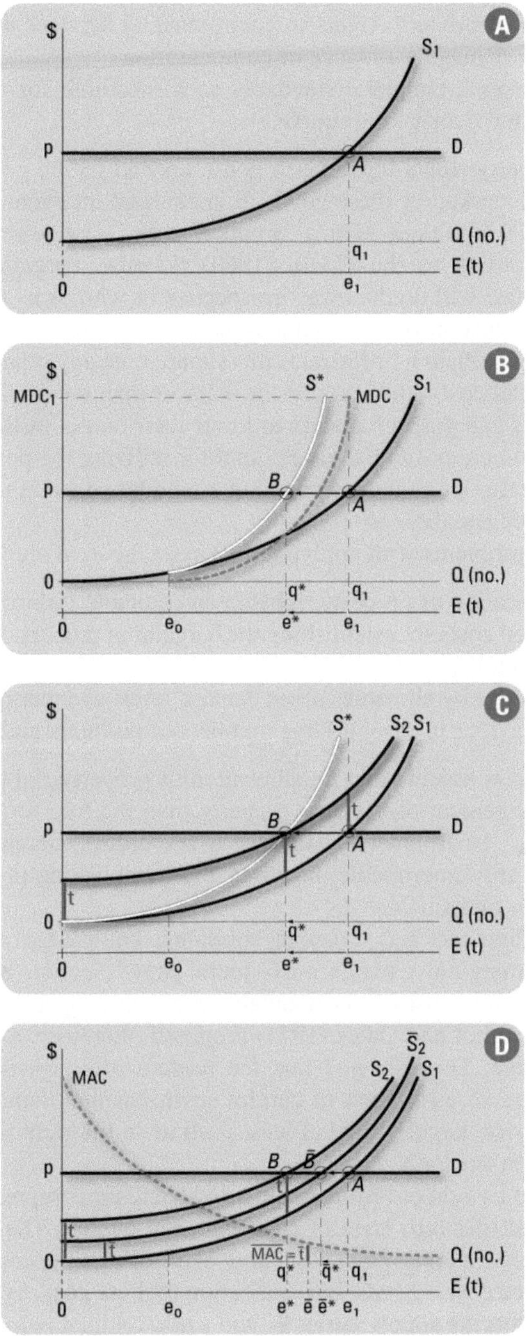

Plate I.1 Internalizing environmental damage (*See Colour Plates*).
Source: Adapted from Turner et al. (1993).

- The allocation of property rights to environmental services and corresponding negotiation for damage avoidance or compensation
- The costing of environmental standards as a substitute for the – difficult-to-assess – marginal damage cost curve.

Establishing property rights for hitherto non-marketed public goods is a first step toward creating a market for these goods. It is the least interventionist solution for reaching the social optimum with q^* production and e^* emission levels through negotiation. According to the *Coase* (1960) *theorem*, bargaining between the polluter and pollutee will do the trick, irrespective of who owns the environmental asset. Negotiation thus prevents or reduces the use of the environment as a sink for discharges. Part B of Plate I.1 illustrates this situation: at any other than the optimal level, marginal total cost either exceeds or is lower than the product price; owners of pollution rights can push producers to lower their output in the former case, or the producer would choose to increase output for utilizing the potential profit margin in the latter case. Ultimately production would be adjusted to the optimum of marginal cost-price equality.
The well-known problems with applying the Coase theorem are

- The initial allocation of property rights in an equitable fashion
- High transaction costs for establishing the bargaining process, which might deter participation in this process
- Lack of knowledge by all parties about damage levels and marginal damage costs
- Free-rider behaviour in case of large numbers of polluters and pollutees.

One can nonetheless assume that the allocation of property rights would induce a more caring management of the new property than the free-for-all situation with common-access to environmental sinks (and resources). Given the above-listed problems, it is highly improbable, however, that establishing property rights will achieve optimality on its own.

In both the Pigovian and Coasean solutions, knowledge and measurement problems about marginal damage costs loom large (Section 8.1). Valuation of environmental damage poses insurmountable problems, especially at national and sectoral levels. Baumol and Oates (1971) proposed, therefore, a practical way out of these difficulties. They showed that the *costing of an environmental quality standard*, based on an estimation of current environmental damage and a derived (maximum) emission target, might at least push us in the right direction – toward the social optimum $B(q^*, e^*)$.

Part D of Plate I.1 sets out from an emission standard $\bar{e}$, representing an exogenously determined (desired) level of environmental quality. The introduction of a marginal avoidance (of emissions) cost curve MAC allows the costing of $\bar{e}$. At $\bar{e}$ the marginal avoidance cost is $\overline{\text{MAC}}$. Internalization of these costs by means of an eco-tax $\bar{t} = \overline{\text{MAC}}$ obtains the supply curve $\bar{S}_2$, and a new optimal output of $\bar{q}^*$ and emission $\bar{e}^*$ ($e_1 > \bar{e}^* > \bar{e} > e^*$). In other words, the envisaged environmental quality standard $\bar{e}$ is not achieved because of the adaptive, cost-absorbing production behaviour of the enterprise. As pointed out by Baumol and Oates (1971), an iterative process of

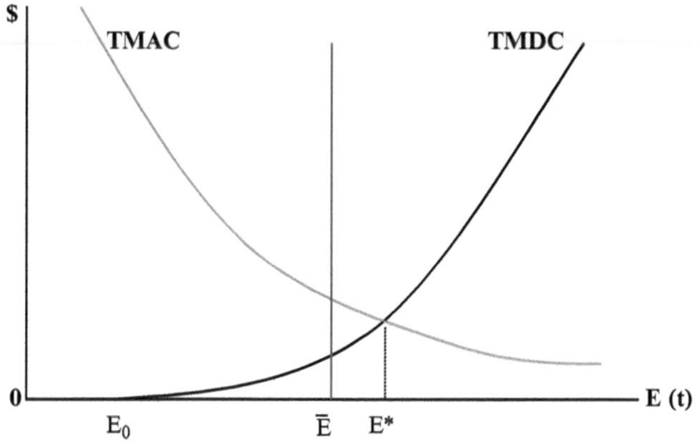

Fig. I.1 Optimal environmental protection

tax-rate and standard adjustments might lead us closer to the social optimum $B(q^*, e^*)$, or at least to cost-efficient solutions.

Note that a similar analysis for environmental costing and standard setting can be made at the national or project/programme level, looking for the *optimal provision of a public good* such as environmental protection. As shown in Figure I.1, optimality in the provision of pollution control requires that total (aggregate) marginal damage cost (TMDC) equals total marginal pollution avoidance cost (TMAC). This is the case at the intersection of the two aggregate marginal cost curves. At this point, an optimal level of emission reduction from some level E to E^* is reached. At E^* the marginal cost of an environmental project or programme equals the marginal benefit (reduction of damage), the key criterion of – optimal – environmental *cost-benefit analysis* (cf. Section 2.3.2).

The figure also illustrates the simplified *cost-efficiency* analysis, avoiding the estimation of damage costs by replacing the TMDC curve with the vertical social damage standard $\bar{E}$. In principle, at $\bar{E}$ one could reduce the marginal efficiency costs that are in excess of the marginal damage costs by allowing more emissions up to their optimal level E^*.

Annex II
Economic Rent and Natural Resource Depletion

The SEEA-2003 introduces the concept of *economic rent* as income or 'benefit to the owner of using all his assets' during an accounting period (United Nations et al., in prep., chs. 7 and 10). In this sense, economic rent is synonymous with a component of value added generated in production, gross operating surplus (GOS). GOS is the residual obtained after deducting labour cost and production taxes from and adding subsidies to gross value added. The term 'gross' refers to the inclusion of capital consumption CC. Economic rent R thus consists of a 'net return to capital' NR and the wear and tear (CC) of capital during an accounting period:

$$R = NR + CC \qquad \text{(II.1)}$$

Deducting produced capital rent R_{pc} from total economic rent = GOS then obtains the economic rent of natural capital R_{nc}:

$$R_{nc} = GOS - R_{pc} \qquad \text{(II.2)}$$

To split total rent into earnings from produced and natural capital one could thus deduct the produced capital rent from total earnings. Alternatively, one could use royalties and taxes on government-owned natural capital use as a direct estimate of natural resource rent.

In analogy to the general rent definition (II.1), the SEEA thus defines R_{nc} as consisting of the net return to natural capital NR_{nc} and natural capital consumption or *depletion* D (= CC_{nc}):

$$R_{nc} = NR_{nc} + D \qquad \text{(II.3)}$$

Depletion is thus the difference between (gross) natural capital rent and the net return to natural capital:

$$D = R_{nc} - NR_{nc} \qquad \text{(II.4)}$$

Note the full consistency of depletion of natural capital with the national accounts concept of capital consumption.

Applying a discount rate r to the flows of rents over the lifetime of a resource obtains a net present value for the opening stock OpSt, which is equal to the discounted values of the closing stock ClSt and rent generated during the accounting period. GOS or rent for natural capital use can then be defined as the difference of the stock values OpSt – ClSt plus the net return (at rate r) received from the use of the natural asset during the accounting period r(OpSt)[2]:

$$R_{nc} = (OpSt - ClSt) + rOpSt = (OpSt - ClSt) + NR_{nc} \qquad \text{(II.5)}$$

Using the definition of (II.4) and applying (II.5), equation (II.6) defines *depletion* more operationally, but still in analogy to produced capital rent, as the change in the (discounted net present) value of the natural capital asset over the accounting period:

$$D = R_{nc} - NR_{nc} = (OpSt - C1St) \qquad \text{(II.6)}$$

As discussed in Section 8.2.2, this change refers only to the use of natural capital in production, excluding price changes of the resource and other 'volume' changes from discovery, revision of estimates, natural regeneration and disaster.

[2] See for a formal derivation United Nations et al. (in prep.), para. 10.26.

Annex III
SEEA Germany – A Pilot Case Study[3]

Figure III.1 is the synoptic presentation of a green accounting study for Germany in the SEEA format of Figure 8.1 (Bartelmus et al., 2003). Time and data constraints prevented the compilation of asset stocks. The asset accounts present therefore stock *changes* only. Asset and flow accounts overlap for capital formation and capital consumption, covering produced and natural capital.

Total environmental cost of natural resource depletion and environmental degradation amount to DM 59.2 billion. The energy sector incurs over 20% of the total environmental cost, followed by agriculture (14%). 'Others', consisting mainly of commercial and private transportation, generate the largest share of environmental cost (45%). Lack of data prevented the further breakdown of this sector. Emission of pollutants accounts for nearly all of the environmental cost. The case study includes most pollutants (except for dust, methane and volatile compounds), whose emissions are measured in the official environmental statistics.

Natural resource depletion is less significant in Germany (0.6% of total environmental cost), as there are few mineral resources and the use of renewables (water and forests) appears to be sustainable at the national level. Some exhaustible resources, notably coal, are subsidized to the extent that they do not show a positive economic value and hence economic value loss. The only depletion costs are for selected fish stocks, and some minerals (mostly oil and gas) and metals. In the absence of usable market prices for these resources, the net price served as a proxy for the net present value.

Actual environmental protection expenditures are part of the conventional accounts. However, the national accounts do not usually present these expenditures separately. Gross capital formation for environmental protection amounted to 0.8% of GDP. In 1999 the national (environmental) accounts recorded total environmental outlays, covering capital and current expenditures at 1.5% of GDP.

Overall environmental cost depends significantly on the level of CO_2 reduction standards since greenhouse gases are the largest environmental cost factor in Germany. The cost calculations of Figure III.1 applied a 40% (from the 1990 level of

[3] From: Bartelmus (2002; with permission by the copyright holder, Springer).

Country: Germany (Monetary unit: million DM)	Agri-culture, forestry, fishery	Mining	Manufacturing	Electricity, gas, hot water, water supply	Public ad-ministration and defence; compulsory social security	Others	Total	Imports	Exports	Final Consumptiom: Govern-ment	Final Consumptiom: Households and NPISHs	Produced assets	Natural 'economic' assets: Land	Natural 'economic' assets: Subsoil assets	Natural 'economic' assets: Fish	Natural 'economic' assets: Water	Environmental assets: Air	Environmental assets: Water
Opening stock																		
Output	72,660	30,760	2,005,610	173,420	519,380	3,205,450	6,007,280											
Intermediate and final uses	35,920	19,370	1,264,060	114,580	266,180	2,061,120	3,761,230	-661,930	778,900	289,100	1,320,710	519,270						
Consumption of fixed capital	11,870	3,550	76,070	16,140	16,660	178,720	303,010					303,010						
Economic use of natural assets	8,298	1,377	12,458	13,303	5,009	18,741	59,186				13,090							
of which:																		
Degradation	8,296	1,244	12,212	13,303	5,009	5,651	45,714				13,090						-46,583	-12,222
Depletion	2	133	246				381							-379	-2			
Shift[a]						13,090	13,090											
Gross value added	36,740	11,390	741,550	58,840	253,200	1,144,330	2,246,050											
Net value added	24,870	7,840	665,480	42,700	236,540	965,610	1,943,040											
Environmentally-adjusted value added	16,572	6,463	653,022	29,397	231,531	946,869	1,883,854											
Other accumulation														23	0			
Other volume changes																		
Revaluation														356	2			
Closing stock																		

Fig. III.1 SEEA application: Germany, 1990

Note: [a] Represents depletion and degradation caused by the government, households and NPISHs.

Source: Bartelmus (2002), table II.3; with permission by the copyright holder, Springer.

emissions) standard to be reached in 2005, using best-available technologies. An alternative reduction scenario of 25%, with different (marginal) reduction cost decreased total environmental cost significantly to about DM 28 billion or 1.4% of NDP.

When considering the much lower depletion cost of natural resource use (0.02% of NDP) one should note Germany's dependence on resource extraction in other countries. In 1990 natural resource imports were about 6% of NDP. Not all of it represents non-sustainable natural capital consumption, but the figure is a first indication of the country's need to 'import' sustainability.

Much of the database stems from the physical input-output tables of the German Federal Statistical Office (Stahmer et al., 1998). Other important data sources are from research and management institutes for fishery, industry and water. The Federal Statistical Office provided cost estimates for meeting greenhouse gas emission standards.

References

Adriaanse, A. (1993). *Environmental policy performance indicators*. Koninginnegracht, the Netherlands: Sdu Uitgeverij.

Adriaanse, A. et al. (1997). *Resource flows: The material basis of industrial economies*. Washington, DC: World Resources Institute.

Akita, T., & Nakamura, Y. (Eds.) (2000). *Green GDP estimates in China, Indonesia and Japan*. Tokyo: The United Nations University.

Albrow, M. (1997). *The global age: State and society beyond modernity*. Stanford, CA: Stanford University Press.

Alfsen, K. H. (1996). Macroeconomics and the environment: Norwegian experience. In V. P. Gandhi (Ed.), *Macroeconomics and the environment*. Washington, DC: International Monetary Fund.

Anielski, M. (2007). *The economics of happiness: Building genuine wealth*. Gabriola Island, BC, Canada: New Society.

Arrow, K. et al. (2004). Are we consuming too much? *Journal of Economic Perspectives, 18*(3), 147–172.

Asian Development Bank (ADB) (2002). *Handbook on environment statistics*. Manila: ADB. Retrieved May 15, 2007, from http://www.adb.org/Documents/handbooks/environment/default.asp.

Auty, R. M., & Mikesell, R. (1998). *Sustainable development in mineral economics*. Oxford: Clarendon.

Ayres, R. U. (1976). *Environment statistics – Draft guidelines for statistics on material/energy balances*, Report of the Secretary-General. United Nations (Statistical Commission, New Delhi, E/CN.3/492).

Ayres, R. U. (1989). Industrial metabolism. In J. H. Ausubel & H. E. Sladovich (Eds.), *Technology and environment*. Washington, DC: National Academy Press.

Ayres, R. U., & Ayres, L. W. (Eds.) (2002). *A handbook of industrial ecology*. Cheltenham, UK: Edward Elgar.

Ayres, R. U., Ayres, L. W., & Martinás, K. (1998). Energy, waste accounting and life- cycle analysis. *Energy, 23*(5), 355–363.

Ayres, R. U., & Kneese, A. (1969). Production, consumption and externalities. *American Economic Review, 59*, 282–297.

Ayres, R. U., & Warr, B. (2002). Economic growth models and the role of physical resources. In P. Bartelmus (Ed.), *Unveiling wealth – On money, quality of life and sustainability*. Dordrecht: Kluwer.

Azar, C., & Holmberg, J. (1995). Defining the generational environmental debt. *Ecological Economics, 14*(1), 7–19.

Barbier, E. B. (1997). Introduction to the environmental Kuznets curve special issue. *Environment and Development Economics, 2*, 357–381.

Barde, J.-P. (1994). *Economic instruments in environmental policy: Lessons from the OECD experience and their relevance to developing countries.* Working Paper No. 92, OECD/GD(93)193. Retrieved May 15, 2007, from http://www.oecd.org/dataoecd/25/36/1919252.pdf.

Barnett, H., & Morse, C. (1963). *Scarcity and growth: The economics of natural resource availability*. Baltimore: Johns Hopkins University Press.

Bartelmus, P. (1979). Limits to development – environmental constraints of human needs satisfaction. *Journal of Environmental Management, 9*, 255–269.

Bartelmus, P. (1980). *Economic Development and the Human Environment*. Munich, Germany: Weltforum Verlag.

Bartelmus, P. (1987). Beyond GDP – New approaches to applied statistics. *The Review of Income and Wealth, 33*(4), 347–358.

Bartelmus, P. (1994a). *Environment, growth and development: The concepts and strategies of sustainability*. London: Routledge.

Bartelmus, P. (1994b). *Towards a framework for indicators of sustainable development.* DESIPA Working Paper Series No. 7. New York: United Nations.

Bartelmus, P. (1997a). Measuring sustainability: Data linkage and integration. In B. Moldan, S. Billharz, & R. Matravers (Eds.), *Sustainability indicators: A report on indicators of sustainable development*. Chichester, UK: Wiley.

Bartelmus, P. (1997b). Whither economics? From optimality to sustainability? *Environment and Development Economics, 2*, 323–345.

Bartelmus, P. (1998). The value of nature – Valuation and evaluation in environmental accounting. In K. Uno & P. Bartelmus (Eds.), *Environmental accounting in theory and practice*. Dordrecht: Kluwer.

Bartelmus, P. (2000). Economic growth, wealth and sustainable development. In R. Kreibich & U. E. Simonis (Eds.), *Global change – Globaler Wandel*. Berlin: Berlin Verlag, Arno Spitz.

Bartelmus. P. (2001). Accounting for sustainability: Greening the national accounts. In M. K. Tolba (Ed.), *Our fragile world: Challenges and opportunities for sustainable development*. Oxford: Eolss.

Bartelmus, P. (2002). Unveiling wealth – Accounting for sustainability. In P. Bartelmus (Ed.), *Unveiling wealth: On money, quality of life and sustainability*. Dordrecht: Kluwer.

Bartelmus, P. (2003). Dematerialization and capital maintenance: Two sides of the sustainability coin. *Ecological Economics, 46*, 61–81.

Bartelmus, P. (2004). Green accounting and energy. In C. Cleveland (Ed.), *Encyclopedia of energy*, Elsevier. Retrieved May 15, 2007, from http://www.sciencedirect.com/science/referenceworks/012176480X.

Bartelmus, P. (2007). SEEA-2003: Accounting for sustainable development? *Ecological Economics, 61*(4), 613–616.

Bartelmus, P., Albert, J., & Tschochohei, H. (2003). Wie teuer ist uns die Umwelt? Zur umweltökonomischen Gesamtrechnung in Deutschland [How much the environment? Integrated environmental and economic accounts in Germany]. *Zeitschrift für Umweltpolitik und Umweltrecht, 3*, 333–370.

Bartelmus, P., Lutz, E., & Schweinfest, S. (1992). *Integrated environmental and economic accounting: A case study for Papua New Guinea*. Environment Working Paper No. 54. Washington, DC: The World Bank.

Bartelmus, P., & Seifert, E.K., (Eds.) (2003). *Green accounting*. Aldershot, U.K.: Ashgate.

Bartelmus, P., Stahmer, C., & van Tongeren, J. (1991). Integrated environmental and economic accounting: Framework for a SNA satellite system. *Review of Income and Wealth, 37*, 111–148.

Baumol, W. J., & Oates, W. E. (1971). The use of standards and prices for protection of the environment. *Swedish Journal of Economics, 73*, 42–54.

Baumol, W. J. (1986). On the possibility of continuing expansion of finite resources. *Kyklos, 39*(2), 167–179.

Bebbington, J., Gray, R., Hibbitt, C., & Kirk, E. (2001). *Full cost accounting: An agenda for action.* ACCA Research Report No. 73. London: The Association of Chartered Certified Accountants.

Beckerman, W. (1992). Economic growth and the environment: Whose growth? Whose environment? *World Development, 20*(4), 481–496.

Beckerman, W. (1994). Sustainable development: Is it a useful concept? *Environmental Values, 3*, 191–209.

Bergman, L. (1990). Energy and environmental constraints on growth: A CGE modelling approach. *Journal of Policy Modeling, 12*(4), 671–691.

Bhagwati, J. (2002). Coping with antiglobalization. *Foreign Affairs, 81*, 2–7.

Bhagwati, J. (2004, 2007). *In Defense of Globalization.* Oxford: Oxford University Press.

Biermann, F., & Bauer, S. (Eds.) (2005). *A world environment organization, Solution or threat for effective environmental governance.* Aldershot, UK: Ashgate.

Birou, A., Henry, P.-M., & J. P. Schlegel (Eds.) (1977). *Towards a re-definition of development.* Oxford: Pergamon

Bossel, H. (1999). *Indicators for sustainable development: Theory, method, applications.* Winnipeg: International Institute for Sustainable Development.

Boulding, K. E. (1966). The economics of the coming spaceship earth. In H. Jarret (Ed.), *Environmental quality in a growing economy.* Baltimore, MD: Johns Hopkins Press for Resources for the Future.

Boutros-Ghali, B. (1995). *Agenda for development.* New York: United Nations sales publication (E.95.V.16).

Boyd, J. (2007). Nonmarket benefits of nature: What should be counted in green GDP? *Ecological Economics, 61*, 716–723.

Bringezu, S. (1993). Towards increasing resource productivity: How to measure the total material consumption of regional or national economies? *Fresenius Environmental Bulletin, 2*, 437–442.

Bringezu, S. (2002). *Towards sustainable resource management in the European Union.* Wuppertal Papers No. 121. Wuppertal Institute for Climate, Environment and Energy, Wuppertal, Germany.

Bringezu, S., & Moriguchi, Y. (2002). Material flow analysis. In R. U. Ayres & L. W. Ayres (Eds.), *A handbook of industrial ecology.* Cheltenham, UK: Edward Elgar.

Bringezu, S., Schütz, H., Steger S., & Baudisch, J. (2004). International comparison of resource use and its relation to economic growth. The development of total material requirement, direct material inputs and hidden flows and the structure of TMR. *Ecological Economics, 51*, 97–124.

Broda, C., & Weinstein, D. E. (2006). Globalization and the gains from variety. *Quarterly Journal of Economics, 121*(2), 541–585.

Brodianski, V. (2001). Earth available energy and the sustainable development of life support systems. In M. K. Tolba (Ed.), *Our fragile world, Challenges and opportunities for sustainable development.* Oxford: Eolss.

Brown, M. T. and Ulgiati, S. (1999). Emergy valuation of the biosphere and natural capital. *Ambio, 28*(6), 486–493.

Caldwell, B. (1982). *Beyond positivism: Economic methodology in the twentieth century.* London: Allen & Unwin.

Canadian Institute of Chartered Accountants (CICA) (1997). *Full cost accounting from an environmental perspective.* Toronto: CICA.

Caribbean Community Secretariat (CARICOM) (2003). *The CARICOM environment in figures 2002.* Georgetown, Guyana: CARICOM Secretariat.

Carnot, N. L. S. (1824). *Réflexions sur la puissance motrice du feu et sur les machines propres à développer cette puissance* [Reflections on the motive power of fire and the machines appropriate for developing this power], reprint 1966. London: Dawson.

Carson, R. (1965). *Silent spring.* London: Penguin.

Casamento, R. (2004). Accounting implications of EU emissions trading directive. In ACCA and Environment Agency, *Advances in environmental accounting* (proceedings of the ACCA/Environment Agency Seminar, January 2004). Retrieved May 16, 2007, from http://www.emawebsite.org/library_detail.asp?record = 125.

Center for Waste Reduction Technologies (CWRT) (1999). *Total cost assessment methodology: Internal managerial decision making tool.* New York: CWRT.
Chiang, A. C. (1984, 3rd ed.). *Fundamental methods of mathematical economics.* New York: McGraw-Hill.
China Council for International Cooperation on Environment and Development (CCICED) (2005). Interim report of the Task Force on Economic Growth and Environment. In *Proceedings, The 4th meeting of the 3rd phase of China Council for International Cooperation on Environment and Development.* CCICED and State Environmental Protection Administration.
Christensen, J. (2005). Are we consuming too much? *Conservation in Practice, 6*(2), 15–19.
Clausius, R. J. E. (1850). Über die bewegende Kraft der Wärme [On the moving power of heat]. In *Annalen der Physik.* Berlin.
Coase, R. (1960). The problem of social cost. *Journal of Law and Economics, 3*, 1–14.
Cobb, C., Halstead, T., & Rowe, J. (1995). If the GDP is up, why is America down? *The Atlantic Monthly*, October, 59–78.
Cohen, J. E. (1995). *How many people can the earth support?* New York: Norton.
Cole, H. S. D., Freeman, C. Jahoda, M., & Pavitt, K. L. R. (1973). *Models of doom, A critique of the limits to growth.* New York: Universe.
Commission of the European Communities (2005). *Thematic strategy on the sustainable use of natural resources*, COM(2005) 670 final. Retrieved June 7, 2007, from http://ec.europa.eu/environment/natres/.
Common, M., & Stagl, S. (2005). *Ecological economics, An introduction.* New York: Cambridge University Press.
Commons, J. R. (1934). *Institutional economics, Its place in political economy.* New York: Macmillan.
Conference of European Statisticians (1973). *Report of the meeting on statistics for environmental studies and policies.* Geneva, 19–23 March 1973 (CES/AC.40/5).
Conrad, K. (1999). Computable general equilibrium models for environmental economics and policy analysis. In J. van den Bergh (Ed.), *Handbook of environmental and resource economics.* Cheltenham, UK: Edward Elgar.
Cooper, C. (1981). *Economic evaluation and the environment.* London: Hodder & Stoughton.
Costanza, R. (1980). Embodied energy and economic valuation. *Science, 210*, 1219–1224.
Costanza, R. (Ed.) (1991). *Ecological economics, The science and management of sustainability.* New York: Columbia University Press.
Costanza, R., Daly, H. E., & Bartholomew, J. A. (1991). Goals, agenda and policy recommendations for ecological economics. In R. Costanza (Ed.), *Ecological economics, The science and management of sustainability.* New York: Columbia University Press.
Costanza, R. et al. (1997a). *An introduction to ecological economics.* Boca Raton, FL: St. Lucie.
Costanza, R. et al. (1997b). The value of the world's ecosystem services and natural capital. *Nature, 283*, 253–260.
Crook, C. (2005). The good company, a survey of corporate social responsibility. *The Economist*, January 22.
Curry-Lindahl, K. (1972). *Conservation for survival: An ecological strategy.* New York: Morrow.
Daly, H. E. (1990). Toward some operational principles of sustainable development. *Ecological Economics, 2*, 1–6.
Daly, H. E. (1991). Sustainable growth: A bad oxymoron. *Grassroots Development, 15*(3).
Daly, H. E. (1996). *Beyond growth.* Boston, MA: Beacon.
Daly, H. E. (1999). Globalization versus internationalization – Some implications. *Ecological Economics, 3*, 31–37.
Daly, H. E., & Cobb, J. B., Jr. (1989). *For the common good: Redirecting the economy towards community, the environment, and a sustainable future.* Boston, MA: Beacon.
Daly, H. E., & Farley, J. (2004). *Ecological economics.* Washington, DC: Island Press.
Darwin, C. (1859, 1951 ed.). *The origin of species.* New York: Dutton.
Das Gupta, A. K., & Pearce, D. W. (1972). *Cost benefit analysis: Theory and practice.* London: Macmillan.

Dasgupta, P. (1994). Optimal versus sustainable development. In I. Serageldin & A. Steer (Eds.), *Valuing the environment*, proceedings of the First Annual International Conference on Environmentally Sustainable Development. Washington, DC: The World Bank.

Dasgupta, P., & Mäler, K.-G. (1991). The environment and emerging development issues. *Proceedings of the World Bank Annual Conference on Development Economics 1990*. Washington, DC: The World Bank.

Dasgupta, P., & Mäler, K.-G. (2000). Net national product, wealth, and social well-being. *Environment and Development Economics, 5*, 69–93.

De Bruyn, S. M. (1997). Explaining the environmental Kuznets curve: Structural change and international agreements in reducing sulphur emissions. *Environment and Development Economics, 2*, 485–503.

De Bruyn, S. M., van den Bergh, J. C. J. M., & Opschoor, J. B. (1998). Economic growth and emissions: Reconsidering the empirical basis of environmental Kuznets curves. *Ecological Economics, 25*, 161–175.

De Groot, R. S. (1992). *Functions of nature*. Groningen, the Netherlands: Wolters Noordhoff.

De Groot, R.S., van der Perk, R. J., Chiesura, A., & van Vliet, A. (2003). Importance and threat as determining factors of criticality of natural capital. *Ecological Economics, 44*, 187–204.

De Haan, M. (2001). A structural decomposition analysis of pollution in the Netherlands. *Economic Systems Research, 13*(2), 181–196.

De Haan, M., & Kee, P. (2004). *Accounting for sustainable development: The NAMEA-based approach*. Retrieved May 16, 2007, from http://www.cbs.nl/NR/rdonlyres/789FC43C-28AC-4A07-A4E1-158745589A50/0/accountingforsustainabledevelopmentthenameabasedapproach.pdf.

Delos Angeles, M. S, & Peskin, H. M. (1998). Philippines: Environmental accounting as instrument of policy. In K. Uno & P. Bartelmus (Eds.), *Environmental accounting in theory and practice*. Dordrecht: Kluwer.

Diamond, J. (2005). *Collapse, How societies choose to fail or succeed*. London: Penguin.

Diefenbacher, H. (1995). Der 'Index of Sustainable Economic Welfare', Eine Fallstudie für die Bundesrepublik Deutschland 1950–1992 [The 'index of sustainable economic welfare', a case study for the Federal Republic of Germany 1950–1992]. Texte und Materialien der Forschungsstätte der Evangelischen Studiengemeinschaft, Heidelberg.

Dietzenbacher, E., & Los, B. (1998). Structural decomposition techniques: Sense and sensitivity. *Economic Systems Research, 10*(4), 307–323.

Dixon, J. A., Fallon Scura, L., Carpenter, R. A., & Sherman, P. B. (1994). *Economic analysis of environmental impacts*. London: Earthscan.

Domingo, E. V. (1998). Philippines: Adaptation of the United Nations system of environmental accounting. In K. Uno & P. Bartelmus (Eds.), *Environmental accounting in theory and practice*. Dordrecht: Kluwer.

Doob, L. W. (1995). *Sustainers and sustainability: Attitudes, attributes, and actions for survival*. Westport, CT: Praeger.

Dorfman, R., Samuelson, P. A., & Solow, R. M. (1958). *Linear programming and economic analysis*. New York: McGraw-Hill.

Drewnowsky, J. (1970). *Studies in the measurement of levels of living and welfare*. Geneva: UNRISD.

Drewnowsky, J. (1974). *On measuring and planning the quality of life*. The Hague: Mouton.

Duesenberry, J. S. (1949). *Income, saving and the theory of consumer behavior*. Cambridge, MA: Harvard University Press.

Durning, A. (1992). *How much is enough?* New York: Norton.

Easterlin, R. (1974). Does economic growth improve the human lot? Some empirical evidence. In P. A. David & M. W. Reder (Eds.), *Nations and households in economic growth*. New York: Academic Press.

Ehrenfeld, J. R., & Chertow, M. R. (2002). Industrial symbiosis: The legacy of Kalundborg. In R. U. Ayres & L. W. Ayres (Eds.), *A handbook of industrial ecology*. Cheltenham, UK: Edward Elgar.

Ehrlich, P. R., & Holdren, J. P. (1971). Impact of population growth. *Science, 171*(3977), 1212–1217.

Eisner, R. (1988). Extended accounts for national income and product. *Journal of Economic Literature, XXVI*, 1611–1684.

Ekins, P., Hillman, M., & Hutchinson, R. (1992). *Wealth beyond measure, An atlas of new economics*. London: Gaia.

Ekins, P., & Simon, S. (2001). Estimating sustainability gaps: Methods and preliminary applications for the UK and the Netherlands. *Ecological Economics, 37*, 5–22.

Ekins, P. et al. (2003). A framework for the practical application of the concepts of critical natural capital and strong sustainability. *Ecological Economics, 44*, 165–185.

Elliot, R. (2001). Ethics and value. In M. K. Tolba (Ed.), *Our Fragile World, Challenges and opportunities for sustainable development*, Oxford: Eolss.

El Serafy, S. (1989). The proper calculation of income from depletable natural resources. In Y. J. Ahmad, S. El Serafy, & E. Lutz (Eds.), *Environmental Accounting for Sustainable Development*. Washington, DC: The World Bank.

England, R. W. (2000). Natural capital and the theory of economic growth. *Ecological Economics, 34*(3), 425–431.

Etzioni, A. (1993). *The spirit of community. Rights, responsibilities, and the communitarian agenda*. New York: Crown.

European Association for Bioeconomic Studies (E.A.B.S.) (1997). *Implications and applications of bioeconomics*. Proceedings of the Second International Conference of the E.A.B.S. (Palma de Mallorca, March 11–13, 1994). Milan: Edizioni Nagard.

European Commission (1993). *Towards sustainability*. Official Journal of the European Commission No. C 138/5 of 17.5.93. Retrieved May 16, 2007, from http://europa.eu.int/comm/environment/actionpr.htm.

Eurostat (2001). *Economy-wide material flow accounts and derived indicators: A methodological guide*. Luxembourg: European Communities.

Faber, M., Petersen, T., & Schiller, J. (2002). Homo oeconomicus and homo politicus in ecological economics. *Ecological Economics, 40*, 323–333.

Factor 10 Club (1994). *Carnoules Declaration*, Wuppertal, Germany: Wuppertal Institute for Climate, Environment and Energy.

Faucheux, S. (2001). Summary principles for sustainable development. In M. K. Tolba (Ed.), *Our Fragile World, Challenges and opportunities for sustainable development*, Oxford: Eolss.

Fergany, N. (1994). Quality of life indicators for Arab countries in an international context. *International Statistical Review, 62*(2), 187–202.

Field, B. C. (2001). *Natural resource economics: An introduction*. Boston, MA: Irwin/McGraw-Hill.

Field, B. C., & Field, M. K. (2002). *Environmental economics: An introduction*. New York: McGraw-Hill.

Fischer-Kowalski, M. (1998). Society's metabolism: The intellectual history of materials flow analysis, part I, 1860–1970. *Journal of Industrial Ecology, 2*(1), 61–78.

Fischer-Kowalski, M. (2002). Exploring the history of industrial metabolism. In R. U. Ayres & L. W. Ayres (Eds.), *A handbook of industrial ecology*. Cheltenham, UK: Edward Elgar.

Fischer-Kowalski, M., & Amann, C. (2001). Beyond IPAT and Kuznets curves: Globalization as vital factor in analysing the environmental impact of socio-economic metabolism. *Population and Environment, 23*(1), 7–47.

Fischer-Kowalski, M., & Haberl, H. (1998). Sustainable development: Socio-economic metabolism and colonization of nature. *International Social Science Journal, 158*, 573–587.

Fischer-Kowalski, M., & Hüttler, W. (1999). Society's metabolism: The intellectual history of materials flow analysis, part II, 1970–1998. *Journal of Industrial Ecology, 2*(4), 107–136.

Food and Agriculture Organization (FAO) (2005a). *Review of the state of world marine fishery resources*, Fisheries Technical Paper 457, section A.1. Retrieved May 16, 2007, from http://www.fao.org/docrep/009/y5852e/Y5852E00.HTM.

Food and Agriculture Organization (FAO) (2005b). *The global forest resources assessment 2005.* Retrieved May 16, 2007, from www.fao.org/forestry/site/fra2005/en.

Frangopoulos, C. (2001). Exergy, energy system analysis, and optimization. In M. K. Tolba (Ed.), *Our Fragile World, Challenges and opportunities for sustainable development*, Oxford: Eolss.

Frank, R. H. (1999). *Luxury fever: Why money fails to satisfy in an era of excess*. New York: Free Press.

Friend, A. (2004). *Valuation of the physical stock-flow accounts through means of ecopricing, Sraffa ecosystem valuation method, SEVM* (Paper presented at the Eighth Biennial Scientific Conference of the International Society for Ecological Economics, Montreal, 11–14 July 2004).

Friend, A., & Rapport, D. (1979). *Towards a comprehensive framework for environmental statistics: A stress-response-approach*. Ottawa: Statistics Canada.

Funtowicz, S. O., & Ravetz, J. R. (1991). A new scientific methodology for global environmental issues. In R. Costanza (Ed.), *Ecological economics: The science and management of sustainability*. New York: Columbia University Press.

Galbraith, J. K. (1986, 4th ed.). *The new industrial state*. New York: Mentor.

Gallopín, G. C. (1997). Part One – Introduction. In B. Moldan, S. Billharz, & R. Matravers (Eds.), *Sustainability indicators*. Chichester, UK: Wiley.

Gardner, G. (2003). Engaging religion in the quest for a sustainable world. In G. Gardner et al. (Eds.), *The state of the world 2003, A Worldwatch Institute report on progress toward a sustainable society*. New York: Norton.

Gawel, E. (1998). Das Elend der Stoffstromökonomie – Eine Kritik [The misery of material flow economics – a critique]. *Konjunkturpolitik, 44*(2), 173–206.

Georgescu-Roegen, N. (1971). *The entropy law and the economic process*. Cambridge, MA: Harvard University Press.

Georgescu-Roegen, N. (1979). Energy analysis and economic valuation. *The Southern Economic Journal, 45*, 1023–1058.

Gilliland, M. W. (Ed.) (1978). *Energy analysis: A new public policy tool*. Boulder, CO: Westview.

Gintis, H. (2000). *Homo economicus*: Evidence from experimental economics. *Ecological Economics, 35*, 311–322.

Goldsmith, E. et al. (1972). *Blueprint for survival*. Boston, MA: Houghton Mifflin.

Goldsmith, E., & Mander, J. (Eds.) (2001). *The case against the global economy and for a turn towards localization*. London: Earthscan.

Gore, A. (1993). *Earth in the balance, Ecology and the human spirit*. New York: Plume, Penguin.

Gottlieb, R. S. (Ed.) (1996). *This sacred earth: Religion, nature, environment*. New York: Routledge.

Goulder, I. H. (1995). Environmental taxation and the double dividend, a readers' guide. *International Tax and Public Finance, 2*, 157–183.

Gowdy, J. M. (1994). *Coevolutionary economics: The economy, society, and the environment*. Boston, MA: Kluwer.

Gray, R. H. (1990). *The greening of accountancy: The profession after Pearce*. London: Association of Chartered Certified Accountants.

Gray, R. H. (1992). Accounting and environmentalism: An exploration of the challenge of gently accounting for accountability, transparency and sustainability. *Accounting Organisations and Society, 17*(5), 399–425.

Grossman, G. M., & Krueger, A. B. (1995). Economic growth and the environment. *Quarterly Journal of Economics, CX*, 353–377.

Guinomet, I. et al. (1997). Approaches to indicators of sustainable development in the European Commission. In B. Moldan, S. Billharz, & R. Matravers (Eds.), *Sustainability indicators*. Chichester, UK: Wiley.

Haberl, H. (2001). The energetic metabolism of societies. *Journal of Industrial Ecology, 5*(1), 11–32.
Haeckel, E. (1866). *Generelle Morphologie der Organismen* [General morphology of organisms], vol. 2. Berlin: Reimer.
Haeckel, E. (1898, 9th ed.). *Natürliche Schöpfungsgeschichte* [History of the natural creation]. Berlin: Reimer.
Hamilton, C. (2004). *Growth fetish*. London: Pluto.
Hankiss, E. (1983). Cross-cultural quality of life research: An outline for conceptual framework. In UNESCO (Ed.), *Quality of life: Problems of assessment and measurement*. Paris: UNESCO.
Hardesty, D. L. (2001). Archaeology. In M. K. Tolba (Ed.), *Our Fragile World, Challenges and opportunities for sustainable development*, Oxford: Eolss.
Hardin, G. (1968). The tragedy of the commons. *Science, 162*, 1243–1248.
Hartwick, J. M. (1977). Intergenerational equity and the investing of rents from exhaustible resources. *American Economic Review, 67*(3), 972–974.
Hartwick, J. M., & Hageman, A. P. (1993). Economic depreciation of mineral stocks and the contribution of El Serafy. In E. Lutz (Ed.), *Toward improved accounting for the environment*. Washington, DC: The World Bank.
Harrison, P., & Pearce, F. (2000). *AAAS atlas of population & environment*. Berkeley, CA: University of California Press.
Hecht, J. E. (2005). *National environmental accounting, Bridging the gap between ecology and economy*. Washington, DC: Resources for the Future.
Heilbroner, R., & Milberg, W. (1995). *The crisis of vision in modern economic thought*. Cambridge: Cambridge University Press.
Hellsten, E., Ribacke, S., & Wickbom, G. (1999). SWEEA – Swedish environmental and economic accounts. *Structural Change and Economic Dynamics, 10*(1), 39–72.
Henderson, D. (2001). *Anti-liberalism in 2000, The rise of new millennium collectivism*. London: The Wincott Foundation.
Henderson, H., Lickerman, J., & Flynn, P. (Eds.) (2000). *Calvert-Henderson quality of life indicators*. Bethesda, MD: Calvert Group.
Hicks, J. R. (1946, 2nd ed.). *Value and capital*. Oxford: Oxford University Press.
Hines, C. (2000). *Localization: A global manifesto*. London: Earthscan.
Hinterberger, F., Luks, F., & Stewen, M. (1999). Wie ökonomisch ist die Stoffstromökonomik? Eine Gegenkritik [How economic is material flow economics? A counter-critique]. *Konjunkturpolitik, 45*(4), 358–375.
Hinterberger, F., Luks, F., Stewen, M., & van der Straaten J. (2000). Environmental policy in a complex world. *International Journal of Sustainable Development, 3*, 276–296.
Hoffmann-Nowotny, H.-J. (1981). *Sozialbilanzierung, Soziale Indikatoren VIII, Konzepte und Forschungsansätze* [Social accounting, Social indicators VIII, Concepts and research approaches]. Frankfurt: Campus.
Hotelling, H. (1931). The economics of exhaustible resources. *Journal of Political Economy, 39*, 137–175.
Huber, J. (2004). Environmental policy shift through technological innovation. In K. Jacob, M. Binder, & A. Wieczorek (Eds.), *Governance for industrial transformation. Proceedings of the 2003 Berlin Conference on the Human Dimensions of Global Environmental Change*. Berlin: Environmental Policy Research Centre. Retrieved May 16, 2007, from http://web.fu-berlin.de/ffu/ffu_e/Publications/bc2003_proceedings.htm.
Hueting, R. (1980). *New scarcity and economic growth*. Amsterdam: North Holland.
Hueting, R. (1993). Calculating a Sustainable National Income: A practical solution for a theoretical dilemma. In A. Franz & C. Stahmer (Eds.), *Approaches to environmental accounting: Proceedings of the IARIW Conference on Environmental Accounting*. Heidelberg: Physica.
Hueting, R., & Bosch, P. (1994). *Sustainable national income in the Netherlands: The calculation of environmental losses in monetary terms* (Paper presented at the London Group meeting on Natural Resource and Environmental Accounting, Washington, DC, 15–17 March 1994).

Hughes, J. D. (Ed.) (2000). *The face of the earth, Environment and world history*. Armonk, NY: Sharpe.

Hughes, J. D., & Thirgood, J. V. (1982). Deforestation in ancient Greece & Rome: A cause of collapse. *The Ecologist, 12*(5), 196–208.

Intergovernmental Panel on Climate Change (IPCC) (2007). *Climate change 2007: The physical base, Summary for policy makers*. Retrieved May 16, 2007, from http://www.ipcc.ch/SPM2feb07.pdf.

International Labour Organization (ILO), International Labour Office (1977). *Employment, growth and basic needs: A one-world problem*. New York: Praeger.

International Union for Conservation of Nature and Natural Resources (IUCN) (2006). *2006 IUCN Red List of Threatened Species, Summary statistics*. Retrieved May 16, 2007, from http://www.iucnredlist.org/info/stats.

International Union for Conservation of Nature and Natural Resources (IUCN), United Nations Environment Programme (UNEP) & World Wildlife Fund (WWF) (1980). *World conservation strategy, Living resource conservation for sustainable development*. Gland, Switzerland: IUCN.

Islam, S. M. N. (2001). Ecology and optimal economic growth: An optimal ecological economic growth model and its sustainability implications. In M. Munasinghe, O. Sunkel, & C. de Miguel (Eds.), *The sustainability of long-term growth: Socioeconomic and ecological perspectives*. Cheltenham, UK: Edward Elgar.

Jacobs, M. (1994). The limits of neoclassicism: Towards an institutional environmental economics. In M. Redclift & T. Benton (Eds.), *Social theory and the global environment*. London: Routledge.

Jesinghaus, J. (1997). Box 2A: Current approaches to valuation. In B. Moldan, S. Billharz, & R. Matravers (Eds.), *Sustainability indicators*. Chichester, UK: Wiley.

Jevons, W. S. (1865, 1965 ed.). *The coal question: An inquiry concerning the progress of the nation, and the probable exhaustion of our coal mines* (Reprint of the 3rd edition). New York: Augustus Kelly.

Jolly, R. (1977). Changing views on development. In J. J. Nossin (Ed.), *Surveys of development: A multidisciplinary approach*. Amsterdam: Elsevier.

Jorgenson, D. W., & Wilcoxen, P. J. (1990). Intertemporal general equilibrium modeling of U.S. environmental regulation. *Journal of Policy Modeling, 12*(4), 715–744.

Kallis, G. (2005). *Developing a coevolutionary analytical approach* (Paper presented at the 6th International Conference of the European Society for Ecological Economics, Lisbon, 14–17 June 2005).

Kapp, K. W. (1950). *The social costs of private enterprise*. Boston, MA: Harvard University Press.

Kates, R. W. (2003). The nexus and the neem tree: Globalization and a transition toward sustainability. In J. G. Speth (Ed.), *Worlds apart, Globalization and the environment*. Washington, DC: Island Press.

Kates, R. W. et al. (2001). Environment and development: Sustainability science. *Science, 292*(5517), 641–642.

Kerr, A. (1997). Box 4A: The development of indicators of sustainability in Canada. In B. Moldan, S. Billharz, & R. Matravers (Eds.), *Sustainability indicators*. Chichester, UK: Wiley.

Keuning, S. J., & de Haan, M. (1998). Netherlands: What's in a NAMEA? Recent results. In K. Uno & P. Bartelmus (Eds.), *Environmental accounting in theory and practice*. Dordrecht: Kluwer.

Keynes, J. M. (1936, 1973 ed.). *The general theory of employment, interest and money*. London: Macmillan.

Kim, S.-W. (1998). *Pilot compilation of environmental-economic accounts, Republic of Korea*. Seoul: Korea Environment Institute.

Komen, M. H. C., Gerking, S., & Folmer, H. (1997). Income and environmental R&D: Empirical evidence from OECD countries. *Environment and Development Economics, 2*(4), 366–367.

Koopmans, T. C. (1973). Some observations on 'optimal' economic growth and exhaustible resources. In H. C. Bos, H. Linnemann, & P. de Wolff (Eds.), *Economic structure and development: Essays in honour of Jan Tinbergen*. Amsterdam: North- Holland.

Kuczynski, M. (1971). Quesnay. In *Ökonomische Schriften* [Economic letters], Bd. 1, Berlin: Akademie-Verlag.

Kunert AG Environmental report. (1994/1995). Immenstadt (Germany): Kunert AG [reprinted in Bartelmus P. & E.K. Seifert (Eds.) (2003). *Green accounting*. Aldershot, UK: Ashgate].

Kuttner, R. (1997). *Everything for sale – The virtues and limits of markets*. New York: Knopf.

Kuznets, S. (1955). Economic growth and income inequality. *American Economic Review, 45*, 1–28.

Lahr, M. L., & Dietzenbacher, E. (Eds.) (2001). *Input-output analysis: Frontiers and extension*. Palgrave Macmillan.

Landefeld, J. S., & Howell, S. L. (1998). USA: Integrated economic and environmental accounting: Lessons from the IEESA. In K. Uno & P. Bartelmus (Eds.), *Environmental accounting in theory and practice*. Dordrecht: Kluwer.

Lange, G.-M. (2004). Wealth, natural capital and sustainable development: Contrasting examples from Botswana and Namibia. *Environmental & Resource Economics, 29*(3), 257–283.

Lange, G.-M., Hassan, R., & Alfieri, A. (2003). Using environmental accounts to promote sustainable development: Experience in southern Africa. *Natural Resources Forum, 27*, 19–31.

Lange, G.-M., Hassan, R., & Hamilton, K. (2003). *Environmental accounting in action, Case studies from southern Africa*. Cheltenham, UK: Edward Elgar.

Leipert, C. (1986). Social cost of economic growth. *Journal of Economic Issues, 20*(1), 109–131.

Leipert, C. (1989). National income and economic growth: The conceptual side of defensive expenditures. *Journal of Economic Issues, 23*(3), 843–856.

Leontief, W. (1951, 2nd ed.). *The structure of American economy 1919–1939, An empirical application of equilibrium analysis*. New York: Oxford University Press.

Leontief, W. (1970). Environmental repercussions and the economic structure: An input- output approach. *Review of Economics and Statistics, 52*, 262–271.

Li, J., Bartelmus, P., Qi, J., & Zhao, J. (Eds.) (2007). *Harmonious xiao kang society and environmental performances in China*. Beijing: China Environmental Science Press [in Chinese].

Lifset, R., & Graedel, T. E. (2002). Industrial ecology: Goals and definitions. In R. U. Ayres & L. W. Ayres (Eds.), *A handbook of industrial ecology*. Cheltenham, UK: Edward Elgar.

Light, A., & Rolston III, H. (Eds.) (2003). *Environmental ethics, An anthology*. Oxford: Blackwell.

Lindblom, C. E. (1959). The Science of 'muddling through'. *Public Administration Review, 19*(2), 79–88.

Lipsey, R. G., & Lancaster, R. K. (1956–1957). The general theory of second best. *The Review of Economic Studies, 24*(1), 11–32.

Lomborg, B. (2001). *The skeptical environmentalist, Measuring the real state of the world*. Cambridge, UK: Cambridge University Press.

Loraine, J. A. C. (1972). *The death of tomorrow*. London: Heinemann.

Lotka, A. J. (1925, 1956 ed.). *Elements of Physical Biology*. New York: Dover.

Lovelock, J. E. (1988, 1995). *The ages of Gaia – A biography of our living earth*. New York: Norton.

Lubbers, R., & Morales, M. C. P. (2001). The Earth Charter: Global ethics for the twenty-first century. In M. K. Tolba (Ed.), Our Fragile World, Challenges and opportunities for sustainable development, Oxford: Eolss.

Lüdeke, M. K. B., & Petschel-Held, G. (1997). Syndromes of global change: An information structure for sustainable development. In B. Moldan, S. Billharz, & R. Matravers (Eds.), *Sustainability indicators*. Chichester, UK: Wiley.

Mäler, K.-G. (1991). National accounts and environmental resources. *Environmental and Resource Economics, 1*, 1–15.

Mäler, K.-G., & Vincent, J. R. (Eds.) (2003, 2005). *Handbook of environmental economics*. Amsterdam: Elsevier.

Malthus, T. (1798, 1963 ed.). *Principles of population*. Homewood, IL: R.D. Irwin.

Mander, J. (2001). Introduction: Facing the rising tide. In E. Goldsmith & J. Mander (Eds.), The case against the global economy and for a turn towards localization. London: Earthscan.

Mander, J. (2003). Intrinsic negative effects of economic globalization on the environment. In J. G. Speth (Ed.), *Worlds apart, Globalization and the environment*. Washington, DC: Island Press.

Markandya, A., & Pavan, M. (Eds) (1999). *Green accounting in Europe – Four case Studies*, Dordrecht: Kluwer.

Martinez-Alier, J. (1987). *Ecological economics, Energy, environment and society*. Oxford and Cambridge: Blackwell.

Martinez-Alier, J. (2002). *The environmentalism of the poor, a report for UNRISD for the WSSD*. Retrieved May 16, 2007, from http://www.foescotland.org.uk/nation/ej_alier.pdf.

Marx, K. (1894). Edited by F. Engels. *Das Kapital, Dritter Band* [reprint 1981: *Capital, Vol. III*. Harmondsworth, UK: Penguin].

Matthews, E. et al. (2000). *The weight of nations, Material outflows from industrial economies*. Washington, DC: World Resources Institute.

Max-Neef, M. (1995). Economic growth and quality of life: A threshold hypothesis. *Ecological Economics, 15*, 115–118.

Max-Neef, M. et al. (1990). *Human scale development: An option for the future*. Uppsala, Sweden: Dag Hammarskjöld Foundation.

Meadows, D. H., Meadows, D. L., Randers, J., & Behrens III, W. W. (1972). *The limits to growth*. New York: Universe.

Meadows, D. H., Meadows D. L., & Randers, J. (1992). *Beyond the limits*. Post Mills, VT: Chelsea Green.

Meadows, D., Randers, J., & Meadows, D. (2004). *Limits to growth, The 30-years update*. White River Junction, VT: Chelsea Green.

Messner, D., & Nuscheler, F. (2000). Politik in der Global Governance Architektur [Policy in the global governance architecture]. In R. Kreibich & U.E. Simonis (Eds.), *Global change – Globaler Wandel, Ursachenkomplexe und Lösungsansätze – Causal structures and indicative solutions*. Berlin: Berlin Verlag, Arno Spitz.

Meyer, B. (1999). Research-statistical-policy co-operation in Germany: Modelling with Panta Rhei. In European Commission (1999), *From research to implementation: Policy-driven methods for evaluating macro-economic environmental performance*. Luxembourg: European Communities.

Meyer, B. (2005). *The economic-environmental model Panta Rhei and its application*. GWS Discussion Paper 2005/3. Retrieved May 16, 2007, from http://www.gws-os.de/Downloads/gws-paper05–3.pdf.

Mill, J. S. (1848, 7th ed. 1909). *Principles of Political Economy*. London: Longmans, Green and Co. Retrieved May 16, 2007, from http://socserv2.socsci.mcmaster.ca/~econ/ugcm/3ll3/mill/prin/index.html.

Millennium Ecosystem Assessment (2005). *Ecosystems and human well-being: Synthesis*. Washington, DC: Island Press. Retrieved May 16, 2007, from: http://www.maweb.org/en/Article.aspx?id = 58.

Mishan, E. J. (1967, rev. ed. 1993). *The costs of economic growth*. London: Weidenfeld and Nicolson.

Mishan, E. J. (1973). *Elements of cost-benefit analysis*. New York: Praeger.

Moldan, B., Billharz, S., & Matravers, R. (1997). *Sustainability indicators: A report on indicators of sustainable development*. Chichester, UK: Wiley.

Müller-Wenck, R. (1978). *Die ökologische Buchhaltung. Ein Informations- und Steuerungsinstrument für umweltkonforme Unternehmenspolitik* [Ecological bookkeeping. An information and control instrument for environmentally sound corporate policy]. Frankfurt: Campus.

Munasinghe, M. (Ed.) (2002). *Macroeconomics and the environment*. Cheltenham, UK: Edward Elgar.
Naess, A. (1976). The shallow and the deep, long-range ecology movement, a summary. *Inquiry, 16*, 95–100.
Neumayer, E. (2000). On the methodology of ISEW, GPI and related measures: Some constructive suggestions and some doubt on the 'threshold' hypothesis. *Ecological Economics, 34*, 347–361.
Neumayer, E. (2001). The human development index and sustainability – a constructive proposal. *Ecological Economics, 39*, 101–114.
Nordhaus, W. D. (1973). World dynamics: Measurement without data. *The Economic Journal, 83*(332), 1156–1183.
Nordhaus, W. D. (ed.) (1998). *Economics and policy issues in climate change*. Washington, DC: Resources for the Future.
Nordhaus, W. D., & Boyer, J. (2000). *Warming the world, Economic models of global warming*. Cambridge, MA: MIT Press.
Nordhaus, W. D., & Kokkelenberg, E. C. (Eds.) (1999). *Nature's numbers – Expanding the national accounts to include the environment*. Washington, DC: National Academy Press.
Nordhaus, W. D., & Tobin, J. (1973). Is growth obsolete? *Studies in Income and Wealth, 38*, 509–564.
Norgaard, R. B. (1994). *Development betrayed, The end of progress and a coevolutionary revisioning of the future*. London: Routledge.
Nováček, P., & Mederly, P. (2002). *Global partnership for development, Sustainable development index*. Olomouc, Czech Republic: Palacky University (for American Council for the United Nations University).
O'Connor, M. (1999). Estimating a 'greened GDP' in line with the GREEN STAMP methodology: An analytical overview. In European Commission (1999), *From research to implementation: Policy-driven methods for evaluating macro-economic environmental performance*. Luxembourg: European Communities.
Oda, K. et al. (1996). *The system of integrated environmental and economic accounting for Japan, trial estimates and remaining issues* (Paper presented at the Special IARIW Conference on Integrated Environmental and Economic Accounting in Theory and Practice, Tokyo, 5–8 March 1996).
Odum, E. P. (1971, 3rd ed.). *Fundamentals of ecology*. Philadelphia, PA: Saunders.
Odum, H. T. (1996). *Environmental accounting, Emergy and decision making*. New York: Wiley.
Odum, H. T. (2002). Emergy accounting. In P. Bartelmus, *Unveiling wealth – On money, quality of life and sustainability*. Dordrecht: Kluwer.
Odum, E. P., & Odum, H. T. (1953, 1st ed.). *Fundamentals of ecology*. Philadelphia, PA: Saunders.
O'Neil, J., Turner, R. K., & Bateman, I. J. (Eds.) (2001). *Environmental ethics and philosophy*. Cheltenham, UK: Edward Elgar.
Opschoor, H., & van der Straaten, J. (1993). Sustainable development: An institutional approach. *Ecological Economics, 7*, 203–222.
Organisation for Economic Co-operation and Development (OECD) (1973). *List of social concerns common to most OECD countries*. Paris: OECD.
Organisation for Economic Co-operation and Development (OECD) (1976). *Measuring social well-being, A progress report on the development of social indicators*, Paris: OECD.
Organisation for Economic Co-operation and Development (OECD) (1985). Treatment of mining activities in the system of national accounts (note by the Secretariat, DES/NI/85.4).
Organisation for Economic Co-operation and Development (OECD) (1989). *Economic instruments for environmental protection*. Paris: OECD.
Organisation for Economic Co-operation and Development (OECD) (1993). *Core set of indicators for environmental performance reviews* (Environmental Monograph No. 83). Paris: OECD. Retrieved May 17, 2007, from http://www.virtualcentre.org/en/dec/toolbox/Refer/gd93179.pdf.

Organisation for Economic Co-operation and Development (OECD) (2003). *OECD environmental indicators – Development, measurement and use* (Reference Paper). Retrieved May 17, 2007, from http://www.oecd.org/dataoecd/7/47/24993546.pdf.

Organisation for Economic Co-operation and Development (OECD), & Statistical Office of the European Communities (Eurostat) (1999). *The environmental goods and services industry, Manual for data collection and analysis*. Paris: OECD.

Osborn, D., & Bigg, T. (1998). *Earth summit II, Outcomes and analysis*. London: Earthscan.

Paris, Q. (1991). *An economic interpretation of linear programming*. Ames, IA: Iowa State University Press.

Pearce, D. W. (1994). Valuing the environment: Past practice, future prospect. In I. Serageldin & A. Steer (Eds.), *Making development sustainable – From concepts to action*, Environmentally Sustainable Development Occasional Paper Series No. 2. Washington, DC: The World Bank.

Pearce, D. W., Markandya, A., & Barbier, E. B. (1990). *Blueprint for a green economy*. London: Earthscan.

Pearce, D., & Ulph, D. (1995). *A social discount rate for the United Kingdom*, CSERGE WorkingPaper GEC 95–01. Retrieved May 17, 2007, from http://www.uea.ac.uk/env/cserge/pub/wp/gec/gec_1995_01.pdf.

Perman, R., Common, M., McGilvray, J., & Ma, Y. (2003, 3rd ed.). *Natural resource and environmental economics*. Pearson Education (Addison-Wesley).

Perrings, C. (1995). Ecology, economics and ecological economics. *Ambio, 24*(1), 60-63.

Perrings, C. (1998). Income, consumption and human development. Environmental linkages. In UNDP (Ed.), *Consumption for human development*. New York: UNDP.

Perrings, C. (2006). Resilience and sustainable development. *Environment and Development Economics, 11*, 417–427.

Pezzey, J. (1989). *Economic analysis of sustainable growth and sustainable development*, Environment Department Working Paper No. 15. Washington, DC: The World Bank.

Pigou, A. C. (1920, 4th ed. 1932). *The economics of welfare*. London: Macmillan.

Pintér, L., Hardi, P., & Bartelmus, P. (2006). *Indicators of sustainable development, Proposals for a way forward* (Paper presented at the United Nations Expert Group Meeting on Indicators of Sustainable Development, New York, 13–15 December 2005). Retrieved May 17, 2007, from http://www.iisd.org/publications/pub.aspx?pno = 769.

Porter, G. (1999). Trade competition and pollution standards: 'Race to the bottom' or 'stuck to the bottom'? *Journal of Environment and Development, 8*(2), 133–151.

Postel, S. (1990). Toward a new 'eco'-nomics. *World-Watch, 3*(5), 20–28.

Powell, M. (1996). *Integrated environmental and economic accounts for Ghana* (Paper presented at the Special IARIW Conference on Environmental and Economic Accounting in Theory and Practice, Tokyo, 5–8 March 1996).

Prescott-Allen, R. (2001). *The wellbeing of nations: A country-by-country index of quality of life and the environment*. Washington, DC: Island Press.

Quesnay, F. (1759). The 'Third Edition' of the *Tableau Economique*. In M. Kuczynski & R. L. Meek (Eds.) (1972), *Quesnay's tableau économique* (facsimile reproduction and English translation). London: Macmillan [reprint in P. Bartelmus & E. K. Seifert (Eds.) (2003). *Green accounting*. Aldershot, UK: Ashgate].

Radermacher, W. (1999). Indicators, green accounting and environment statistics – information requirements for sustainable development. *International Statistical Review, 67*(3), 339–354.

Redefining Progress (1994–2004). *Sustainability indicators programme, Ecological footprint accounts*. Retrieved May 17, 2007, from http://redefiningprogress.org/programs/sustainability indicators/ef/methods/calculating.html#assumptions, http://redefiningprogress.org/publications/2004/footprintnations2004.pdf.

Redman, C. L. (1999). *Human impact on ancient environment*. Tucson, AZ: University of Arizona Press.

Rees, M. (2003). *Our final hour: A scientist's warning: How terror, error, and environmental disaster threaten humankind's future in this century – on earth and beyond*. Cambridge, MA: Basic.

Rees, W. E. (2000). Eco-footprint analysis: Merits and brickbats. *Ecological Economics, 32*, 371–374.

Rennings, K., Koschel, H., Brockmann K. L., & Kühn, I. (1999). A regulatory framework for a policy of sustainability: Lessons from the neo-liberal school. *Ecological Economics, 28*, 197–212.

Repetto, R. et al. (1989). *Wasting assets, Natural resources in the national income accounts*. Washington, DC: World Resources Institute.

Ricardo, D. (1817, 1963 ed.). *The principles of political economy and taxation*. Homewood, Ill.: Irwin.

Rodrik, D. (1997). *Has globalization gone too far?* Washington, DC: Institute for International Economics.

Rørmose, P., & Olsen, T. (2005). *Structural decomposition analysis of air emissions in Denmark 1980–2002* (Paper submitted to the 15th International Conference on Input-Output Techniques, Beijing, 27 June to 1 July 2005). Retrieved May 17, 2007, from http://www.iioa.at/pdf/15th%20Conf/roermose_olsen.pdf.

Rose, A., & Casler, S. (1996). Input-output structural decomposition analysis: A critical appraisal. *Economic Systems Research, 8*(1), 33–63.

Rothman, D. S. (1998). Environmental Kuznets curves – Real progress or passing the buck? A case for consumption-based approaches. *Ecological Economics, 25*(2), 177–194.

Russel, C. S. (2001). *Applying economics to the environment*. Oxford: Oxford University Press.

Sachs, I. (1976). Environment and styles of development. In W. H. Matthews (Ed.), *Outer limits and human needs*. Uppsala, Sweden: Dag Hammarskjöld Foundation.

Sachs, I. (1980). *Stratégies de l'Ecodéveloppement* [Strategies of eco-development]. Paris: Editions Ouvrières.

Sachs, J. D. (2005). *Investing in development, A practical plan to achieve the millennium development goals*. London: Earthscan. Retrieved May 17, 2007, from http://www.unmillenniumproject.org/reports/fullreport.htm.

Sachs, W. (1995). From efficiency to sufficiency. *Resurgence, 171*, 6–8.

Sachs, W. (2000). *Development. The rise and decline of an ideal*, Wuppertal Papers No. 108. Wuppertal Institute for Climate, Environment and Energy.

Sachs, W. et al. (1998). *Greening the North, Post-industrial blueprint for ecology and equity*. London: Zed.

Samuelson, P. A., & Nordhaus, W. D. (1992, 14th ed.). *Economics*. New York: McGraw-Hill.

Sandel, M. (1982). *Liberalism and the limits of justice*. Cambridge, MA: Cambridge University Press.

Santarius, T., Dalkmann, H., Steigenberger, M., & Vogelpohl, K. (2004). *Balancing trade and environment, An ecological reform of the WTO as a challenge in sustainable global governance*, Wuppertal Papers No. 133e. Wuppertal Institute for Climate, Environment and Energy.

Schaltegger, S., & Burritt, R. (2000). *Contemporary environmental accounting. Issues, concepts and practice*. Sheffield, UK: Greenleaf.

Schor, J. B. (1998). *The overspent American*. New York: Basic.

Scitovsky, T. (1976). *The joyless economy*. New York: Oxford University Press.

Seers, D. (Ed.) (1981). *Dependency theory: A critical reassessment*. London: Oxford University Press.

Segal, J. M. (1999). *Graceful simplicity. Towards a philosophy and politics of simple living*. New York: Holt.

Seppälä, T., Haukioja, T., & Kaivo-oja, J. (2000). *The EKC hypothesis does not hold for material flows! Environmental Kuznets curve hypothesis tests of direct material flows in some industrial countries* (Paper presented at the 3rd Biennial Conference of the European Society for Ecological Economics, Vienna 3–6 May 2000).

Sessions, G. (Ed.) (1995). *Deep ecology for the 21st century: Readings on the philosophy and practice of the new environmentalism*. Boston, MA: Shambhala (distributed by Random House).

Shogren, J. F. (2001). Natural resource economics. In M. K. Tolba (Ed.), (2001). *Our Fragile World, Challenges and opportunities for sustainable development*, Oxford: Eolss.
Siebenhüner, B. (2000). *Homo sustinens* – Towards a new conception of humans for the science of sustainability. *Ecological Economics, 32*, 15–25.
Simon, H. (1982). *Models of bounded rationality*. Cambridge, MA: MIT Press.
Simonis, U. (2005). Global environmental governance: Why we need a World Environmental Organisation. In G. Allan & M. Allshouse (Eds.), *Nature, truth, and value – Exploring the thinking of Frederick Ferré*. Lanham, MD: Lexington.
Slesser, M. (1975). Accounting for energy. *Nature, 254*, 170–172.
Smith, A. (1776, 1991 ed.). *Wealth of nations*. Amherst, NY: Prometheus.
Soddy, F. (1933). *Wealth, virtual wealth and debt: The solution of the economic paradox*. New York: Dutton.
Söderbaum, P. (1999). Values, ideology and politics in ecological economics. *Ecological Economics, 28*, 161–170.
Solórzano, R. et al. (1991). *Accounts overdue: Natural resource depreciation in Costa Rica*. San José, CA: Tropical Science Center and Washington, DC: World Resources Institute.
Solow, R. (1974a). Intergenerational equity and exhaustible resources. *The Review of Economic Studies, 41*, 29–45.
Solow, R. (1974b). 'The economics of resources or the resources of economics', *The American Economic Review*, *64*(2), 1–14.
Solow, R. (1992). *An almost practical step toward sustainability*. Washington, DC: Resources for the Future.
Solow, R. (2000, 2nd ed.). *Growth theory, An exposition*. Oxford: Oxford University Press.
Spangenberg, J. H., Femia, A., Hinterberger, F., & Schütz, H. (1999). *Material flow-based indicators in environmental reporting*, Environmental Issues Series of the European Environment Agency, No. 14. Luxembourg: European Communities.
Speth, J. G. (Ed.) (2003). *Worlds apart, Globalization and the environment*. Washington, DC: Island Press.
Stahmer, C., Kuhn, M., & Braun, N. (1998). *Physical input-output tables for Germany, 1990*, Eurostat Working Papers No. 2/1998/B/1. European Commission.
Statistisches Bundesamt (2002). *Umwelt, Umweltökonomische Gesamtrechnungen 2002* [Environment, Environmental-economic national accounts 2002]. Wiesbaden: Statistisches Bundesamt.
Steinbrink, A. (2001). Zero Emissions: Der Weg zur Vereinigung von Ökologie und Ökonomie als Wirtschaftssystem dieses Jahrtausends [Zero emissions: Towards unification of ecology and economy as the economic system of this millennium]. *Wasser und Boden, 53*(11), 8–12.
Stern, N. (2007). *The economics of climate change: The Stern review*. New York: Cambridge University Press. Retrieved May 20, 2007, from http://www.hm-treasury.gov.uk/independent_reviews/stern_review_economics_climate_change/stern_review_report.cfm.
Steurer, A. (1992). *Stoffstrombilanz Österreich 1988* [Material flow balance Austria 1988], Schriftenreihe Soziale Ökologie, Vol. 26. Vienna: Institut für interdisziplinäre Forschung und Fortbildung der Universitäten Innsbruck, Klagenfurt und Wien.
Stiglitz, J. (1974). Growth with exhaustible natural resources: Efficient and optimal growth paths. *Review of Economic Studies, 41*, Special Symposium Issue, 123–137.
Stiglitz, J. E. (2003). *Globalization and its discontents*. New York: Norton.
St. James, E. (1996). *Living the simple life*. New York: Hyperion.
Strassert, G. (2000). Stoffflüsse und Systempreise, produktionstheoretische Zusammenhänge von monetärer und physischer Input-Output-Rechnung [Material flows and system prices, some aspects of production theory in monetary and physical input-output systems]. In S. Hartard, C. Stahmer, & F. Hinterberger (Eds.), *Magische Dreiecke, Band 1, Stoffflussanalysen und Nachhaltigkeitsindikatoren*. Marburg, Germany: Metropolis.
Strassert, G. (2001). Eigenwerte der Natur [Eigen values of nature]. In S. Hartard & C. Stamer (Eds.), *Magische Dreiecke, Band 2, Bewertung von Nachhaltigkeitsstrategien*. Marburg, Germany: Metropolis.

Sustainability Now (2006). Wellbeing of Nations, Background. Retrieved May 17, 2007, from http://www.sustainability.ca/index.cfm?body = SourceView.cfm&ID = 422.

Sylvan, R., & Bennet, D. (1994). *The Greening of ethics*. Cambridge, UK: The White Horse Press and Tucson, AZ: The University of Arizona Press.

Szargut, J. (2001). Exergy analysis of thermal processes and systems with ecological applications. In M. K. Tolba (Ed.), Our Fragile World, Challenges and opportunities for sustainable development, Oxford: Eolss (on compact disc).

Szargut, J., Morris, D. R., & Steward, F. R. (1988). *Exergy analysis of thermal, chemical, and metallurgical processes*. New York: Hemisphere.

Tainter, J. A. (1988). *The collapse of complex societies*. Cambridge: Cambridge University Press.

Tainter, J. A. (2001). Complexity, collapse, and sustainable problem-solving. In M. K. Tolba (Ed.), Our Fragile World, Challenges and opportunities for sustainable development, Oxford: Eolss.

Tapscott, D., & Ticoll, D. (2003). *The naked corporation, How the age of transparency will revolutionize business*. New York: Free Press.

Tietenberg, T. (2005, 7th ed.). *Environmental and natural resource economics*. Boston and others: Addison-Wesley.

Todaro, M. P. (1977). *Economic development in the third world: An introduction to problems and policies in a global perspective*. London: Longman.

Tolba, M. K. (Ed.) (2001). *Our fragile world, Challenges and opportunities for sustainable development*. Oxford: Eolss.

Tulbure, I. (2001). An approach to define sustainable development indicators. In M. Matthies, H. Malchow, & J. Kriz (Eds.), *Integrative systems approaches to natural and social dynamics, Systems science 2000*. Berlin: Springer.

Turner, R. K., Pearce, D., & Bateman, I. (1993). *Environmental economics, An elementary introduction*. Baltimore, MD: Johns Hopkins University Press.

Ullman, A. A. (1976). The corporate environmental accounting system: A management tool for fighting environmental degradation. *Accounting Organizations and Society, 1*, 71–79.

United Nations (1973). *Report of the United Nations Conference on the Human Environment, Stockholm, 5–16 June 1972*. New York: United Nations (sales no. E.73.II.A.14).

United Nations (1975). *Towards a system of social and demographic statistics*. New York: United Nations (sales no. E.74.XVII.8).

United Nations (1977a). *Statistical Commission, Report of the nineteenth session*. New York: United Nations.

United Nations (1977b). *The feasibility of welfare-oriented measures to supplement the national accounts and balances: A technical report*. New York: United Nations (sales no. E.77. XVII.12).

United Nations (1979). *Studies in the integration of social statistics: Technical report*. New York: United Nations (sales no. E.79.XVII.4).

United Nations (1982). *Survey of environment statistics: Frameworks, approaches and statistical publications*. New York: United Nations (sales no. E.82.XVII.4).

United Nations (1984). *A framework for the development of environment statistics*. New York: United Nations (sales no. E.84.XVII.12).

United Nations (1988). *Concepts and methods of environment statistics: Human settlements statistic*. New York: United Nations (sales no. E.88.XVII.14).

United Nations (1991). *Concepts and methods of environment statistics: Statistics of the natural environment*. New York: United Nations (sales no. E.91.XVII.18).

United Nations (1993). *Integrated environmental and economic accounting*. New York: United Nations (sales no. E.93.XVII.12).

United Nations (1994). *Earth summit, Agenda 21, the United Nations programme of action from Rio*. New York: United Nations (sales no. E.93.I.11).

United Nations (1996). *Indicators of sustainable development, Frameworks and methodologies*. New York: United Nations (sales no. E.96.II.A.16).

United Nations (1999). *Handbook of input-output table compilation and analysis*. New York: United Nations (sales no. E.99.XVII.9).

United Nations (2000a). *Integrated environmental and economic accounting, An operational manual*. New York: United Nations (sales no. R.00.XVII.17).

United Nations (2000b). *Links between business accounting and national accounting*. New York: United Nations (sales no. E.00.XVII.13).

United Nations (2001a). *Environmental management accounting, Procedures and principles*. New York: United Nations (sales no. E.01.II.A.3).

United Nations (2001b). *Indicators of sustainable development: Guidelines and methodologies*. New York: United Nations (sales no. E.01.II.A.6).

United Nations (2002a). *Environmental managerial accounting, Policies and linkages*. New York: United Nations (sales no. E.02.II.A.3).

United Nations (2002b). *Use of macro accounts in policy analysis*. New York: United Nations (sales no. E.02.XVII.5).

United Nations (2003). *Johannesburg declaration on sustainable development and plan of implementation of the world summit on sustainable development*. New York: United Nations.

United Nations (2004). *National accounts: A practical introduction*. New York: United Nations (sales no. E.04.XVII.4).

United Nations (2005). *Population, resources, environment and development databank*. New York: United Nations (on compact disc). Retrieved May 17, 2007, from http://www.un.org/esa/population/publications/predbank4/predbank.htm.

United Nations (2006). *The millennium development goals report*. New York: United Nations. Retrieved May 17, 2007, from http://mdgs.un.org/unsd/mdg/default.aspx.

United Nations Conference on the Human Environment (1972). *Development and environment*, report and working papers of a panel of experts. Mouton, France: United Nations and Ecole Pratique des Hautes Etudes.

United Nations Development Programme (UNDP) (2003a). *Human development report*. Oxford: Oxford University Press.

United Nations Development Programme (UNDP) (2003b). *Making global trade work for people*. London: Earthscan.

United Nations Development Programme (UNDP), United Nations Environment Programme (UNEP), World Bank, & World Resources Institute (WRI) (2003). *World resources* 2000–2004. Washington, DC: World Resources Institute. Retrieved May 17, 2007, from http://pubs.wri.org/pubs_description.cfm?PubID = 3764.

United Nations Environment Programme (UNEP) (1975). The Proposed Programme (UNEP/GC/30), Nairobi.

United Nations Environment Programme (2002). *Global environmental outlook 3*. London: Earthscan. Retrieved May 17, 2007 from http://www.unep.org/GEO/geo3/.

United Nations Environment Programme (UNEP/GRID-Arendal) (2005). *Vital climate change graphics*. Arendal, Norway. Retrieved May 17, 2007, from http://www.vitalgraphics.net/_documents/clmate_change_update.v15.pdf.

United Nations Environment Programme (UNEP) (2006). *GEO yearbook 2006*. London: Earthscan.

United Nations Environment Programme (UNEP), & United Nations Conference on Trade and Development (UNCTAD) (1974). Symposium on 'Patterns of resource use, environment and development strategy', *The Cocoyoc Declaration*. Retrieved May 17, 2007, from http://www.southcentre.org/publications/conundrum/conundrum-06.htm.

United Nations et al. (1993). *System of national accounts 1993*. New York: United Nations (sales no. E.94.XVII.4).

United Nations et al. (in prep.). *Integrated environmental and economic accounting 2003*, final draft circulated for information prior to official editing. Retrieved February 1, 2008, from http://unstats.un.org/unsol/envaccounting/seea.asp.

Uno, K., & Bartelmus, P. (Eds.) (1998). *Environmental accounting in theory and practice*. Dordrecht: Kluwer.

Upadhyay, S. (2004). *Tribal self-rule law and common property resources in scheduled areas of India: A new paradigm shift or another ineffective sop?* (Presented at 'The Commons in an Age of Global Transition: Challenges, Risks and Opportunities', the Tenth Conference of the International Association for the Study of Common Property, Oaxaca, Mexico, August 9–13). Retrieved from http://dlc.dlib.indiana.edu/archive/00001553/.

van den Bergh, J. (Ed.) (1999). *Handbook of environmental and natural resource economics*. Cheltenham, UK: Edward Elgar.

van der Voet, E. et al. (2005). *Policy review on decoupling: Development of indicators to assess decoupling of economic development and environmental pressure in the EU-25 and AC-3 countries*. Institute of Environmental Sciences (Leiden University) and others. Retrieved May 17, 2007, from http://www.leidenuniv.nl/cml/ssp/projects/dematerialisation/policy_review_on_decoupling.pdf.

van Dieren, W. (1995). *Taking nature into account*. New York: Copernicus, Springer.

van Kooten, G. C., & Bulte, E. H. (2000). The ecological footprint – useful science or politics? *Ecological Economics, 32*(3), 385–389.

van Tongeren, J. et al. (1991). *Integrated environmental and economic accounting: A case study for Mexico*, Environment Working Paper No. 50. Washington, DC: The World Bank.

Veblen, T. (1899, 1967 ed.). *The theory of the leisure class*. New York: Macmillan.

Vemuri, A. W., & Costanza, R. (2006). The role of human, social, built, and natural capital in explaining life satisfaction at the country level: Toward a national well-being index (NWI). *Ecological Economics, 58*, 119–133.

Venetoulis, J., & Cobb, C. (2004). *The genuine progress indicator 1950–2002* (2004 update). Oakland, CA: Redefining Progress. Retrieved May 17, 2007, from http://www.rprogress.org/publications/2004/gpi_march2004update.pdf.

Venetoulis, J., Chazan, D., & Gaudet, C. (2004). *Sustainable indicators program*. Redefining Progress. Retrieved May 17, 2007, from http://redefiningprogress.org/publications/2004/footprintnations2004.pdf.

Vitousek, P. M., Ehrlich, P. R., Ehrlich, A. H., & Matson, P. A. (1986). Human appropriation of the products of photosynthesis. *Biometrics, 36*(6), 368–373.

von Carlowitz, H. C. (1713). *Sylvicultura oeconomica*. Leipzig, Germany: Braun.

von Weizsäcker, E.U., Lovins, A., & Lovins, H. (1997). *Factor four: Doubling wealth, halving resource use*. London: Earthscan.

Wackernagel, M., & Rees, W. (1996). *Our ecological footprint, Reducing human impact on earth*. Gabriola Island, BC: New Society.

Waggoner, P. E., & Ausubel, J. H. (2002). A framework for sustainability science: A renovated IPAT identity. *Proceedings of the National Academy of Sciences, 99*(12), 7860–7865.

Walker, B. H., & Pearson, L. (2007). A resilience perspective of the SEEA. *Ecological Economics, 61*, 708–715.

Wall, G. (2001a). Exergetics. In M. K. Tolba (Ed.) (2001). *Our Fragile World, Challenges and opportunities for sustainable development*. Oxford: Eolss (on compact disc).

Wall, G. (2001b). The use of natural resources in society. In M. K. Tolba (Ed.) (2001). *Our Fragile World, Challenges and opportunities for sustainable development*. Oxford: Eolss.

Ward, M (2004). *Quantifying the world, UN ideas and statistics*. Bloomington and Indianapolis: Indiana University Press.

Weber, J.-L. (2007). Implementation of land and ecosystem accounts at the European Environment Agency. *Ecological Economics, 61*, 695–707.

Weterings, R., & Opschoor, P. H. (1992). *The ecocapacity as a challenge to sustainable development*. Rijswijk, the Netherlands: Netherland Advisory Council for Research on Nature and Environment.

Wheeler, D. (2002). Beyond pollution havens. *Global Environmental Politics, 2*(2), 1–10.

White, L. (1967). The historical roots of our ecological crisis. *Science, 155*, 1203–1207.

Wicke, L. (1993). *Umweltökonomie: Eine praxisorientierte Einführung* [Environmental economics: A practical introduction]. Munich, Germany: Franz Vahlen.

Wohlfahrt, W. (1999). *Der Weg zum Umweltmanagementsystem: Gegenüberstellung von Öko-Audit-Verordnung und DIN ISO 14001* [The way to the environmental management system: Comparing the environmental management and audit scheme with DIN ISO 14001]. Berlin: Beuth Verlag.

Wolfe, A. (1989). *Whose keeper?* Berkeley, CA: University of California Press.

World Bank (1982). *Tribal peoples and economic development: Human ecologic considerations*. Washington, DC: The World Bank.

World Bank (1992). *World development report 1992: Development and the environment*. Washington, DC: The World Bank.

World Bank (1997). *Expanding the measure of wealth, Indicators of environmentally sustainable development*. Washington, DC: The World Bank.

World Bank (2003). *World development report 2003, Sustainable development in a dynamic world, Transforming institutions, growth, and quality of life*. New York: Oxford University Press.

World Bank (2006). *Where is the wealth of nations? Measuring capital for the 21st century*. Washington, DC: The World Bank.

World Commission on Environment and Development (WCED) (1987). *Our common future*. Oxford: Oxford University Press.

World Resources Institute (WRI) (1992). *World resources 1992–93*. New York and Oxford: Oxford University Press.

World Resources Institute (2006). *Earth trends, Environmental information*. Retrieved May 17, 2007, from http://earthtrends.wri.org/#.

World Watch Institute (WWI) (2000). *State of the world 2000*. New York: Norton.

World Watch Institute (WWI) (2003). *State of the world 2003*. New York: Norton.

Yale Center for Environmental Law and Policy, & Center for International Earth Science Information Network (1997–2006). *Environmental sustainability index 2002, 2005; Environmental performance index 2006*. Retrieved May 17, 2007, from http://sedac.ciesin.columbia.edu/es/esi/.

Index*

---

* Page numbers in italics refer to the main descriptions of key terms and definitions.

F

Colour Plates

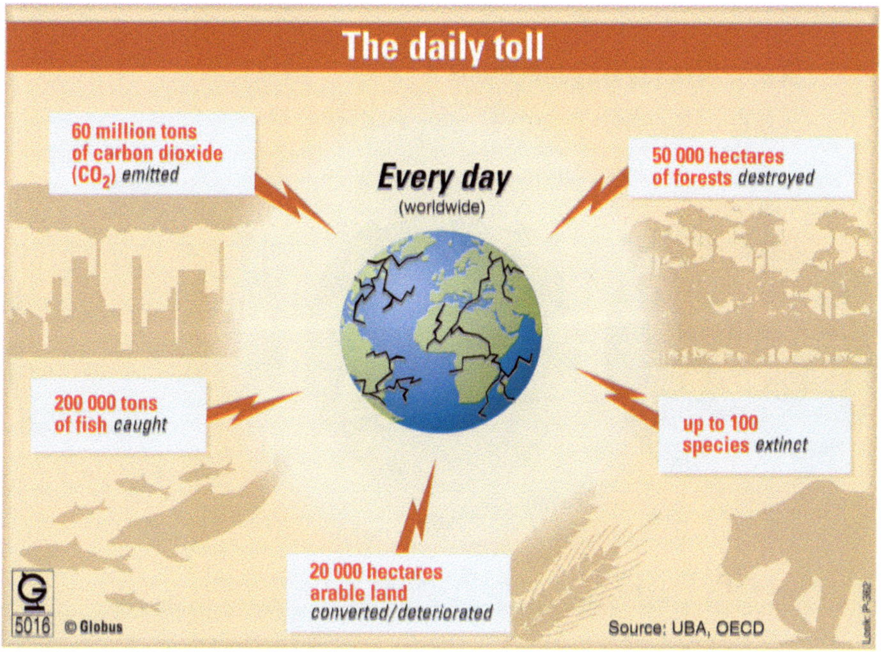

Plate 1.1 Environmental indicators.
Source: Globus Infografik GmbH.

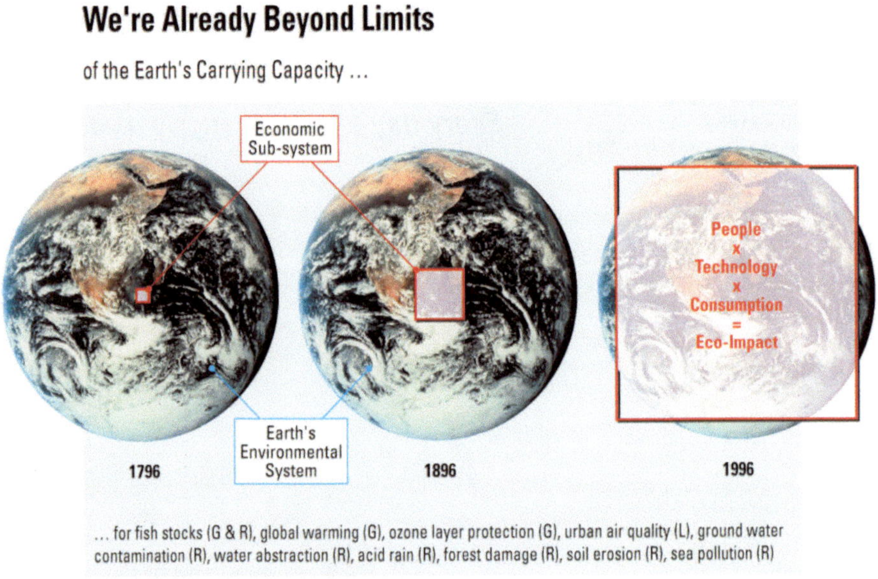

Plate 1.2 Full World?
Source: Based on Daly (1996); copyright VisLab/Wuppertal Institute for Climate, Environment and Energy; with permission by the copyright holder.

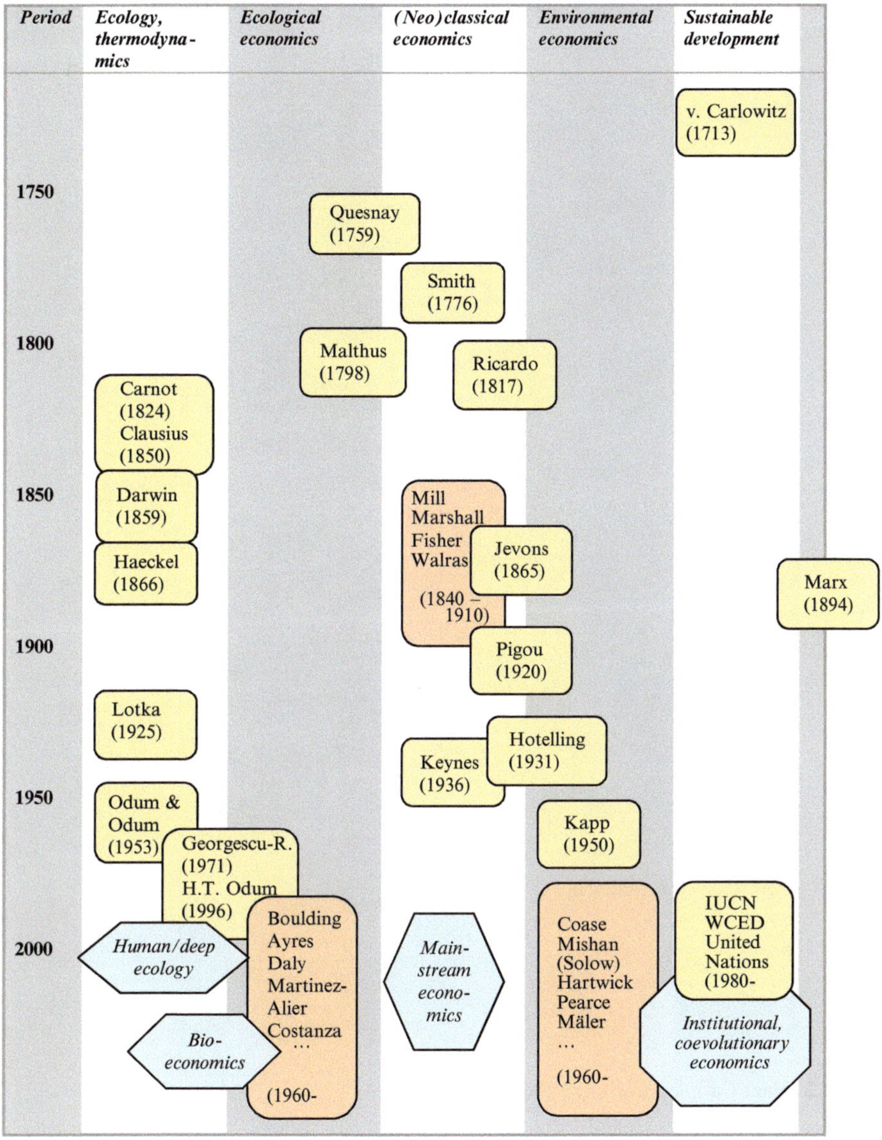

Plate 2.1 Historical perspective of *eco*–nomics

Plate 2.2 Hans Carl von Carlowitz (1645–1714)

Plate 2.3 Getting physical or monetary?
Copyright VisLab/Wuppertal Institute for Climate, Environment and Energy; with permission by the copyright holder.

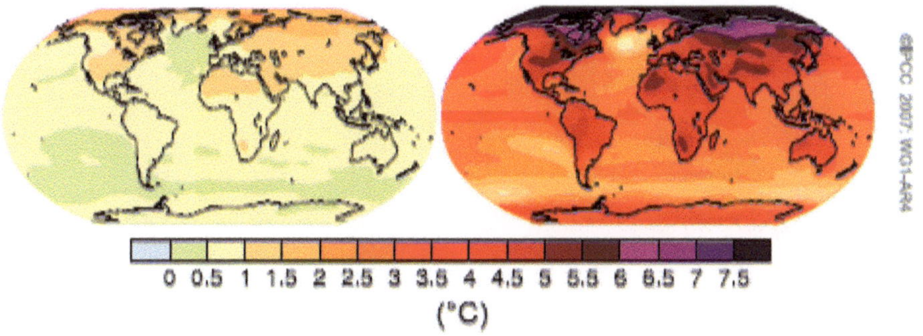

Plate 4.1 Projected surface temperature increase in the 21st century[a]
Note: [a]"Best estimates" for the high-impact scenario, compared to 1980–1999.
Source: IPCC (2007) – Climate Change 2007: The Physical Science Basis, Summary for Policymakers. Intergovernmental Panel on Climate Change.

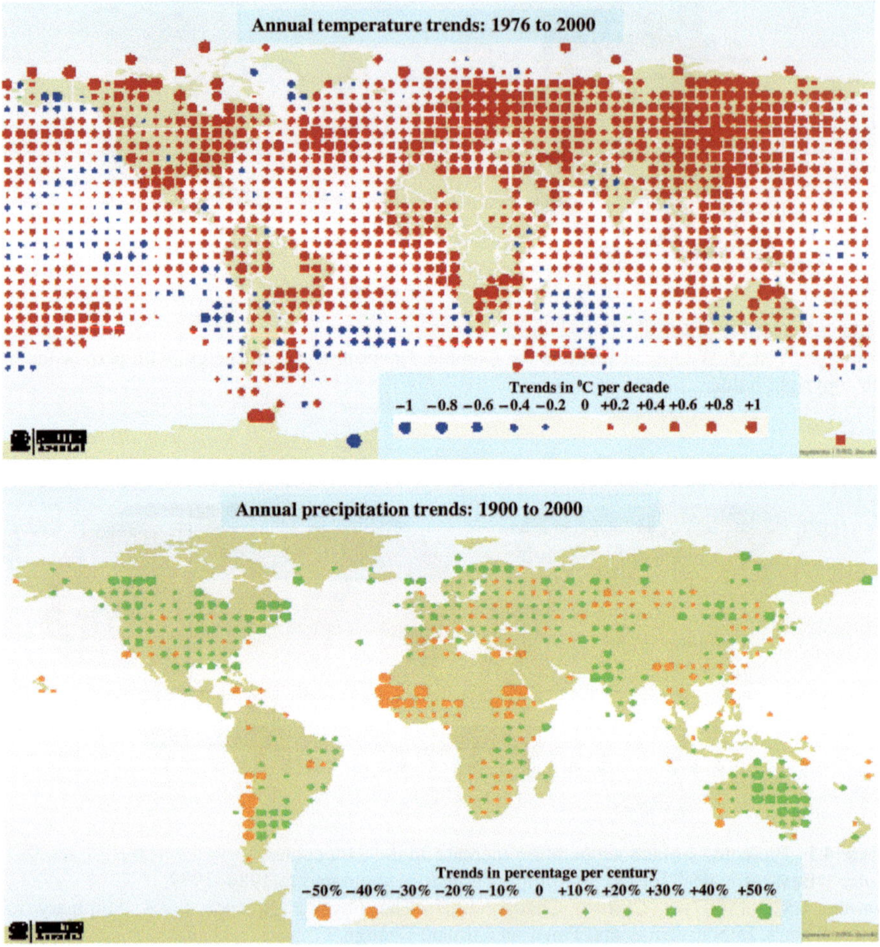

Plate 5.1 Overlay mapping: global warming and precipitation effects
Source: UNEP/GRID-Arendal (2005), Vital Climate Change Graphics.

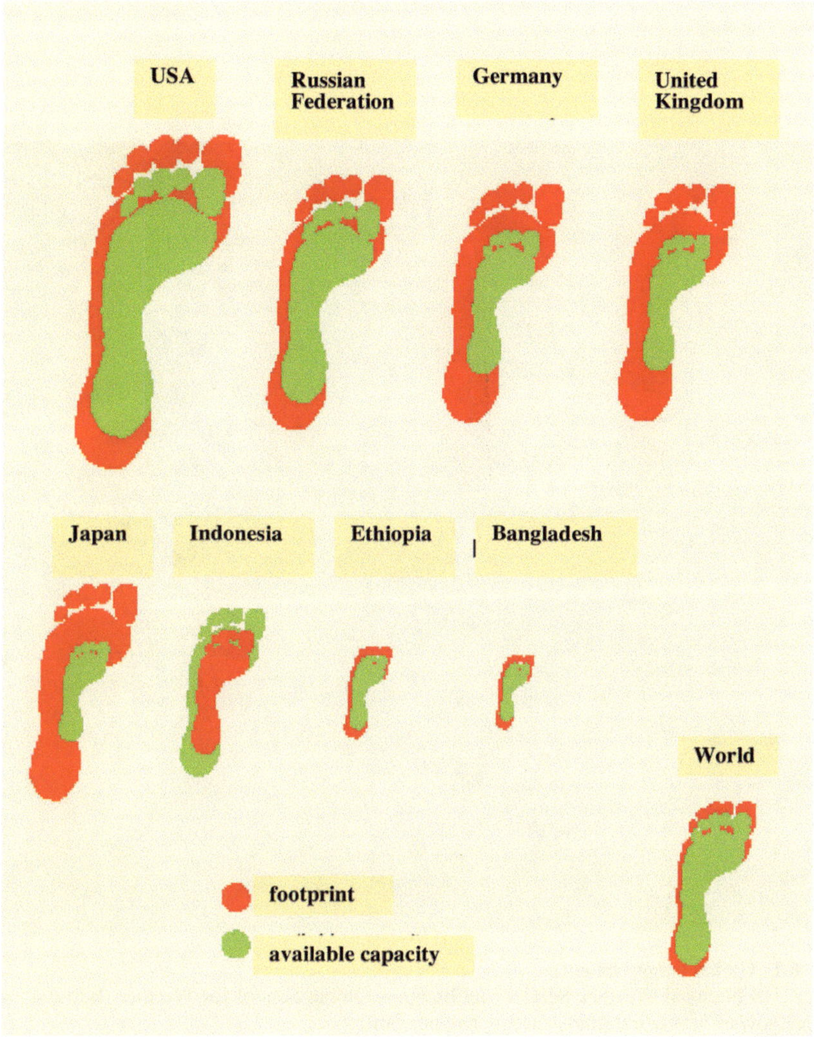

Plate 5.2 Ecological footprint
Source: Copyright VisLab/Wuppertal Institute for Climate, Environment and Energy; with permission by the copyright holder.

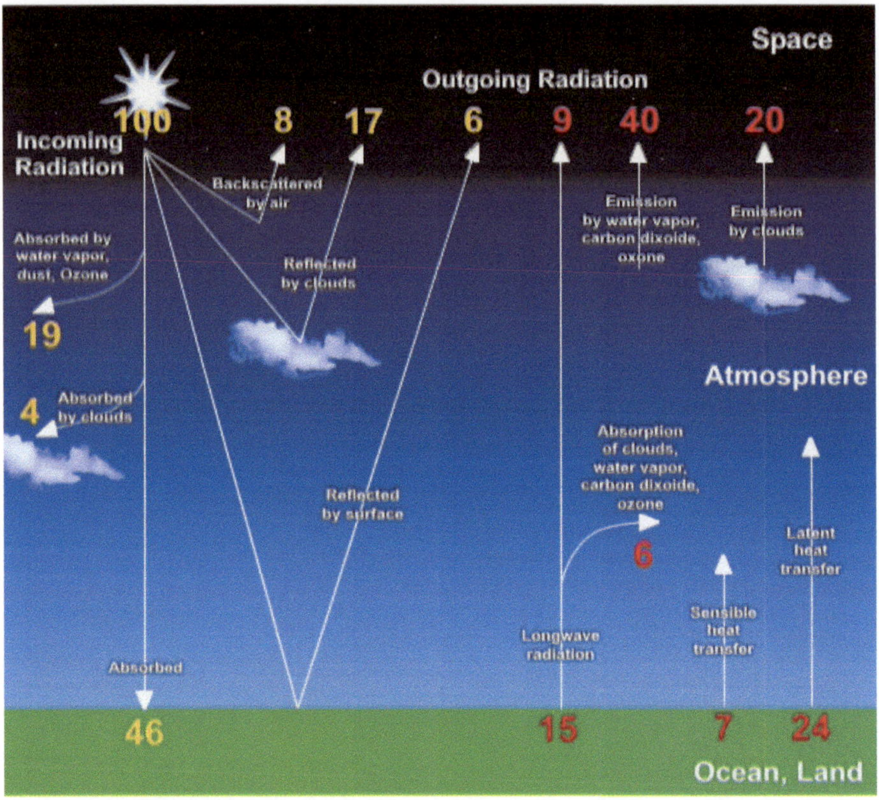

Plate 6.1 Global energy balance
Source: US National Weather Service, JetStream – Online School for Weather (http://www.srh.weather.gov/srh/jetstream/atmos/energy_balance.htm).

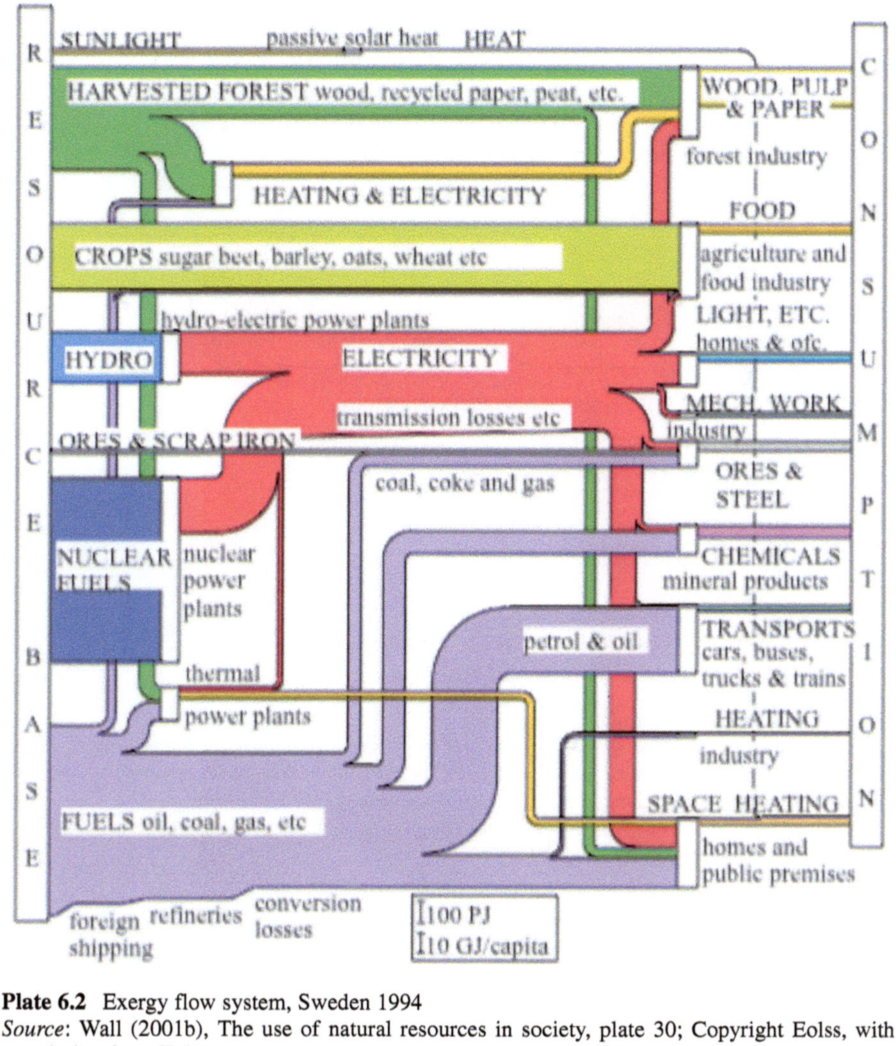

Plate 6.2 Exergy flow system, Sweden 1994
Source: Wall (2001b), The use of natural resources in society, plate 30; Copyright Eolss, with permission from Eolss.

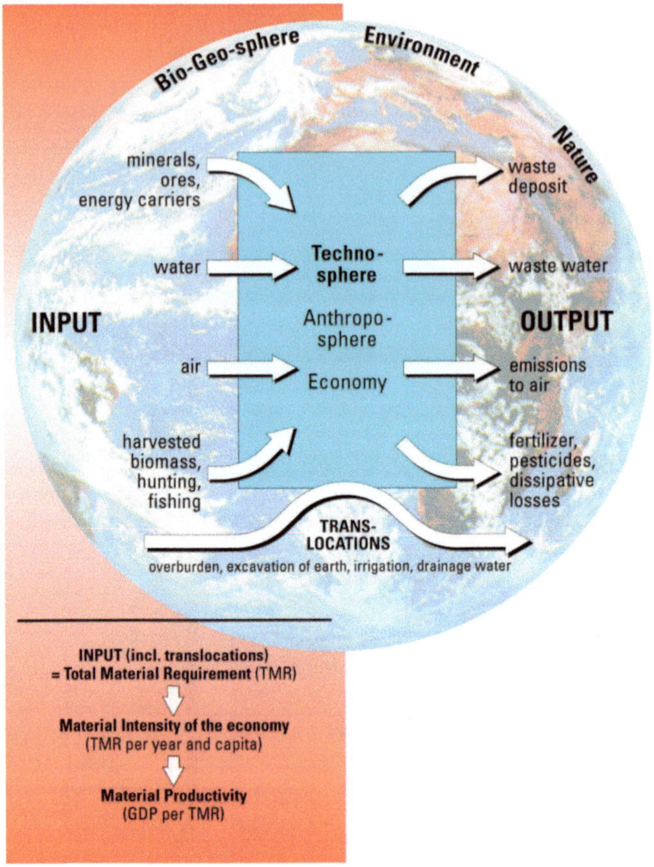

Plate 6.3 Material flows through the economy
Source: S. Bringezu (2000). *Ressourcennutzung in Wirtschaftsräumen*. Berlin: Springer, cover page (translated by the author); with permission by the author, VisLab/Wuppertal Institute for Climate, Environment and Energy, and Springer Science and Business Media.

Plate 6.4 Ecological rucksack of a wedding band: 'too heavy to marry?'
Source: Seppo Leinonen, with permission by the artist.

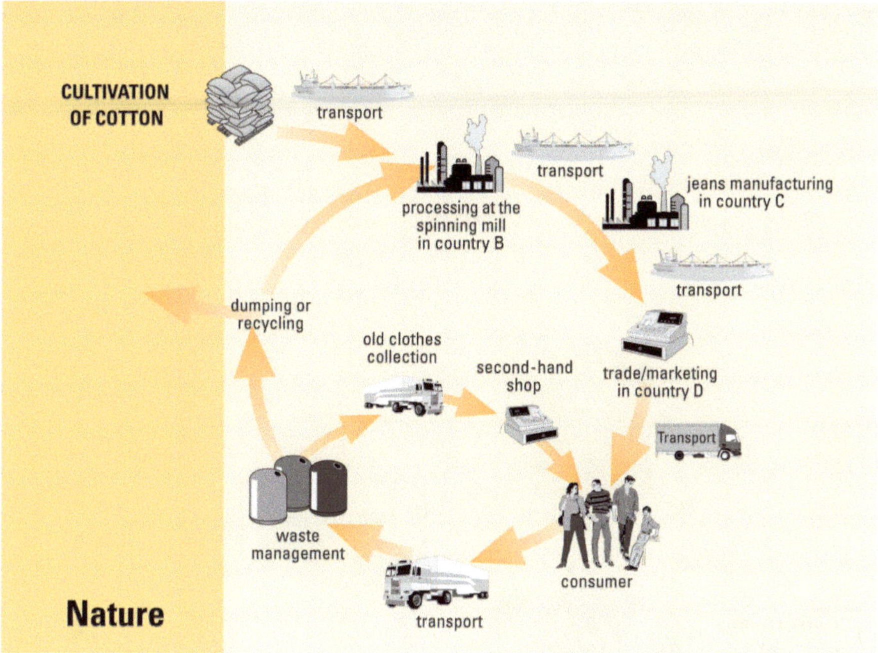

Plate 9.1 Life cycle of jeans
Copyright VisLab/Wuppertal Institute for Climate, Environment and Energy; with permission by the copyright holder.

Plate 9.2 EMAS logo
Source: http://europa.eu.int/comm/environment/emas/index_en.htm; with permission by the copyright holder, Stora Enso Kabel Mill, Germany.

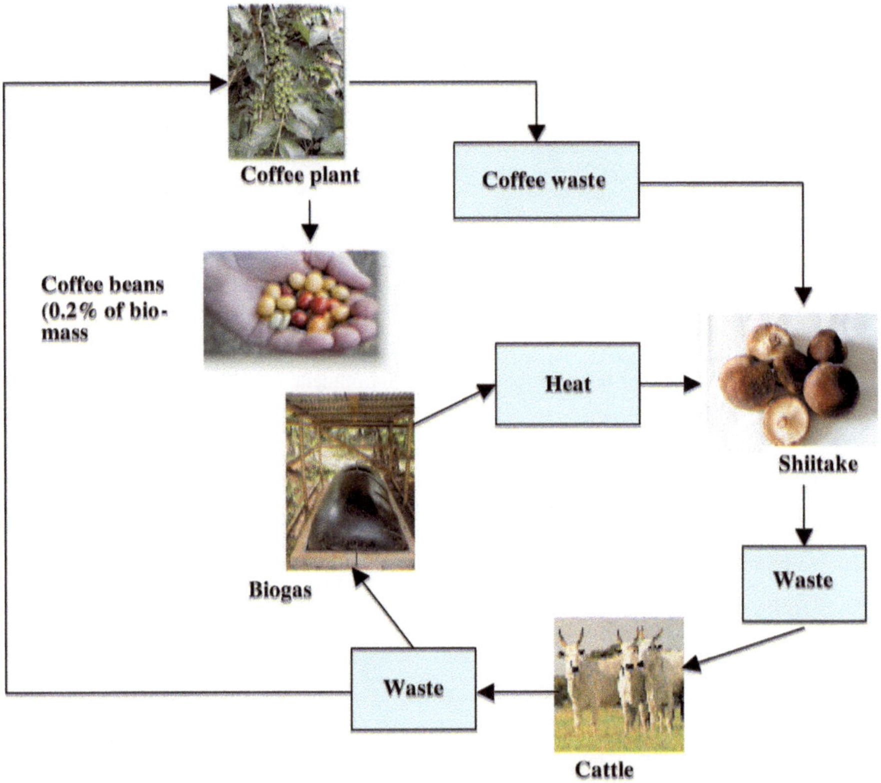

Plate 13.1 Metabolic consistency: coffee and mushroom production
Source: Based on Steinbrink (2001), fig. 2; with permission by the copyright holder, Zero Emission Research Initiative, ZERI

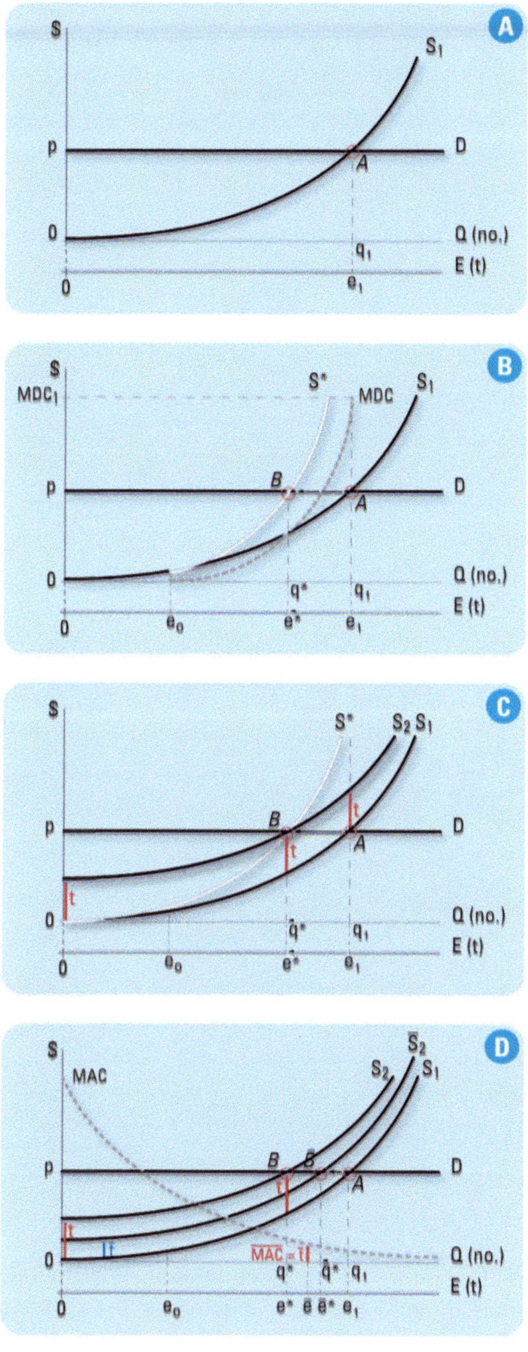

Plate I.1 Internalizing environmental damage
Source: Adapted from Turner et al. (1993).

MIX
Papier aus verantwortungsvollen Quellen
Paper from responsible sources
FSC® C105338

If you have any concerns about our products,
you can contact us on
ProductSafety@springernature.com

In case Publisher is established outside the EU,
the EU authorized representative is:
Springer Nature Customer Service Center GmbH
Europaplatz 3, 69115 Heidelberg, Germany

Printed by Libri Plureos GmbH
in Hamburg, Germany